PROFESSOR PORSCHE'S WARS

PROFESSOR PORSCHE'S WARS

THE SECRET LIFE OF LEGENDARY ENGINEER
FERDINAND PORSCHE
WHO ARMED TWO BELLIGERENTS
THROUGH FOUR DECADES

Pen & Sword
MILITARY

First published in Great Britain in 2014 by
PEN & SWORD MILITARY
an imprint of
Pen and Sword Books Ltd
47 Church Street
Barnsley
South Yorkshire S70 2AS

Copyright © Karl Ludvigsen, 2014

ISBN 978 1 78303 019 4

The right of Karl Ludvigsen to be identified as the author of this work has been asserted by him in accordance with the Copyright, Designs and Patents Act 1988.

A CIP record for this book is available from the British Library.

All rights reserved. No part of this book may be reproduced or transmitted in any form or by any means, electronic or mechanical including photocopying, recording or by any information storage and retrieval system, without permission from the Publisher in writing.

Printed and bound in India by
Replika Press Pvt. Ltd.

Typeset in Sabon by
CHIC GRAPHICS

Pen & Sword Books Ltd incorporates the imprints of
Pen & Sword Archaeology, Atlas, Aviation, Battleground, Discovery, Family History, History, Maritime, Military, Naval, Politics, Railways, Select, Social History, Transport, True Crime, and Claymore Press, Frontline Books, Leo Cooper, Praetorian Press, Remember When, Seaforth Publishing and Wharncliffe

For a complete list of Pen and Sword titles please contact
Pen and Sword Books Limited
47 Church Street, Barnsley, South Yorkshire, S70 2AS, England
E-mail: enquiries@pen-and-sword.co.uk
Website: www.pen-and-sword.co.uk

Suffolk County Council	
30127 08267841 3	
Askews & Holts	Nov-2014
623.7409	£30.00

CONTENTS

Introduction vii
Chapter 1   Porsche Meets the Military..1
Chapter 2   Power to the Dual Monarchy..13
Chapter 3   Austria in the Air 28
Chapter 4   Land Trains and Mortar Movers...................................45
Chapter 5   Advancing Aviation Power 60
Chapter 6   Mercedes Makes Military..74
Chapter 7   Army Tanks and Navy Twelves....................................88
Chapter 8   Porsche and Hitler Meet..103
Chapter 9   Third Reich Assignments..113
Chapter 10   Kübelwagen Takes the Road 131
Chapter 11   Masterpiece Schwimmwagen......................................147
Chapter 12   Leopards at Nibelung..160
Chapter 13   Fate of the Ferdinands...177
Chapter 14   Versatility and Variety..197
Chapter 15   The Mouse that Roared 214
Chapter 16   Cherrystone in Production 231
Chapter 17   Dispersal Under Pressure..245
Chapter 18   Occupation and Resolution..259

Acknowledgements..275
Bibliography..277
Index..280

To the memories of
Dean Batchelor
and
L. Scott Bailey
who backed my first explorations
into the world of Porsche

INTRODUCTION

When I asked Ernst Piëch where his grandfather was based during the Second World War his answer surprised me: 'Well, mainly in sleeping cars in trains between Stuttgart, Berlin and Vienna. He more or less slept in the train and worked during the day. Sometimes he stayed in a hotel, but he was more and more on trains than in hotels. He was very, very busy.'

I knew anecdotally that Ferdinand Porsche was well occupied during the Second World War but this was a fresh perspective from a man who knew. Thanks to Ernst Piëch, Porsche's oldest grandson, I had a good grasp of Porsche's work in the First World War. He backed research into this period that led to my book *Porsche – Genesis of Genius*, published by Bentley. This covers the Porsche career from the earliest days to the establishment of his own engineering company at the beginning of the 1930s.

Also for Bentley Publishing I researched some aspects of the pre-war and wartime activities of Porsche and his designers when they were evacuated from Stuttgart to the Austrian town of Gmünd for safekeeping. In spite of these ructions they and their leader, the Professor, kept most of their many platters spinning. Porsche's son Ferry played a progressively more important role.

In the course of these researches I learned much that has hitherto been inadequately recorded about one of the most prolific and productive engineers ever to tread the world's stage. Ferdinand Porsche is known for his cars, especially the Volkswagen, the Auto Union racing cars and, at the end of his career, the first Porsches. Knowledgeable readers will also be aware of the cars he designed for Lohner, Austro Daimler, Mercedes, Mercedes-Benz, Steyr and Wanderer, among which are some of the most respected automobiles of all time.

His career as a designer and builder of automobiles marks Ferdinand Porsche as one of the most creative and forward-thinking of his ilk. His achievements in that field are rivalled by very few. Yet Porsche had another vocation altogether, one that ranks virtually as a secret career. His work for the military in and between both world wars is little known and poorly understood. This remains the case although his contributions to military technology were astonishingly significant and pathbreaking. Often – and this is one reason for the dearth of attribution to Porsche – they only bore fruit after he left the company for which he was working.

No member of the Porsche/Piëch dynasty has done more than his oldest grandson to keep alive the memory of Ferdinand Porsche and his work. His career and its consequences are subjects of the contents of a remarkable museum, fahr(T)raum, opened in 2013 at Mattsee, Austria, by the Family Piëch-Nordhoff. They do not exaggerate by saying of its exhibits and displays that 'All of them are the work of an unrivalled visionary and pioneer of his time and beyond.'

In our conversations Ernst Piëch has brought to life the events of his grandfather's later years. 'My grandfather was very, very open with me,' said Piëch. 'He'd take me to every discussion. He said, "Look, this is top secret but you should see it." At that time I knew a lot of things which were absolutely top secret.

'We were driving at St Valentin with a tank chassis on the test track,' Piëch continued, 'up a slope of 45 degrees. When we were half-way up grandfather said "Stop" so I stopped. Then he said "Start again" and it didn't start. It just sprayed stones behind us. Then he said "Stop again" and then "Can't you use a softer clutch?" They made some changes and then it went up. But in a tank, at 45 degrees, with both of us in the tank – it was quite something.'

Ernst Piëch viewed at close range the techniques that made and kept Ferdinand Porsche a leader in his field when working under two diametrically different ideologies in two world wars. 'What I found brilliant,' he said, 'and

The author, right, and Ernst Piëch present their book on the early work of Ferdinand Porsche next to Piëch's superbly restored 1910 Prince Heinrich Austro Daimler racing car.

so different to the other engineers, was that my grandfather was very fair to other ideas.

'After the war,' Piëch continued, 'we went to an automobile show in Paris. We were standing in front of a car – a very famous car – and he said, "Look at this. They have a brilliant idea how to do it." He was quite without any jealousy. He really said, "This is just brilliant." Throughout his life he kept a big team together in this way by accepting other ideas and combining ideas to achieve a good result.'

As well, Porsche was daring. Although he sometimes gazed sideways at the work of others, Porsche was happiest when creating from first principles. This implied risk but he knew that without risk no gains could be achieved. His daring showed in his willingness not only to test-drive his creations but also to race them in open competition. He shared this trait with only a few leading engineers in the early days of the auto. Porsche continued personally test-driving his creations right through the Second World War to the first sports cars that bore his name in 1948.

Burnished by his daring was Porsche's confidence. As this record shows, he was not one to shrink from a challenge. Backed by his first-class team, he was confident of being able to cope with any requirement. Without pomp or artifice he manifested this confidence in his dealings with his colleagues and, most importantly, with his customers. They could only be impressed by the clarity, logic and passion with which Porsche set out his stall. No one would have picked this short, often dumpy, balding and moustachioed hard-core engineer as a superb salesman, but his record shows that Ferdinand Porsche's confidence helped make him one of the very best.

I am tempted to make the claim on his behalf that Ferdinand Porsche surpasses all other engineers in the depth and breadth of his contributions to military technology when taken over the entire scope of the years from 1914 to 1945. Challengers doubtless exist; the reader is encouraged to identify them. But they will have to go some to match the achievements of Porsche, who exploited in both world wars the Mixte drive system that he first invented in 1901. That alone is a distinctive hallmark of the life and work of this phenomenally creative engineer and outstanding personality.

Karl Ludvigsen
Hawkedon, Suffolk
April 2014

Reviewing maps of the manoeuvres in Austria-Hungary's Sárvár region, Crown Prince Franz Ferdinand was seated next to his driver Ferdinand Porsche in the latter's latest creation, which drove its front wheels through hub motors.

CHAPTER 1

PORSCHE MEETS THE MILITARY

September of 1902 found the armed forces of Austria-Hungary staging their autumn manoeuvres. They marshalled near Sárvár in the west of Hungary, close to its border with Austria, where foothills and the Rába River, winding its way north to the Danube at Györ, offered challenging countryside. Commanding the Western Army was Archduke Franz Ferdinand, heir-apparent to the Habsburg throne and Royal Prince of Hungary and Bohemia.

Important though traditional horse-drawn cavalry was to the Archduke and his noble colleagues, Franz Ferdinand wanted to be seen as welcoming the motor vehicle to his armies. Accordingly he was chauffeured on the manoeuvres by a reserve infantryman in the Imperial and Royal Deutschmeister Regiment, who in fact had designed the radical new car in which they rode. This was Ferdinand Porsche, just turned 27, at the wheel of his Lohner-Porsche.

Porsche's creation was radical, driven through its front wheels by electric motors. Their power came from an engine-driven generator that gave great flexibility to what the engineer called his 'Mixte' power train. This all-Austrian innovation was welcomed on parade by the high and mighty. They came no higher and mightier than Crown Prince Franz Ferdinand, who settled comfortably into the Lohner-Porsche Mixte to be driven, with his retinue, by Porsche from his headquarters to and around their military manoeuvres.

Afterward the Archduke's adjutant wrote to Porsche, saying 'how satisfied His Most Serene Highness was in every respect' with the services of both man and machine. Naturally enough, Porsche had worn the uniform appropriate to his lowly rank in the reserve. This was part and parcel of his campaign to interest the dual monarchy's military in his company's products. Indeed, the Sárvár manoeuvres launched Porsche's career of more than 40 years in the service of the military. Never more, however, would he see the need to wear a uniform.

Germany's forces staged their own manoeuvres a few months earlier. Their

A proud Ludwig Lohner caressed the bonnet of the Lohner-Porsche Mixte with which Porsche, at its wheel, won the Large Car Class of 1902's Exelberg hillclimb. Later in the year the same car was used for military manoeuvres, as at left.

nation having pioneered horseless carriages, thanks to Gottlieb Daimler and Karl Benz, Germany's General Staff officers took pride in rolling entirely on wheels – bicycles, motorcycles and automobiles. Exploiting a new age of motive power, Germany greeted the new century with a wholehearted commitment to the motorisation of its military.

Neighbouring ally Austria-Hungary could not afford to tarry. Ending almost 350 years of rule by occupiers, Franz Josef I formally established the dual monarchy of Austria-Hungary in July 1867 in response to the strident nationalism of Bismarck's Prussia. He ruled both as emperor of Austria – a title dating from 1806 – and as king of Hungary through a cabinet, common to both, that dealt with finance, defence, an internal customs union and external relations. Franz Josef allied with Germany in 1879 and added Italy, Serbia and Romania as alliance partners by 1883.*

The emperor-king and his kindred Habsburgs had their hands full with the aspirations for independence of their many member states. They ruled through an extensive bureaucracy from their capital in and near Vienna, which thus gained primacy over their empire's other major cities Budapest and Prague. Popular reforms contributed to educational advances and rapid industrialisation. Vienna was the cultural and commercial hub for Austria-Hungary's 50 million citizens, for whom their nobility and military were important sources of pride and a sense of unification.

Leading the charge to give motive power to the dual monarchy was Captain Robert Wolf, heading the army's research and development in his domain at Klosterneuburg on Vienna's north side.† Strikingly open to new ideas, Wolf saw motor vehicles as a means of giving his artillery greater mobility at lower cost. Austria-Hungary was in advance of many other nations in recognising and exploiting the potential of powering its military, though it was a step resisted by traditionalists reluctant to give up the glory and versatility of the horse.

As early as 1897 Wolf ordered a 5-ton truck from Germany's Daimler, dealing through its Vienna agents Bierenz & Hermann. In 1898 he received it, an oak-framed machine with a twin-cylinder 6.0-litre engine giving 10hp. It was named the 'Dromedary', a tradition followed with two more German-built vehicles, the 4-ton 'Hyena' and 3-ton 'Kangaroo'.‡ It was Wolf's idea to fit his trucks with a powered winch whose long cable could be used to extract a heavy vehicle from any situation.

When it came time to order a 2-tonner, Wolf could get it from a home producer. Daimler's Austrian agent became a vehicle maker on 11 August 1899 when the Daimler-Motoren-Kommanditgesellschaft Bierenz, Fischer & Co. was founded. Its lengthy cognomen acknowledged the role in the company of Josef Eduard Bierenz, who had been Gottlieb Daimler's man in Austria since 1890. It also marked participation in the new company by Eduard Fischer, whose family-owned Fischer Brothers machine works was founded in 1848 and had been based in Wiener Neustadt, 26 miles south of Vienna, since 1866. Fischer placed his facilities at the disposal of this pioneering effort to build Daimler-patent cars and trucks for the Austro-Hungarian market.

Initially structured as a partnership, the new company marked time until

---

* Italy would prove an unreliable ally. It chose neutrality in the early phases of the First World War and then joined the Entente Powers to fight Austria-Hungary.

† Wolf would soon be promoted to Major and later to General.

‡ The naming of its motor vehicles was a custom continued by the Austro-Hungarian military until 1913. The names of various predatory hounds were given to workshop and support vehicles, while tugs were named after large beasts and 2- and 3-tonners after small and domesticated animals.

its designated chief engineer, Gottlieb Daimler's eldest son Paul, was freed of his responsibilities in Stuttgart to move to Wiener Neustadt. The appointment was made by Paul's father before his untimely death in March 1900. In January 1902 young Daimler finally took up his post. Effective 24 June of the same year, the company was reformed as the Österreichische Daimler-Motoren-Gesellschaft, a corporation.

Branded as Daimlers, the new business's products offered little competition to those from the mother company in Stuttgart. Some passenger cars were made, based on Wilhelm Maybach's designs for the Germany company. A Paul Daimler design with a transverse twin-cylinder engine failed to catch the public's fancy. Concentrating on commercial and military vehicles, the company was pleased to receive Robert Wolf's order for a 2-tonner. Delivered in time for the 1902 manoeuvres, this was named *Ilitis* or 'Polecat'.

Ferdinand Porsche's Mixte was a brash interloper among the motor vehicles deployed near Sárvár in the dual monarchy's autumn 1902 exercises. It was a product of another enterprise eager to make its mark in Austria-Hungary's vehicle market. The Royal and Imperial Court Carriage Factory Jakob Lohner & Co. was no callow newcomer to vehicle manufacture. Tracing its origins to 1821, as a coachbuilder Lohner supplied 'by special appointment' the courts of Sweden, Norway and Romania as well as Austria-Hungary from its workshops in Vienna's Floridsdorf. It proudly celebrated its 20,000th carriage in 1890.

A third-generation head of Lohner, at the age of 34 Ludwig Lohner took command of the company on his father's death in 1892. Lohner soon viewed his inheritance as a poisoned chalice, for the market for fine horse-drawn carriages was commencing its decline. Four-figure annual volumes soon contracted to the low three figures – although celebratory 1900 was a good year with 508 carriages.

An engineer by training and young enough to be open to new ideas, Ludwig Lohner was easily seduced by the exciting new world of the motor vehicle. However, it was, wrote Lohner historian Erwin Steinböck, 'a breathtaking leap into cold water', for this new technology was to bring as many risks as opportunities at a time when petrol- and battery-powered vehicles were vying for marketplace supremacy with steam power an outside bet.

A warm personality with a lively sense of humour, seldom without a smile, Ludwig Lohner had an outgoing nature that was an asset when he travelled abroad in 1896 to assess the state of Europe's auto industry. He hoped above all to forge an alliance with Cannstatt's Daimler, but he declined a proposal that would have made Bierenz a partner as well. Nor did he find Rudolf Diesel ready to build auto engines. Experiments with French engines were disappointing.

Finally Lohner decided to explore opportunities for electric cars, whose smooth, silent progress and easy operation seemed a perfect fit with his upper-class clientele. For power Lohner turned to Austria's number one firm in electrical equipment, United Electrical Corporation, formerly Béla Egger &

Company. VEAG, its German acronym, kept its links with Béla Egger because the Hungarian had made a great name in Vienna as Thomas Edison's representative.

With the VEAG's help Ludwig Lohner's own team built its first electric car. Dubbed an Egger-Lohner, it united two great Viennese names. When first tested, in June 1898, 'We were initially unable to climb the gradient on Börsegasse,' recalled one of the VEAG's men who was assisting the trials, 'but after replacing many of the lead fuses we even reached the top of Ringstrasse.

'I was then given the job of strengthening the motor,' the Egger man continued. 'I don't need to say how much pleasure it gave me to undertake this task. After this augmentation we even managed to climb Berggasse and its 10 per cent incline with only one lead fuse. These events explain the mysterious, nocturnal journeys that took place at that time in Vienna's 9th district, which constantly plunged the automobile quarter of the city into turmoil.'

Enter, his coveralls deeply stained as usual, the unprepossessing figure of Ferdinand Porsche. Scarcely 5 feet 6 inches tall, so slender his mother would always worry that he wasn't eating, the 22-year-old Porsche wore his hair medium-length and, like his contemporaries, was growing a moustache. As overseer of test activities at the VEAG it fell to his lot to assist in the trials of the motors they had made for an important customer. This held the potential of significant business for the VEAG, as Ludwig Lohner pointed out in his negotiations with them.

Ferdinand Porsche had been with the VEAG since 1893. It was his university. He had arrived as an 18-year-old trainee whose main duties were cleaning the workshop and greasing the equipment. Fascinated by the potential of electricity, he lost not a moment in exploring the works Béla Egger had founded. In his spare time the youngster audited classes that interested him at Vienna's Technical University. Recognising his potential, Ernst Egger gave Porsche fresh responsibilities that he soon mastered. When Ludwig Lohner arrived with his commission, Ferdinand Porsche was in charge of the VEAG's test laboratories and an assistant in the calculation office.

From his earliest days Porsche was intimate with machinery and making things. Behind the family home at number 38 on the right bank of the Niesse River in Maffersdorf, Bohemia was the tinsmithery owned by his father Adolf. With its staff two-dozen strong, it was an important supplier of plumbing, carpentry and metal products to businesses in the area and in Reichenberg, the city less than 3 miles to the west. The region's economy was uplifted during the nineteenth century by the boom in wool and textiles.

At the empire's northernmost fringe, a scant 10 miles from the border with Germany, Ferdinand Porsche was born on 3 September 1875. He was the second son of Anton and Anna Porsche.* Anton and Anna, née Ehrlich, who had married in 1871, were 30 and 25 respectively at the time of Ferdinand's birth. The newcomer was the third to arrive. Eldest was a brother, Anton, followed by Hedwig. Trailing Ferdinand into the Porsche family were Oscar and Anna in that order. When Anton sadly died after an accident in the workshop's machinery, Ferdinand as the next eldest son was expected to train to take over the management of the tinsmithery.

---

* His birth was not registered until the 8th. Also born in 1875 were Walter Chrysler, Harry Miller, Alfred P. Sloan Jr., Maurice Ravel, Albert Schweitzer and Rainer Maria Rilke. Born the following year was Charles F. Kettering, an American inventive genius. Bizet's *Carmen* was first performed in 1875 and Mark Twain's *The Adventures of Tom Sawyer* was published the same year.

The Roman Catholic Porsche family could trace its origins to 1672. The name had evolved from 'Borso' centuries earlier, derived from the male name 'Borislav', through 'Boresch' and 'Porse'; in the seventeenth century some were spelling it 'Pursche'. Ferdinand was named after his paternal grandfather, who was a tenant farmer and master tailor. Great-grandfather Antonius mixed carpentry with hydraulic engineering, looking after his region's brooks and canals. Earlier antecedents made their living on the farm. Weavers were among Anna Ehrlich's forebears.

Although the family enjoyed reasonable prosperity, life in the Porsche household was no bed of roses. 'Anton Porsche was a very strict and harsh father and taskmaster,' wrote Porsche biographer Richard von Frankenberg. 'He ran his family in a patriarchal manner.' He reserved the more genial side of his character for his civic activities, serving as vice-mayor of Maffersdorf, an official in the fire department and a member of a banking board. Young Ferdinand learned early that the best policy at home was to keep his head down and get on with his work.

In 1886, while Porsche was an 11-year-old, electricity came to Maffersdorf. First to exploit its labour-saving potential was the textile mill of the Ginzkey brothers, Willy, Ignatz and Alfred, employing 2,000 of the town's almost-6,000 population. Workmen brought from Vienna to electrify Ginzkey's soon got used to the curious youngster under their feet, probing their installations and peppering them with questions.

'Ferdinand was a thinker and a tinkerer,' a classmate recalled. 'Otherwise he was taciturn. He was punctual and orderly. He brightened up in choir practice and joined in cheerfully. In his gymnastics a certain ambition emerged.' His grades were middling but adequate. What fascinated him most was electricity. An enthusiastic skater, he rigged battery-powered electric lamps for his skates. A career in electricity for his son, however, was not on Anton Porsche's agenda. He discouraged such fantasies and prepared the boy for an apprenticeship in tinsmithing.

Thus Anton was all the more enraged when he ventured into the lad's attic hideaway and discovered the crude batteries his son had made in secret to generate his own electricity. Furious, he stomped them into smithereens. To his surprise and dismay the battery acid ravaged his shoes, trouser and shins. That mattered not to Anton, who was sure he had shattered his son's aberrant yearnings.

With acid dripping through her bedroom ceiling, it was time for Anna to intervene. She could see the positive side of Ferdinand's

By the age of 13 Ferdinand Porsche had electrified both his home and his family's workshop in Maffersdorf. A shaft from the workshop's machinery drove a high-speed dynamo, its output controlled from a master board.

Young Porsche commuted by rail to evening classes in electrical engineering at Reichenberg, the town of which Maffersdorf is an appendage to the south-east. In the Czech Republic they are Liberec and Vratislavice respectively.

shenanigans to which his father was blind. Anna persuaded Anton that they should allow their youngster to take some evening classes at the Staatsgewerbeschule in Reichenberg. That way he would either be inspired or – as Anton hoped – discouraged from pursuing his rebellion from tinsmithing. Ferdinand commuted after his long days at work to study electrical engineering under Professor Joseph Pechan. To the young Porsche this was the opportunity of a lifetime. He took full advantage of it.

The decisive initiative came from the Ginzkey brothers, who urged that the young Porsche's evident genius be given a chance to flourish. Oscar, not Ferdinand, would have to train to take over the family business. The Ginzkeys were also influential in finding a place in Vienna at Béla Egger's firm for the 18-year-old Porsche. Catching the eye of his superiors at Egger's, as the company was still known, impressing them with his quick comprehension, initiative and problem-solving ability, the slender, unassuming Ferdinand Porsche readily mastered the tasks set him by the foreman in charge of the test department.

Although electrical engineering was Ferdinand Porsche's profession, he couldn't help being fascinated by the motor vehicle. It was exploding into raucous life all around him in the news if not in actuality. Intrigued by the potential of wheeled power, Porsche built an electrocycle for his commute to the works from his digs at 30–32 Metzleinsdorferstrasse. Harshly sprung, it often let him down. Nevertheless an eight-sided electric motor flanked its rear wheel, a design he called the 'Octagon'.

Thinking about vehicle design, Ferdinand Porsche was intrigued by the potential of front-wheel drive. He foresaw the movement of engines from the rear, as in most nineteenth-century cars, to the front, where they and their radiators could most easily be cooled. In his mind it followed logically that the front wheels should be driven to create a coherent power train.

Ludwig Lohner had been impressed by the way young Porsche had dealt with the problems thrown up by his first electrics. The Bohemian seemed to have a knack for visualising solutions to knotty problems. In the meantime Porsche had been thinking his own thoughts about the way an electric car should be propelled. It seemed to him sensible that if the motors were at the wheels – 'where they belong', he said – their tractive effort could be applied directly without the weight, bulk and cost of gears and differentials. As well it would be a cinch to provide drive through the steering front wheels, since only electric cables had to link wheels to vehicle.

In a conventional electric motor an armature rotates inside a magnetic field. Constant reversals of polarity of the current in the armature, caused by successive contacts from the brushes delivering power to it, rotate the armature because its changing polarity reacts against that of the magnets that surround it. Ferdinand Porsche's concept was the reverse of this. He chose to hold the magnetic field stationary within the wheel and to make the armature rotate around it and, with it, the wheel rim.

Thus by definition, because wheel and motor rotated together, Porsche's was a low-speed motor. While traction motors of the time were running at 300–500rpm, his would revolve in the 100–200rpm range for car speeds of 15–25mph. To gain power and efficiency under these conditions he provided numerous poles in his magnetic field, as many as 12 or 14 instead of the four in a conventional motor. Around them electromagnetic forces pushed the windings of the armature, and with them the wheel.

Combined with other features, such as an ingenious design for the vital brushes and commutator, this was the essence of the idea that Ferdinand Porsche presented to Ludwig Lohner. Its engaging simplicity appealed to the industrialist. Equally appealing was that these were new conceptions that were patentable. In these early freebooting years of the motor industry, industrialists like Lohner were eager to sequester promising technologies that they could then license at home and abroad. What Daimler, de Dion and others were doing, Lohner could do as well. Into the bargain, here was an indigenous

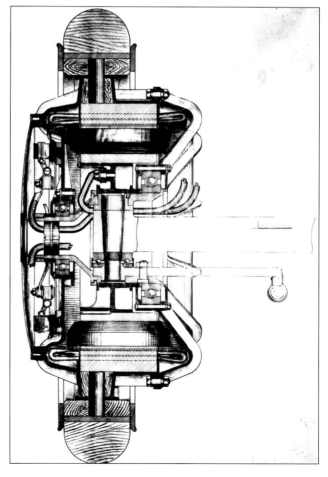

Shown here with hard-rubber tyres for a heavy-vehicle application, Lohner-Porsche wheel motors had commutators on their outer faces where they were accessible for service and immune to vertical motions.

In the Vienna-Floridsdorf workshop of Jakob Lohner & Co. Ferdinand Porsche found skilled tools and workmen to create a bewildering variety of vehicles driven by his wheel motors, powered both by batteries and by engine-driven dynamos.

Austrian invention for the use of which Lohner would not have to pay royalties to others.

The upshot was that near the end of 1899 Ferdinand Porsche left the VEAG to start working for Ludwig Lohner. At the factory in Floridsdorf, across the Danube in Vienna's north-west quarter, he found 400,000 square feet of workshop space with a large assembly hall and 120 machine tools powered by a traditional steam-driven overhead-belt system. He also found skilled craftsmen who could cope with his demands – although he was about to test their metalworking skills. Porsche still planned to rely for electrical components on the VEAG, but he would make his motor housings and other parts at Floridsdorf.

Granted the opportunity of his young life, the 24-year-old Ferdinand Porsche grasped it with both hands. He set out to design in detail the motors he promised his patron. He and Lohner manufactured their unique powered wheels in three versions, Types I, II and III, before they had sold or even publicly displayed a System Lohner-Porsche vehicle in 1900. This suggested the immense confidence that Porsche's design engendered in Ludwig Lohner. 'Three times we've tried to get to grips with making cars,' Lohner told a friend, 'but this time with the right result.'

Their first Lohner-Porsches were displayed and sold in 1900 as pure electric cars, finding custom with Vienna's elite as well as its taxi operators. One, equipped with a pair of de Dion engines driving generators that topped up its batteries, was a forerunner of the modern hybrid named the *Semper Vivus* ('Always Alive'). This gave Porsche an idea which he was quick to exploit.

The year 1901 found Porsche at work on his new concept. He hit on a dramatic way to demonstrate the capabilities of his wheel motors and front-wheel drive. Instead of powering them with batteries, he installed a petrol engine to drive a generator that would feed current to the wheels without a battery's intervention. This was not the work of a moment. Knotty problems in controlling the engine, generator and motors waited to be solved. But they had found the right man to solve them.

From the Austrian Daimler company Porsche obtained a 5.5-litre four-cylinder engine. He placed this at the front of a slender channel-steel chassis with solid axles front and rear sprung by semi-elliptic leafs. A straight shaft from the engine passed under the front floor to drive, under the seat, a powerful generator. Under the floorboards at the left was the master controller, which was rotated from one position to the next by the main control lever at the left of the right-hand steering wheel. At the extreme of its travel this lever applied the rear band brakes, cutting out the drive current at the same time and, indeed, short-circuiting the motors so they provided braking of the front wheels.

In 1901 Porsche created his first Mixte, a vehicle whose Daimler engine drove a dynamo under its front seats that sent electricity to its front-wheel motors. Its rear tonneau was removable.

Here was a big step forward from the *Semper Vivus* of 1900. Instead of a hybrid, Porsche had a car with an electric transmission. It did have a small battery, but this was only used to power accessories and to start the engine, using the generator as a motor. Starting, a major bugbear of the early cars, especially those with big engines, was a snap with this new species of Lohner-Porsche. To describe it, its creators adopted a French expression, *mixte*, as it was a blend of petrol and electric power trains. 'Mixte' would characterise a new generation of Lohner-built vehicles.*

Working all the hours God sent, as was his wont, Ferdinand Porsche not only designed, built and developed this prototype but also produced four more Mixtes of the same design in 1901. With experience he improved the steering action of his motor wheels for a new generation introduced in 1902. What did Porsche have in mind as a suitable debut for his new concept? Competition, of course, for he had already raced one of his pure electric cars at Vienna's demanding Semmering hill-climb.

His new car's chassis was not unlike the 1901 models, with a cable-braced straight channel-steel frame and a petrol tank low at the extreme rear. Steering gear was improved, following a Mercedes design. A substantial generator was

* In his writings Porsche preferred a German version of the term, Mixt, but Mixte was the public style.

Neighbours gathered to marvel when Ferdinand Porsche drove his first Mixte to Maffersdorf in 1901. Father Anton and brother Oscar were in the tonneau for the 26-year-old's demonstration of his new creation.

Entrepreneur Emil Jellinek bought five of Porsche's lively 1902-model Mixtes but his interest failed to evolve into regular orders. However, he would not forget the precocious Porsche's design talent.

underneath the front seats. Thanks to the use of a more advanced Mercedes 28 HP four-cylinder engine its hood was lower with bevelled edges behind a honeycomb radiator with matching bevels.

More wieldy and better balanced than his 1901 Mixte, Porsche's 1902 model was an appealing package. It could ask for no better public introduction than France's wealthy Côte d'Azur where the Nice Week of motoring competitions was held in March 1902. Two were entered, for Porsche himself

Porsche's stern gaze at his fiancée Aloisia Kaes as she took the wheel of his 1902 Lohner-Porsche was typical, said Ernst Piëch, who experienced the same scrutiny in drives with his grandfather.

and veteran Lorraine Barrow. In terms of sheer speed the new Lohner-Porsches were overmatched by the latest Mercedes, although they had clear advantages in controllability with their 15 forward speeds.

'There is no doubt that the car is extremely quiet,' *The Autocar* reported, 'and runs with every possible variation of speed without gearing of any kind.' The British weekly warned, however, that 'there remains the old objection, that the owner must not only be thoroughly acquainted with his petrol motor, but must also be an electrician.' According to one observer at Biarritz, who saw Lorraine Barrow in difficulty, the Mixte was reluctant to restart on a very steep hill.

Undaunted, Lohner and Porsche presented their new racer at Austria's Exelberg hill-climb in April. In a nature park on Vienna's western fringe, this was a short, sharp climb of 2.6 miles from Neuwaldegg up a 9 per cent gradient on loose surfaces to Exelberg. Taking the wheel himself, the inventor won the large-car class and set a new record for the hill. This was a sensational success that offered rich pickings for Lohner's advertising.

As if to emphasise the versatility of this Lohner-Porsche Mixte, its creators fitted a rear tonneau and flowing wings to make it a four-seater for road use. In this very guise Ferdinand Porsche had driven his Mixte the 70 miles south-east to Sárvár to take part in the dual monarchy's 1902 manoeuvres. Fitting right in, Porsche disguised his youth by flaunting a moustache that rivalled those of some of the royals.

At the close of the military exercises their key participants gathered in Vienna at the Emperor Franz Josef's summer palace for a debriefing. A report

At the close of 1902's manoeuvres young Porsche, here at the wheel, was closely questioned about his versatile Mixte by the elite of Austro-Hungarian aristocracy, up to and including the emperor himself.

Although his successful 1902 demonstration at Sárvár failed to lead to military orders for his Mixte, Porsche kept in touch with Austria-Hungary's officers. In 1903 he explained a more ambitious Mixte with a 70hp Panhard engine.

of 24 September observed that 'the monarch and Archduke Franz Ferdinand spent quite some time conversing with a member of the lowest rank of the Deutschmeister Regiment who looked anything but parade-ready, standing by his automobile. As well the German crown prince joined the group and conversed very deeply with this man. Foreign attachés engaged him in lengthy discussions.'

Here was recognition at the highest level, not only for the 27-year-old engineer's achievements but also for his maturity and aplomb. Illuminated was a character trait that would stand Ferdinand Porsche in good stead in years to come. He possessed *complete confidence*. He manifested this confidence in overcoming the resistance of a dominant father to his chosen career. He showed it in his move to Vienna and Béla Egger at the age of 18 and again – even more so – in his acceptance of the challenge of designing and developing a completely new technology of his own conception at Lohner.

This self-assurance, backed up by Porsche's unequalled knowledge of his own field and interest in many others, was a captivating asset. It enabled him to deal easily with royalty, politicians, the military and his industrial allies. On his ground Porsche was peerless. It also brought his colleagues with him, even into such trackless jungles as the motor-wheel adventure. This seemed to hold vast promise, both for passenger vehicles and for military applications.

CHAPTER 2

POWER TO THE DUAL MONARCHY

To Ludwig Lohner's disappointment the dual monarchy failed to follow up Ferdinand Porsche's successful 1902 demonstration with orders for Lohner-Porsche Mixtes.* Lohner and Porsche would continue to invite members of the military to Floridsdorf to see more powerful Mixtes under development and witness them in action on the company's test circuit. With demand for both Mixtes and electrics modest, however, not least because of their high prices, Ludwig Lohner began streamlining his business and phasing out their production.

Lohner's withdrawal from the motor industry in 1906 coincided with fresh interest taken by diplomat and entrepreneur Emil Jellinek in what he called the 'antiquated' Austro Daimler firm. His description was apt; the Fischer Brothers machine shop had not been transformed into an automobile company of even moderate series-production capabilities. It was ripe for the entrepreneur's attentions.

To meet his aim of dominance of the electric-vehicle market Emil Jellinek funded two new machining and assembly halls for the Austro Daimler works at Wiener Neustadt. They were valuable assets when war broke out.

---

* Lohner's man in Germany was Dr A. Isbert, with offices in Frankfurt. In 1905 he drove German generals in a handsome Mixte during manoeuvres 'to their fullest satisfaction'. The German army didn't order any either.

The very model of a modern motor magnate, the 35-year-old Porsche posed at his new villa on the grounds of the Austro Daimler works on Pottendorfstrasse in 1910. It would be a triumphant year for him.

Key players in the pre-war success of Austro Daimler were chief engineer Porsche (right), *and managing director Eduard Fischer* (centre). *With Count Heinrich Schönfeld* (left) *they made up the 'Iron Team' of the company's competition drivers.*

Jellinek, famous for his backing of new Daimler models that bore the name of his daughter Mercedes, had seen his personal coffers enhanced in 1905 since the acquisition by German's Daimler of 60 per cent of his interest in the Mercedes brand he had created, now controlled by a Paris company in which he retained a 40 per cent share. For this concession to Daimler he received one million francs, some $195,000 – a fortune at the time. With bank backing as well, Jellinek took control of Austro Daimler.

Wiener Neustadt had to step up to its role in the Jellinek Grand Plan. Two spacious new halls were erected on open land adjacent to the existing buildings, one for vehicle assembly. Lofty and amply skylit, they were of modern steel-reinforced concrete. One hall, for machine tools, was described by a visitor as 'snow white and leaves nothing to be desired in terms of cleanliness and brightness'.*

Nor was the machine park overlooked. Hundreds of new machines were acquired, many from America, including the latest lathes and gearcutters. Emil Jellinek made good on his pledge 'to transform the antiquated Austrian Daimler works in Wiener Neustadt into a modern automobile factory', at least by the standards of the early twentieth century.

With Bierenz out of the picture since the 1902 restructuring and staff on loan from Daimler returning to Germany, Austro Daimler was led in all respects by the charming, technically educated and professionally capable Eduard Fischer. Its engineering was rudderless, however, for in 1905 Paul Daimler was recalled to Stuttgart to assist in design work there; he became the German Daimler company's engineering chief in 1907.

---

* This hall was the subject of local controversy when, after the concrete forms were removed prematurely, part of it collapsed. The authorities threatened to have it completely torn down and rebuilt, but common sense prevailed and the necessary repairs were made.

The canny Jellinek settled on the man he wanted to take charge of his new expanded factory and its products. This, he decided, was a job for Ferdinand Porsche. The brilliant and confident young engineer who had carved out a distinctive niche with his designs for Lohner was his nominee for the job. Porsche affixed his signature to an Austro Daimler contract on 19 July 1906. Also joining was 24-year-old design engineer Otto Köhler, who was to become a strong right arm for Porsche.

With the help of Lohner's technology and new Porsche designs with rear-mounted motor wheels, Jellinek's aim was two-fold: to dominate the market for electric vehicles and to launch a new auto range named after another of his daughters, Maja. An important advantage was that this new design would liberate the company from the payment of licence fees to Daimler in Stuttgart.

Neither initiative scored success. Electrics had had their day and the Maja of 1907 foundered on a mediocre design – Porsche's first petrol-engined vehicle – and high warranty costs. Before the end of 1907 the venture tumbled into bankruptcy and liquidation. This was the end of his automotive career for the disillusioned Emil Jellinek, who saw his gasoline-engine venture expire along with his hoped-for hegemony in the electric-vehicle market.

Although depressed business conditions contributed heavily to the collapse of both his ventures, Jellinek was quick to place the blame on product problems and on the man he had put in charge of engineering. When in later years Jellinek reminisced about his automotive entrepreneurship, Ferdinand Porsche was among those he denigrated. The talent was there; he had spotted it. But he demanded too much of it too soon.*

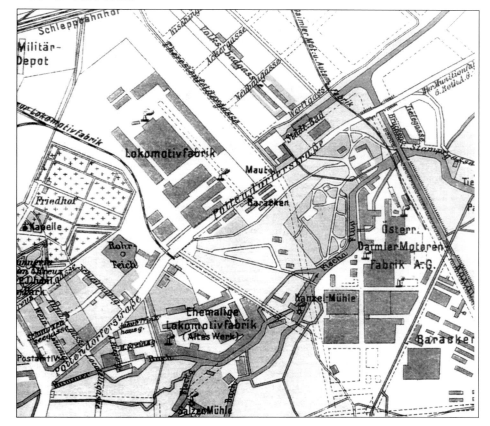

In the north-east quarter of Wiener Neustadt the Austro Daimler works was surrounded not only by locomotive factories but also by military installations whose officers often inspected Porsche's latest ideas.

---

* Emil Jellinek-Mercedes left Nice in 1914 and in 1915 settled with his family in Switzerland, away from the war. A stroke ended his life on 21 January 1918.

Porsche's predecessor at Austro Daimler, Paul Daimler, created an ambitious and ingenious four-wheel-driven armoured car in 1904. However, its 1905 demonstrations failed to win military orders.

By the end of 1908 Eduard Fischer and Ferdinand Porsche found themselves bereft of business partners. The deep-pocketed distributors who promised to buy so many vehicles from their expanded Austro Daimler workshops had melted away. Even worse, they wallowed in the trough of a pan-European business slowdown that discouraged purchase of such luxury items as motorcars. Rescue came at the end of the year in the form of an order for a dozen mortar-towing tugs for an increasingly important customer, the Austro-Hungarian army.

From its founding, Austro Daimler attracted the attention of the dual monarchy's military, not least by virtue of its location. Wiener Neustadt had been the choice of the Austrian Empress Maria Theresa as the site of a military academy, which opened its doors on the town's south-east periphery in 1752 as the first institution of its kind.* Ultimately the Austro Daimler works was surrounded by barracks and training grounds that swarmed with officers and non-coms. They would take full advantage of Austro Daimler's capabilities.

Paul Daimler had already given the military option a good try. In 1904 he and senior engineer Otto Stahl designed and built an armoured car that they hoped would catch the army's fancy at a time of Central European rearmament. This was a dedicated undertaking that included full-time four-wheel drive. Although its 4.4-litre four of 30hp was antiquated, its chassis showed new thinking.

Instead of adapting a conventional transmission to drive all four wheels, Daimler designed an all-indirect four-speed gearbox that took the drive downward through two selectable pinion pairs that gave a choice of ratios for eight forward speeds in total. An advantage of this layout was that a rearward continuation of the mainshaft could be used to drive a generator, winch or other equipment.

---

* After its 1938 annexation of Austria, Germany installed Erwin Rommel as the Academy's commandant.

Drive to the front and rear wheels was through a differential that could be locked out in difficult going. A similar lockout could bypass the rear differential. Patented by Daimler was a rear axle whose ratio reduction was entirely in a pinion and internal gear at each artillery wheel that allowed the rear wheels to have positive camber, the better to cope with the road surfaces of the day.

Drive to the front wheels of Paul Daimler's 1904 armoured car was also innovative. To permit the wheels to steer without the use of universal joints, he and Stahl placed a pair of bevel gears at each wheel to turn shafts, around vertical king pins, that then drove ring gears built into the faces of the steel-disc front wheels. Later Daimler patented a more elaborate system of frame-pivoted gears and shafts that gave the front axle more mobility on rough terrain. All the tyres were solid rubber.

Steel shrouded the engine and a two-man driving compartment whose seats could be raised and lowered to allow the crew to be either exposed or concealed behind viewports. At the rear of Daimler's vehicle a rotatable turret housed a water-cooled Maxim machine gun and space for two soldiers and their ammunition. Considered the first turreted armoured car, the final version was armoured to a 3mm thickness with 4mm for the turret. This brought its weight to just over 3 tons.

Capable of speeds up to 23mph, Paul Daimler's offering was mustered for the German Imperial Army's manoeuvres of 1905. The following year, updated with a pair of Maxims, it featured in the trials of the Austro-Hungarian military. Deputed to demonstrate the machine was a young lieutenant friendly to Austro Daimler, Count Heinrich Schönfeld. As related by historian Martin Pfundner, when Schönfeld fired up its raucous four the racket spooked a nearby horse which reared, dislodging its noble rider. This happened to be a friend and contemporary of the Kaiser, who witnessed the incident. 'Something like that can't be used for military purposes' was the emperor's spontaneous reaction, putting an end to any hope of a career with his army for Paul Daimler's creation. In the wake of Daimler's departure the armoured car was disposed of to the French military, which scrapped it after its evaluations.

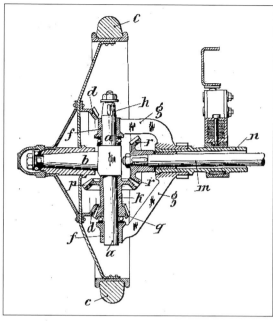

In his patent Paul Daimler showed how he carried his armoured car's drive to the front wheels through bevel gears, some of which rotated around the vertical king pin. Porsche made use of this practical solution.

A variant of this four-wheel-drive chassis would come to the rescue of Ferdinand Porsche and Eduard Fischer. With its wheelbase stretched from 2,860 to 3,200mm and a bigger 8½-litre engine producing 50bhp at 1,000rpm, it was built as a tug for field artillery. To meet Robert Wolf's requirements it also had an engine-driven cable winch at its nose. 'Its drive ratio', said Otto Stahl of the winch,

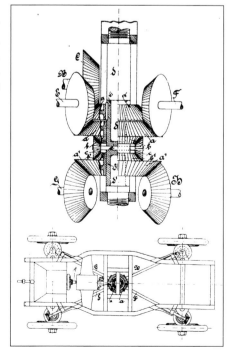

Archduke Salvator's patented four-wheel-drive mechanism took power from a central bevel-gear cluster to the machine's four corners. Although powerful, a vehicle built on these lines by Skoda in 1908 failed to satisfy.

Elements of Paul Daimler's 1904 four-wheel-drive system contributed to Austro Daimler's first military tug, the M 06. It was one of the first 1906 projects of Porsche, testing its winch here with Eduard Fischer (right).

In the works courtyard Porsche and Fischer (left) stood next to one of the dozen M 08 artillery tugs built for the army's Robert Wolf in 1908. With its 'Seals' Austro Daimler was at last gaining traction with the military.

'was chosen so that the vehicle could attain forward motion almost anywhere.'

Wolf considered this a piece of kit that his Austro-Hungarian forces needed. However, he could only propose, not dispose. He had to convince the dual monarchy's inspector general of artillery, Archduke Leopold Salvator. With a technical background that would lead to an honorary doctorate of engineering from the Vienna Technical Institute, Salvator was an innovator in his own right. In 1907 the Archduke patented a four-wheel-drive system of his own. At its heart in the centre of the chassis was a single complex combining a differential with bevel pinions from which shafts splayed out diagonally to power all four wheels. Using hub-mounted bevel gears akin to those of Paul Daimler, four-wheel steering was another feature.

In 1908 Salvator succeeded in having a sample built as an artillery tug, using a chassis made by Skoda in Pilsen. A feature was a vast steering wheel so that two men could attack it simultaneously – hinting at the problems experienced in handling the 'Lion', as it was named. Although credited with a 12-ton towing

capacity on level ground, this vehicle proved too cumbersome for its 40hp Gräf & Stift engine.*

Predisposed to the merits of four-wheel drive, Salvator was open to the ideas of Robert Wolf. 'Wolf was able to interest Archduke Leopold Salvator so much in this new form of drive,' wrote Otto Stahl, 'that in fact from this time until the end of the war Wiener Neustadt remained the main producer of tugs for artillery weapons for the Austro-Hungarian army.' It would now be up to Porsche to design and produce tugs that could cope with increasingly heavy weaponry.

Ferdinand Porsche's first major project for the military was his M 08 of 1908, given the name *Robbe* or 'Seal'. A four-wheel-drive tug, it was powered by a new 80hp six-cylinder engine of 13,854cc. Following Robert Wolf's philosophy, each of the dozen made had a winch under its nose that could be operated from the driver's seat. Though built for trailer towing, each M 08 had a rear platform for added gear.

One of the vehicles, intended to accompany others in action, mounted a vast winch from a specialist supplier on its rear deck. Carrying some 1,000ft of cable, it could extract another machine and its cargo from any desperate situation. Considered both baulky to handle and too heavy at first, the Seal's four-wheel drive was far from perfected at this early date.

In his designs Porsche had to keep wartime conditions very much in mind, for in 1908 tensions were increasing both inside and outside the empire. Thirty years earlier the Treaty of Berlin had granted Austria-Hungary the right to occupy Bosnia-Herzegovina to the south. It annexed those territories outright in October 1908, attaching them to neither half of the dual monarchy but instead directly to the Habsburgs' central command. This triggered unrest both internal and external, especially displeasing Serbia to the south. Rising temperatures encouraged rearmament by all nations in the region – especially the variegated empire that was at its heart.

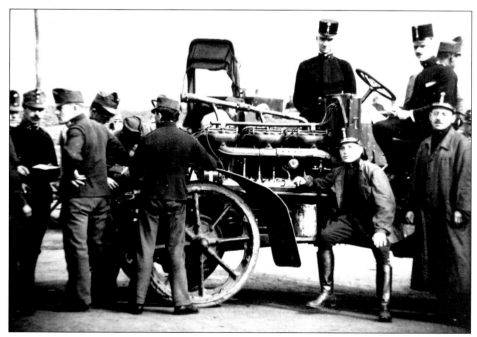

By 1910 troops in the field were getting to know the intricacies of the four-wheel-driven M 08 tug, including its impressive 13.9-litre six-cylinder engine developing 80hp.

* In 1914 Salvator would patent a pair of caterpillar-tracked bogies to be carried beneath a vehicle. Driven by chain from the rear wheels, they were to be lowered into position when the going was poor, as in mud or snow.

On 3 March 1910 the dual monarchy's Inspector General of Artillery, Archduke Leopold Salvator, visited Austro Daimler. He is fourth from right, Fischer second from right and the begoggled Porsche in a typical pose.

Pictured on the bonnet of an Austro Daimler 9/20 were (left to right) Porsche's children Louise and Ferry and their cousin Ghislaine Kaes. Kaes would become Ferdinand Porsche's secretary and, later, a keeper of his flame.

Domestically Ferdinand Porsche had already secured his succession. In 1896 at the VEAG in Vienna he had met Aloisia 'Louise' Johanna Kaes. On a sunny Saturday, 17 October 1903, the two were married in the church in Maffersdorf where Ferdinand had been baptised. True to Porsche's workaholic character, their honeymoon in Austria, Italy and France included several business meetings. Their first-born, Louise Hedwig Anna Wilhelmine Marie Porsche, came into the world on 29 August 1904. Following her was Ferdinand Anton Ernst Porsche, who was born on 19 September 1909. His name celebrated his father, his grandfather Anton and Aloisia's brother Ernst, who was also godfather to young 'Ferry'. Both would play important roles in the founding of a Porsche dynasty.

On 3 March 1910 Austro Daimler directors Ferdinand Porsche and Eduard Fischer took time away from their normal duties to pick up Archduke Leopold Salvator at the Steinfeld Airfield for a visit to their works. Pleased by what he found at Wiener Neustadt, the Archduke was especially intrigued by the modified racing car that Porsche was using to test airship propellers.

Salvator also witnessed work under way on a project that would make Ferdinand Porsche famous throughout the world of automobiles. In 1909 Germany's Prince Heinrich first sponsored a demanding trial of production cars that ranged through Germany and Austria. After a mediocre performance that year, Porsche determined to do better in 1910, when the Prince Heinrich Trial would cover 1,200 miles with heavy penalties for faults and delays.

Its mandatory four-passenger bodywork was misleading, for his 1910 'Prince Henry' Austro Daimler was the first pure racing car from the pen of Ferdinand Porsche. He designed its every part for speed with reliability. Revving to 2,300rpm, its 5.7-litre overhead-cam four gave 95hp. In 1910 no engine developed more power per litre. Shaping the body to reduce drag, said Porsche, 'I have gone so far as to shroud every single nut so the air has nothing to catch on.'

Seeking safety in numbers, Porsche and Austro

Daimler managing director Eduard Fischer laid down ten of these radical new cars. Among their friends and business partners they found seven who would take up Prince Heinrich entries in addition to the three works cars that were allocated to Porsche, Fischer and the same Count Heinrich Schönfeld who had demonstrated Daimler's armoured car four years earlier.

Crushing entries from all Europe's leading producers, Austro Daimlers dominated the 1910 Trial. Porsche himself drove the winner, with Eduard Fischer second and Schönfeld third. By sweeping the top three places the 'Iron Team' – as it came to be known – put the team prize far out of reach. The Austrians won half the 16 prizes offered. To rub it in, Porsche also won outright Prince Heinrich's magnificent perpetual trophy, a solid-silver model of a touring car, trumping his rivals from Benz and Opel in a drawing at the prize-giving dinner on 9 June. Porsche thus added good luck to his formidable array of attributes.

Of most interest to Archduke Salvator on his 1910 visit to Wiener Neustadt was Porsche's progress in improving the M 08 tug. 'Our first attempts were directed more towards a cannon-carrying truck,' Porsche explained, 'but there was no valid connection between motor vehicle and gun. We came closer to the matter as we sought to negotiate the steep incline near Liesing at Hochrotherd with a mortar. A similar attempt with a mortar followed whereby we ascended the Kranichberg, which we managed quite well.

'We already recognised that it was important to enable the scaling of steep inclines through a special type of drive,' Porsche continued. 'Our trials experienced significant encouragement through the considerable interest which the Archduke Leopold Salvator took in the development.' His work successfully evolved into the M 09 by 1911. This became the staple pairing with Skoda's 240mm howitzer, type M 98, firing a 410lb round.

Both howitzers and mortars suited the appetites of the dual monarchy, encircled as it was by mountain ranges from the Tatras and Erzgebirge to the north and west, the Carpathians to the east and the Dolomites and Alps to the south. Such artillery was ideal for this troublesome terrain, on which Austria-Hungary expected her future battles to be fought.

In 1910 Ferdinand Porsche posed with some of the loot that he and his team-mates took away from that year's Prince Heinrich Trial. Austro Daimler's success in this competition elevated Porsche to the top of his profession.

Its power increased to 80bhp, by 1911 the M 08 tug evolved into the M 09, eight of which awaited collection by the army in Austro Daimler's courtyard at Wiener Neustadt.

Known as the 'Titan', the M 09 artillery tug was most often paired with Skoda's 240mm howitzer. From 1911 to 1913 a business transition made Skoda the controlling shareholder in Austro Daimler.

Another important mission for these heavy weapons was the demolition of forts. To the south in Italy and to the west in Belgium and France faith was placed in massive fortifications as fixed points of defence. Their formidable nests of cannon and machine guns posed daunting challenges to any would-be invader. Only heavy siege howitzers and mortars had the heft to take on and neutralise these emplacements.

The AD-Skoda pairing soon took on more concrete form. Allied though the dual monarchy was with Germany, the news that by 1912 Daimler of Stuttgart no longer held any shares in the Austrian company was welcomed by the factory's military customers. They wanted to keep their secrets safe at home. In that year, in fact, encouraged by the army, another Austro-Hungarian company started taking an interest in Austro Daimler.

In 1869 Emil Skoda, then 29, took over ownership of an expanding metalworks at Pilsen, west of Prague, that had been established in 1856. With its 150-strong staff the Skoda Works soon became renowned for its skill in casting and machining huge and complex components and assemblies. After the death of his knighted father in 1888, Emil became Baron von Skoda. The title passed to his son Karel when Emil died suddenly in 1900. Having completed his studies, Karel von Skoda took command of the company in 1909 at the age of 30. With its headquarters moved to Vienna, closer to military decision-makers, Skoda was Austria-Hungary's prime contractor for its heaviest weapons.

It was one thing to build huge mortars and cannon, but another to move them from place to place to cope with the exigencies of war. Here Karel von Skoda found a willing partner in Austro Daimler, producer of powerful tugs. Their co-operation soon led to consolidation. In 1911 the two companies declared a relationship, while behind the scenes Ferdinand Porsche acted on Baron von Skoda's behalf by quietly buying up parcels of shares in Austro Daimler.

Effective 8 July 1913 Skoda stepped front and centre as the largest and controlling shareholder of the Wiener Neustadt company. This marked the end of the influence of Eduard Fischer.* While Baron von Skoda placed his general secretary, Herr Burmann, in the same role at Austro Daimler, henceforth Ferdinand Porsche was the ranking operating executive, responsible only to his supervisory board. Just short of his 38th birthday he was in complete charge of one of Austria-Hungary's most prominent producers of both vehicles and war materiel with a staff some 1,000 strong.

The next challenge for Wiener Neustadt was Skoda's 305mm mortar, the M 11. With a muzzle velocity of 1,080ft/sec., this formidable weapon could hurl an 840lb shell at targets up to 6 miles away. Its development dated from 1906, when Skoda began work on a mortar capable of destroying the latest Italian emplacements at the request of General Franz Conrad von Hötzendorf, chief of the Austro-Hungarian general staff. After exhaustive testing two dozen such mortars were ordered at the end of 1911.

Designed by Skoda's Oswald Dirmoser and his brother Richard, the M 11 exploited Skoda's latest technology for heavy weapons. It used breech loading and recoil recuperation. Weighing as much as 55 tons when emplaced, the mortar broke down into three units for transport: the barrel, the gun carriage and the base, a steel box. This liberated the weapon from dependence on the railways, allowing it to be delivered and erected from roads in the most difficult terrain – provided that it had a suitable tug.

For this purpose Porsche's team began work in March 1912 on a new larger vehicle. This resulted in the M 12, nicknamed the 'Hundred' for the output of its new 20.3-litre six-cylinder engine, fitted with dual ignition and twin updraft carburettors. Like the M 08/09, the M 12 carried over the system of bevel gears on the kingpin, to drive its steerable front wheels, that had been introduced to Wiener Neustadt by Paul Daimler. The Hundred's centre differential could be open or locked, to choice, depending on the terrain.

Weighing some 17,600lb, the M 12 differed visibly from its predecessor in having double-tyred rear wheels that were much larger than those at the front. This helped it achieve a rating for towing of between 30 and 36 tons. Said to have been paid for by an army slush fund rather than a formal budget appropriation, the first such tugs as well as the mortars were ready in 1912. Forty such Hundreds were ultimately built to haul the equipment of eight artillery batteries deploying two mortars apiece.†

'Working hand in hand with the great cannon factory Skoda Works,' Porsche explained, 'and enthusiastically supported by the influential gentlemen of the Austro-Hungarian army administration, today's motorised batteries – which perform so brilliantly – gradually emerged. Our test drives were kept a total secret.

Enlarged and paired rear wheels were hallmarks of Austro Daimler's M 12, which was ready for action in 1913. Known as the 'Hundred' for the power of its 20.3-litre six-cylinder engine, the M 12 saw major service in the First World War.

---

* During the war Fischer took a fast Austro Daimler – possibly a Prince Heinrich – to the northern battlefields where he provided a courier service as a member of the Volunteer Automobile Corps.

† With three tugs nominally needed to tow the elements of a single mortar, the complement should theoretically have been 48 M 11s. Some of the weapons would have been deployed by rail and others sited in relatively static positions.

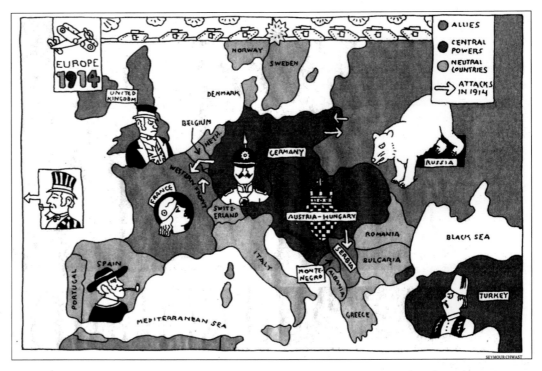

Combined, pre-war Germany and Austria-Hungary were the 800lb gorilla of Central Europe. Although substantial in size, the dual monarchy was weakened by the ramshackle governing structure of the Habsburgs.

During the German advance of August 1914, Skoda 305mm mortars brought rapidly to the western front by Austro Daimler M 12 tugs contributed to the utter devastation of the Liège forts, thought indestructible.

Thus it was possible for the motorised batteries to be tested and ready two years before the outbreak of the war.'

Secrecy had also been the watchword in the creation of the Skoda mortar. Only officers were allowed to accompany it to the test-firing ground at Hajmasker near Lake Balaton and to attend its trials. Indeed, these mortars and their tugs turned out to be a significant secret weapon when war broke out.

The fuse lit by Austria-Hungary's annexation of Bosnia-Herzegovina in 1908 exploded its bomb on 28 June 1914 when Archduke Franz Ferdinand was assassinated in Bosnia's capital Sarajevo by Bosnian Serb militants. Anticipating a Serb-inspired revolt and backed by their alliance partner Germany, the Habsburgs presented Serbia with a lengthy list of humiliating demands. It took only Serbia's inevitable rejection of these to trigger the war, which by August was well under way with fighting on the eastern and western fronts of what became known as the Central Powers.

In August 1914 both the Skoda mortars and their AD tugs played decisive roles in Germany's battles in the west. In accordance with the precepts of Field Marshal Alfred von Schlieffen, Germany attacked France through the neutral states of Holland and Belgium. The Belgians had committed their defence to an array of concreted forts around its key cities

Namur and especially Liège, which was encircled by a dozen forts deploying 400 guns. On 5 August the German forces began their effort to reduce these obstacles.

When the forts proved tougher than expected, even with 420mm Krupp weapons, the Germans swallowed their pride and called on Austria-Hungary for support. M 11/M 12 mobility was such that eight of them, as requested, were quickly on the scene. To the astonishment of international observers as well as the hapless Belgians, they soon made short work of the heavily armoured emplacements in what was one of the first great sensations of the war. Said *The Times History of the War*,

> Nothing which the Belgians could have hoped to do could have been of any avail against the overwhelming German numbers and the great guns which slowly lumbered up into position and to which the Belgians had no artillery that could hope to reply effectively, nor any fortifications that could offer resistance. According to eyewitnesses, nothing so terrible had ever been seen in war as the effect of the great shells fired into the Liège forts. Men were not simply killed or wounded; they were blackened, burnt and smashed. No wonder that three of the forts, although they had been expected to hold out for at least a month, surrendered within the week when the real bombardment began.

Porsche created a new 13.5-litre four-cylinder engine for his M 17 tug, using vee-inclined overhead valves to produce 80bhp at 800rpm. Appropriately dubbed 'Goliath', it entered service during the First World War.

Austro Daimler's works at Wiener Neustadt offered a challenging proving ground for an early version of the M 17 tug. On the heights Porsche, to the right in the group, oversaw its tests.

In front of one of the works billets at Austro Daimler in 1917 an M 17 'Goliath' towed a Skoda M 11 305mm mortar. Cleated wheels gave it traction to reach speeds of up to 9mph.

Both the weapons and their transport came as a stunning surprise to Belgium as well as to France and her allies. Tank historian John Glanfield called their impact 'the first tactical surprise of the war, the destructive power of German howitzers which demolished the Belgian fortresses of Liège and Namur in late August 1914. The 42cm monsters threw a half-ton shell, reducing in days defences intended to survive for months.'

Austro Daimler's contribution continued to be built through 1915. Improved variations of the tug were designated M 12/14 and M 12/15. The M 12/16, laid out in September 1917, was a dedicated cable-reel carrier intended for the Italian front. It scaled 13¼ tons at the kerb, two-thirds of which rested on its wide rear wheels, 59 inches in diameter.

An M 12 was challenged by a new-fangled caterpillar-tracked vehicle, an American Holt, at a war department trial in May 1914.* Towing an artillery piece on 'semi-swampy' terrain, the Austro Daimler vehicle's wheels sank into the mire after 500ft, while the tracked Holt completed the 750ft crossing. Austro-Hungarian officers, who had been similarly impressed by a Holt success against an Austro Daimler tractor near Vienna in October 1912, negotiated a deal for local assembly of caterpillar-tracked vehicles. This collapsed when the onset of war blocked shipments of needed components from the United States.†

Learning from feedback from the field, Ferdinand Porsche designed and produced another multi-purpose tug during the conflict. The M 17's emphasis was on efficiency, with power from a four-cylinder engine of 13½ litres using technology being developed in parallel for aero engines in the form of vee-inclined overhead valves. Producing 80bhp at 800rpm, it was governed to a maximum of 1,050rpm.

Weighing 10 tons on a wheelbase of 118.1 inches, the M 17 had four steel wheels 57½ inches in diameter with wide cleated surfaces to cope with the mucky conditions that the Holt had managed so well in 1914. Appropriately nicknamed 'Goliath', Porsche's M 17 could achieve 9.0mph in its topmost of four gears and was capable of towing the complete Skoda M 11 mortar single-handed. This was but the dawn of Porsche's contributions to First World War transportation.

His designs in co-operation with Skoda made Ferdinand Porsche one of the dual monarchy's leading engineers of ordnance. On the right at this gathering on the eastern front in 1915, Porsche commanded respect with his knowledge.

---

* The initiative was that of a Hungarian engineer and farmer, Dr Leo Steiner, who learned of Holt's machines and imported one from Stockton, California. Having become a Holt dealer in July 1912, Steiner negotiated exclusive selling rights for both Germany and Austria-Hungary.

† As a result of these negotiations initial supplies of Holt tracks and equipment would remain in Austria at the outbreak of war. They would have a role to play in the future of German tank design.

CHAPTER 3

AUSTRIA IN THE AIR

Eduard Fischer (left) *was an investor with Camillo Castiglioni in a producer of fabrics for the balloons and airships that were becoming interesting to the military. Castiglioni would be both an ally and an adversary for Porsche.*

Porsche was a board member of the company that produced Austria's first semi-rigid airship to the designs of August von Parseval. Austro Daimler made its 60bhp four-cylinder propulsion engine and the structure of its control car.

---

* The Österreichisch-Amerikanischen Gummi-Fabrikations-Gesellschaft.

Ferdinand Porsche was aloft early, not with powered craft but with an Excelsior gas balloon. In 1909 he had a number of flights, indeed adventures, learning about navigation in this new medium. Once he found himself blown almost to Hungary, while on another occasion he opened his luncheon package to discover he had taken his wife's newly purchased bodice instead. Struggling to gain altitude in another flight, Porsche had to throw everything overboard, including lunch, to avoid clouting a tower. His worst experience was a flight that threatened to soar too quickly and burst the balloon because a relief valve jammed. Porsche clambered up into the rigging and managed to free it before disaster struck.

The engineer's aviation interest was encouraged by Austro Daimler's Eduard Fischer, who joined him in these flights. Fischer was an investor in the Austrian-American Rubber Factory,* which made special rubberised fabric for balloons and airships. Its principal shareholder and investor was the short, round-faced Camillo Castiglioni, who was born in 1885 in Italian-populated Trieste on the Adriatic, then part of Austria-Hungary. At the age of only 25 in 1910 the free-wheeling Castiglioni was the principal of a venture that could take him into the automotive and aviation industries at a time of hectic growth for both.

Castiglioni's company supplied the fabric for Austria's first fully manoeuvrable airship, built to the designs of Major August von Parseval. Working in Berlin on non-rigid airships since 1906, the Bavarian von Parseval saw his first Austrian ship completed in 1909. Commissioned by the military, it was built at Fischamend, on Vienna's east side along the Danube, by the Austrian Motor Aircraft Company.* This was effectively a joint venture between the Castiglioni enterprise and Austro Daimler, with Ferdinand Porsche among its board members. Thus Austro Daimler had a major role in its construction, including not only its engine but also the tubular structure of its four-passenger control car.

For the 11½ft pusher propeller of the Parseval, as the 160ft semi-rigid airship was known, Porsche provided a unique four-cylinder T-head engine weighing, with its radiator and accessories, 880lb. Its cylinders were individual castings, spaced to allow the crankshaft to have five main bearings. Each crank throw inhabited its own sump so lubricant would be available whenever the airship and its engine were at odd angles. With dual ignition its output was 70bhp. The four's flywheel served as the fan for its own radiator. It also drove, through a belt, a blower that helped inflate the hydrogen-filled gas bag. Great care was taken to isolate the engine from the inflammable gas, including triple enclosures of both inlet and exhaust piping. A chain coupled the engine to the slower-running propeller.

The 1911 semi-rigid Lebaudy airship's two propellers were driven by a single 100hp Austro Daimler four. Porsche handled its engine's controls during its first major overland flight.

The Parseval's first venture from its vast construction hall at Fischamend was on 26 November 1909. Ferdinand Porsche himself tended its engine during a 20-minute jaunt. Two days later, on a Sunday, it caused great excitement by rumbling above the capital, rounding the spire of St Stephan's Cathedral. In September of the following year it was one of the stars of Wiener Neustadt's aviation meeting, fighting a headwind to cover the 26 miles from Fischamend. The Parseval was in service until 1914, when the cost of making needed repairs was judged uneconomic.

The Austrian Motor Aircraft Company built two more large airships. The first was to a French design originated in 1902 for the Lebaudy brothers. Driving two twin-blade propellers, its Austro Daimler engine was akin to the Parseval's but uprated to 100bhp. Measuring 230ft in length in its final version, the Lebaudy machine was distinctly sharp-nosed. In March 1911 it made a successful flight to Linz with Ferdinand Porsche at its engine's controls. Also built to the order of the military, the Lebaudy airship was mustered out of service in 1913.

The third Austrian airship, the largest of all, was an ambitious private venture by engineer Hans Otto Stagl and First Lieutenant Franz Mannsbarth. Its 300ft length was subdivided into four interconnected gas chambers. Two gondolas with a walkway between them carried 150hp Austro Daimler engines, still four-cylinder units with individual T-head cylinders. Each drove a pair of 13ft propellers. From each engine a shaft and bevel gears drove another propeller, whose angle of attack could be adjusted in flight for control purposes.

Dubbed the *Austria* and built with the aim of commercial service, the Stagl-Mannsbarth airship could carry 25 passengers. Although capable of altitudes in

* The Österreichische Motor-Luftfahrzeug Gesellschaft.

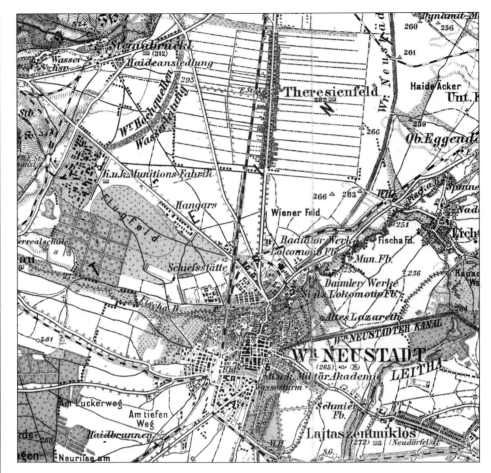

North-west of the centre of Wiener Neustadt was the Flugfeld (flying field), established near a government munitions factory in mid-1909. It could scarcely have been sited more advantageously for Austro Daimler.

excess of 8,000ft and a 20-hour mission at a speed of 37mph, no commercial contracts or military interest came its way. These non-rigid airships had proven difficult to control and prone to hydrogen leakage. In addition, the army quickly realised, they were an attractive target for ground fire. From 1912 the military sourced only heavier-than-air craft. In 1914, after 56 flights, the Stagl-Mannsbarth machine was dismantled.

Defects in these machines, rather than lack of interest in aviation's potential, marked the attitude of the dual monarchy's General Staff. Its chief since 1906 was General Franz Conrad von Hötzendorf, who as early as 1907 advocated the addition of airborne forces to the Austro-Hungarian arsenal. All he lacked was successful native aviators. Thanks in part to Conrad's encouragement, these soon stepped forward.

Among Austria's pioneer flyers Igo Etrich stood out. Etrich, whose father Ignatz was a textile manufacturer in Bohemia's Trautenau, began essaying flight with a Lilienthal glider in 1898. He took up the idea of a fellow Austrian, Franz Wels, who held that the uniquely stable gliding properties of the seed of Java's zanonia vine would provide a suitable basis for the design of an aircraft. Starting in 1904, their experiments with gliders and powered models were promising. In February 1906 they jointly applied for a patent on a zanonia-seed flying wing that would be steered by varying the speeds of its two propellers.

Igo Etrich was the first to rent hangars along the north side of the new flying field. Bowler-hatted Porsche looked on as a monoplane powered by one of his new four-cylinder engines was readied for flight.

Planning as they were for powered flight, Etrich and Wels needed an airfield. They identified an open area of the Steinfeld, on the north-west periphery of Wiener Neustadt, as eminently suitable. Its open spaces were already well known to the military, which used some of its acreage as an artillery test range. Their request for approval went as high as General Conrad and even to the parliament of Lower Austria. Lodged in May 1909, a formal request from the military to the mayor and city council resulted in the latter's 11 June decision to build and rent hangars for what would be the dual kingdom's first official field for heavier-than-air craft.*

For the engine of his victorious 1910 Prince Heinrich racer Porsche, driving here, explored new and advanced design concepts that contributed to the merits of his aero engines.

---

* The field at Wiener Neustadt anticipated by four months the first flight on the Simmering Meadow on Vienna's eastern periphery. This took place on 23 October 1909 when Louis Blériot, fresh from his historic first crossing of the English Channel on 25 July 1909, baptised the new field in front of the Emperor and 300,000 onlookers during his celebratory tour of European capitals.

First to rent a double hangar was Igo Etrich, who moved his manufacturing operations there from Vienna. By the end of the year four more of the seven hangars built were taken by the newly formed Military Aeronautical Institute. With the loan of early aircraft by private donors, aviation in the dual monarchy was up and running. Fortuitously Austria-Hungary's new air age flew directly to the doorstep of Ferdinand Porsche.

Franz Wels and Igo Etrich parted company in 1909, when the latter made a short, straight hop with his first man-carrying aircraft on the Steinfeld in July. He replaced its Anzani engine with a Clerget and on 29 October 1909 made a proper controlled flight, the first ever for an Austrian aircraft.

Igo Etrich pioneered aeroplanes with the monoplane layout at a time when most inventors still used multiple aerofoils. He called his 1909 monoplane *Sperling* or 'Sparrow'. For 1910 he undertook a major redesign in a larger aircraft he called the *Taube* or 'Pigeon', whose elegant design was the subject of a patent applied for on 11 September 1909. Assisting him in its erection was Karl Illner, a skilled mechanic and locksmith who had an aptitude for flight.

In April 1910 Illner qualified for Austria's third pilot's licence and in May he made a 68-minute flight in the *Taube*.* This was achieved with the French Clerget engine, but Etrich – in addition to aiding Porsche with the streamlining of his Prince Heinrich entries – had already been in touch with the Austro Daimler engineer about more powerful and reliable engines to power his aircraft.

Etrich and Illner were eager to have an Austro Daimler engine for their new *Möwe* or 'Gull', a smaller monoplane built expressly for racing. The creation of an all-new aviation engine was yet another task for Ferdinand Porsche in the hectic early months of 1910 when he was readying his Prince Heinrich Trial entries.

The engine he created for these successful competition cars was a major advance not only for Porsche but for power units in general. Measuring 5.7 litres in capacity, it used a shaft-driven single overhead camshaft to open inclined overhead valves, in hemispherical combustion chambers, through rocker arms. Its cylinders were individual iron castings with integral heads while its crankcase was aluminium. It produced 86bhp at 1,900rpm and reached its maximum output of 95bhp at 2,100rpm.

One of these Prince Heinrich engines, weighing 660lb, was built as a flight version for Igo Etrich's entry in a June meeting at Budapest when Porsche's dedicated aero engine was still incomplete. Already equipped with dual ignition, it was revised to produce 40bhp at 1,450rpm. This performed well in the air but the *Möwe* suffered an accident on the ground in Hungary and had to be withdrawn.

Finally at the end of August the first pure-bred Austro Daimler aero engine was ready for installation. Like Porsche's Prince Heinrich engine, it was an appealing and ingenious power unit that showed his meticulous attention to detail and acute awareness of the state of the current engine art.

His first airborne four-cylinder displaced 3.8 litres and weighed a svelte 180lb. Although originally planned for 35hp, it ultimately produced 48 at 1,600rpm. Closely following the first engine was a larger four measuring 120 x 140mm for

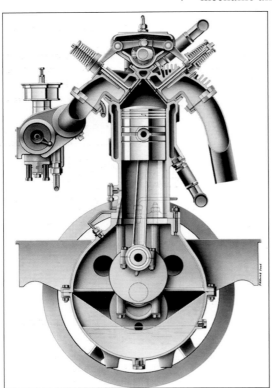

Wolfgang Franke's drawing shows the rocker arms operating the valves of Porsche's Prince Heinrich four, unusually with valve springs above them. Cylinders are offset from the crankshaft centre in the desaxé manner.

---

* First to be granted a licence was Adolf Warchalowski, who learned to fly in France and had his own Farman. Second was Alfred Ritter von Pischof, licensed on 24 April 1910 after a flight in an aircraft of his own design.

Porsche placed drives to the camshaft-driven water pump, visible, and magneto of his 1910 four-cylinder aero engine at the centre where they would not interfere with its installation in a variety of airframes.

Desaxé cylinder placement is seen in the cross-section typical of Porsche's aero fours and sixes, as are the ultra-thin water jackets and single-pushrod valve gear with cam lobes that gave positive control of the mechanism.

6.3 litres. This developed 65bhp at 1,350rpm and a maximum of 70 at 1,500rpm. It weighed 210lb. Initial production was chiefly of the larger engine.

Novelties abounded in Ferdinand Porsche's first heavier-than-air engines, which were similar in design. He gave them four individual cylinders on an aluminium crankcase, with six bearers for attachment to the aeroplane, in which white-metalled main bearings were held by individual caps. This allowed the aluminium bottom cover to be a simple oil pan that could be removed for inspection, in the aircraft, without disturbing the bearings.

The main and big-end bearings were small, in the order of 40mm, to allow the crankshaft to be exceptionally light. Pistons were thin-wall cast iron. Porsche offset the cylinders from the crankshaft centreline towards the downstroke, right-hand side of the engine by 18mm, in the manner known as *désaxe*, to reduce piston side thrust and encourage smoother running. Adjacent to the centre main bearing Ferdinand Porsche arranged a spiral drive from the crankshaft to a cross shaft which drove the water pump on the right and the ignition on the left. His reason for this was to get these accessories out of the way so that the aeroplane designer could take a drive from either or both ends of the engine as he desired.

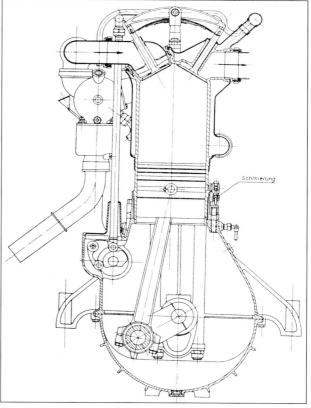

Ignition was a combined magneto and coil unit to spark two plugs per cylinder. A single camshaft ran inside the crankcase on the right, with its gear drive taken at the normal propeller end of the four. A bevel drive from the tail of the camshaft turned the Friedmann lubricator – a favourite Porsche accessory – on the engine's left rear quarter.

At its top end Porsche's first heavier-than-air engine resembled his 1910 Prince Heinrich design only in having inclined overhead valves in a hemispherical combustion chamber. A direct antecedent of the cylinder head of Porsche's four was seen in Fiat's Grand Prix racing car of 1905. This placed two large vee-inclined valves in the head, equally disposed at a 60-degree included angle.

Valve stems could be short and light because no closing coil spring was needed. Instead a single leaf spring was mounted atop the head, curving down at both sides to effect the closing of both valves. Above this, on a central pivot, was a single long rocker arm that tilted back and forth to open both valves. Unlike Fiat, Porsche ingeniously anchored the leaf spring to the bottom of the rocker arm in such a way that its effect was stronger on the closed valve and lesser on the valve that was being opened.*

The rocker arm was operated by a vertical rod attached to a pivot on the engine's right or inlet side. The vertical rod functioned in both tension and compression. When it was pushed up by the tappet at its bottom end the rocker opened the exhaust valve, and when pulled down it opened the inlet valve. In Fiat's patented design the pulling down, against the leaf spring's pressure, was achieved by an even stronger coil spring, at the tappet, that pressed the latter against the cam lobe. Here too Porsche found a much more elegant solution. He controlled the push-pull rod with two cam lobes side-by-side. With this desmodromic control one cam lobe pushed the rod upward while another, working through a bell crank, pulled it down.

To take advantage of its good bore-surface properties, fine cast iron was used for the cylinders. While the exhaust-valve port and guide were integral with the head, the inlet valve was carried by a screwed-in cage. When this was removed, the exhaust valve could be dismantled without having to tear down the cylinder.

During the evolution of subsequent versions of this design Ferdinand Porsche found that the iron that was ideal for the bores was brittle enough that the flanges at the cylinder bases could break away. He invented and patented the idea of a screwed-on steel collar at the cylinder's base that provided a more secure anchorage. Lest the screw threads be a source of failure, he bathed them in molten tin or zinc before assembly to provide a cushioning effect.

The iron cylinders were cast without water jackets. Porsche's process for forming the jacket began with the casting on the cylinder's outer surface of a volume that represented the planned water capacity. The cast-on material was metallic, an alloy that melted at less than water's boiling point.† Then the cylinder was suspended in an electrolytic bath of copper until a thickness of 1½mm had been deposited (about 1/16 of an inch). Finally in boiling water the internal metallic former melted away. The copper water jacket remained.

Icing of the inlet manifold was prevented by a warm-water jacket around it at the carburettor inlet. The carburettor itself was a special Porsche design for aviation. Instead of a conventional float to control fuel flow, he used a ring-shaped float in a bowl that surrounded the jet. This ensured a more consistent supply of

---

* Austro Daimler was granted a patent on this innovation.

† The author has not been able to identify the specific alloy used. It may well have been a variant of Wood's metal, a mixture of 50% bismuth, 25% lead, 12.5% tin and 12.5% cadmium that melts at a temperature of 158° Fahrenheit (70° Celsius).

fuel with the plane at different attitude angles and g loadings. Ahead of the cockpit an inverted-vee radiator also served as a pylon for guy wires to an aeroplane's wings.

Here was an engine prepared for its mission with great care. When Vienna's *AAZ* visited Wiener Neustadt in September it also saw the gondolas and engines of the Stagl-Mannsbarth airship under construction and the first elements of a new electric-drive vehicle train being completed. Elsewhere were conventional vehicles, of course, including a wheel-motor chassis for the Vienna fire department.

Small wonder that the *AAZ* reporter was moved to remark, 'One can see that progress has set out its stall in the Daimler works. Rather than get bogged down by routine, one keeps abreast of the times.' This was a tribute to the company's uncommonly versatile and forward-looking technical chief.

Unlike some of his aero-engine rivals, who produced what were known as 'five-minute wonders', Ferdinand Porsche placed heavy emphasis on durability. Before committing his new units to the air he ran them with the load of a propeller for 10 hours. In pole position to receive one was Igo Etrich, who fitted it to his *Möwe* racer. Its first test hops began at 6:00 a.m. on 28 August 1910 with Karl Illner as pilot.

Illner and Etrich were happy with 28 minutes of air time, but this was not enough for Porsche. He needed his engine to be tested more thoroughly. Illner went up again for 31 more minutes. A final hop at the end of the day was cut short by a trail of smoke. Illner landed safely, to find that a mechanic had forgotten to top up his engine's water supply. The Aëro-Daimler showed its mettle by completing the flight undamaged.

A few days later another flyer, Adolf Warchalowski, received his engine and fitted it to his biplane. Previously, he said, six men could easily hold his plane against its engine's full power. This was no longer possible. Both his aircraft and the *Möwe* – though its propeller was not yet adapted to the new engine – reached 50mph speeds. Warchalowski flew successfully with a passenger, the intrepid Count Schönfeld.

Powered by Porsche's 3.8-litre four-cylinder engine, this Etrich Type IV was the first completed by the Jacob Lohner company. Karl Illner (right) piloted the yellow Taube on its first overland flights in October 1910.

Although still in their baby shoes, Porsche's aero engines were prominent at Wiener Neustadt's 17–19 September 1910 flying meet. On the 19th the gathering was graced by a visit by Kaiser Franz Josef, who examined the machines with great interest and then met the flyers. Seeing Porsche among the dignitaries he asked,

'You now have a great deal to do with aviation?'

'Yes indeed, your majesty,' the engineer answered. 'The four best pilots at Wiener Neustadt are using our engines, and the orders are so numerous that we can only fill them with difficulty.' This was putting it mildly, because the crankshaft in a competitor's engine on the Sunday had only been machined the Wednesday before.

So comprehensively did Aëro-Daimler-powered aircraft sweep the board at the three-day Wiener Neustadt meeting that the *AAZ* concluded that 'since the engine is the heart of an aeroplane, one can well say: these aircraft have won thanks to their Daimler engines. One can speak of a glittering success for Austro Daimler in Wiener Neustadt's flying meet without being in the least guilty of overstatement.'

Messrs Illner and Warchalowski were the fastest overland flyers. Warchalowski reached the highest altitude of 1,500ft and Illner made the longest flight at 3 minutes

Porsche drove to Horn on 10 October 1910 to oversee engine checks before Karl Illner's return flight to Vienna. It was the date – 10/10/10 – on which practical aviation arrived in Austria-Hungary.

short of 2 hours. Both height and duration were new Austrian records. In May 1910 Karl Illner made Austria's first overland passenger-carrying flight.

Confirmation, if it were needed, of the merit of Ferdinand Porsche's new engine came on 10 October when Karl Illner accepted the city of Vienna's challenge of a 20,000-crown (£800) prize for a flight from its Simmering Heath to Horn, 50 miles to the north-east, and return. This went flawlessly, taking 74 minutes for the outbound run and 69 for the return. Illner, in a yellow Etrich Taube, set a new Austrian altitude record of 3,300ft on the way.*

Although he missed the start, Ferdinand Porsche arrived at Horn in his Prince Heinrich tourer in good time to supervise the checking of his big four. Back in Vienna the Taube had already landed when the engineer arrived at Simmering to find a crowd still celebrating the feat. From it erupted the burly figure of Illner, who embraced Porsche with some emotion and said, 'Herr Director, I thank you. Your engine made this performance possible. My life depended on its robust operation.' A surprised Porsche was visibly moved by these unexpected yet sincere plaudits. He and his engine had conquered the new medium of the air – and in style.

With twin magnetos, one of Austro Daimler's aviation engines was on outdoor test in 1911 attended by Otto Köhler, chief designer under Porsche. Köhler would be an important ally through the 1920s and in the Second World War as well.

Easy to build and relatively safe to Fly, Igo Etrich's Taube was the most popular pre-war aeroplane in Germany and Austria-Hungary. In the latter the military authorities set demanding standards for acceptance. Engines had to prove an ability to run continuously for 6 hours under load as well as a minimum of 2 hours airborne with a 300lb payload. At least 45mph had to be attained. Airframes had to be dismantled within an hour for ground transport and reassembled in no more than 2 hours. Etrich and Illner took their Taube apart in 8 minutes and erected it in 30. A flight of 2½ hours proved the credentials of the Austro Daimler four.

Berlin's Edmund Rumpler was a prominent German licensee for the Taube, while in Austria Etrich arranged with none other than Jakob Lohner & Co. to produce his aircraft. To guide his aviation programmes Ludwig Lohner engaged an autodidact engineer, Karl Paulal, who had been at Lohner in the Porsche years. In 1901 he had raced an electric at Semmering and in 1905 he accompanied Count Schönborn when he drove his Lohner-Porsche from Vienna to Breslau and return.

Under Paulal's guidance Lohner developed its own swept-wing biplane, the *Pfeilflieger* or 'Arrow-Flyer', with Aëro-Daimler power. This became the successor to the Taube as the preferred service aeroplane for Austria-Hungary. Lohner also became Austria-Hungary's leading propeller manufacturer.

Lohner-built Taubes were sold and serviced by the Austrian Motor Aircraft Company, entering this new field in addition to its airship projects. As its base for this activity the latter company took over another aeroplane works on the Wiener Neustadt airfield, complete with some of Etrich's earliest planes. On 2 April 1911 Austria-Hungary bought its first Taube for military use, to be shared among the four soldiers who had qualified as pilots.† It was powered by Austro Daimler's larger 6.3-litre four.

---

* An obelisk memorialising this flight still stands in Horn. Illner had originally attempted the round trip on 3 October, but the weather closed in and forced him down near Krems. Some damage on landing had to be made good before he could return to Vienna to make his second and successful attempt.

† This is thought to have been the Lohner-built Etrich IX.

In his characteristic bowler Porsche observed flight preparations with one of his engines on the Steinfeld in 1911. In the craft's rearmost seat was diminutive aviatrix Lilli Steinschneider.

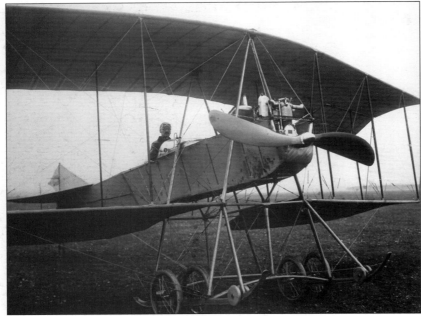

The handsome biplane of Austrian pioneer Adolf Warchalowski gained improved performance with the installation of an Aëro-Daimler four in 1910. Here in 1911 it was piloted by Captain Hans Umlauff von Frankwell.

To drum up interest among potential pilots, the dual monarchy's first aeroplanes were demonstrated in the empire's southern reaches at Görz, near Udine, and the airfield at Pula, principal port for the Austro-Hungarian navy. Early incentives included a joining reward of 1,000 crowns plus 600 crowns for necessary flying garb, followed by 15 crowns a month for aircraft maintenance. After making a cross-country flight of at least 62 miles at an altitude of more than 1,640ft the flyer received 2,000 crowns and a 'campaign pilot' rating.

After the successful proving of its first Etrich Taube, the War Department bought seven more of the Lohner-built monoplanes at 25,000 crowns apiece. Although advances in aircraft design during the First World War were so rapid that the Taube was soon outdated as a combat plane, it was still valuable in a reconnaissance role. It had one signal military achievement to its credit. Several Aëro-Daimler-powered Taubes were sold to Italy, which in September 1911 opened hostilities with Turkey over its African holdings.

Turkish troops at Ain Zara and the Oasis of Jagiura in Libya, then Tripolitania, were startled on 11 November 1911 to see and hear a Taube overhead. They were even more discomfited when its pilot, Commander Gavotti, dropped four 2kg Cipelli grenades in their midst. Further raids by the Lohner-built Taube with Cipellis and Swedish hand grenades followed this, history's first aerial bombing in wartime.

Porsche's 13.9-litre six of 1911 was a landmark prime mover in his career and in the history of aviation power. Viewed here on its exhaust side, it had twin updraft carburettors. The favoured Friedmann lubricator was above its magnetos.

A cutaway view of the 1911 Austro Daimler six showed not only its Friedmann oiler's internals but also the appealing simplicity and lightness of its valve gear with a single rocker arm and leaf spring above each cylinder.

Another pioneer using four-cylinder Porsche power was Alfred Ritter von Pischof, trained as an engineer, who designed his own gliders and, in 1910, his 'Autoplan'. This was an ingenious high-winged monoplane with its AD four in front driving a mid-placed propeller behind the pilot. He installed a clutch between engine and propeller to eliminate the need for helpers to hold the Autoplan before take-off. Von Pischof made several noteworthy flights in his twin-rudder Autoplan, which reached modest production.

Ferdinand Porsche's aeronautical four found other applications in these pioneering years. In 1910 it powered France's Deperdussin monoplane, driving a six-bladed Rapid-patent propeller. The four continued to be offered into 1914, with its power increased to 68bhp at 1,400rpm and 66bhp at its cruising speed of 1,350rpm. This 255lb engine was available in the UK for £495. The lesser figure of £315 obtained the 180lb 40hp four, still in peacetime production.

Responding to the obvious need for higher power in military aero engines, Ferdinand Porsche's next step took him into aviation's hall of fame. First he added two cylinders to his larger aeronautical four to produce a six. Several engines of exactly this type were built, sharing the same cylinder dimensions with the big four to give 9½ litres and 90bhp at 1,300rpm. This was an interim development along the road to Porsche's larger six of 1911, which was ready astonishingly quickly. From its capacity of 13.9 litres, the larger 419lb six initially produced 120bhp at 1,200rpm, soon raised to 130bhp.

All aspects of the six's construction were akin to those of the successful four. For greater stiffness to suppress torsional vibration, the crankshaft's bearings were enlarged to 45mm and given hollow journals for lightness. Bevel gears at the back end of its crankshaft drove two magnetos at the sides, a water pump at the bottom and, at the top, the Friedmann oiler. The latter had a new task, which was to meter oil to the bottom of each bore on the thrust side of its piston. Gradually introduced at Wiener Neustadt were less-time-consuming techniques for forming the water jackets that used thin sheet-metal fabrications, initially of copper and later steel.

Still exposed atop the cylinders, Porsche's single-rocker valve gear was unchanged in principle. Though a disadvantage of the design was that no overlap could occur at top dead centre between exhaust-valve closing and inlet-valve opening, the generous apertures of the 65mm valves gave ample gas flow. Front and rear trios of cylinders were fed by separate manifolds, each with its own updraft carburettor.* As with the fours, coolant flow in and out of the cylinders was concentrated on the exhaust side.

Experts in the field endow this pioneering Austro Daimler six with landmark status. 'Reliable and efficient,' wrote authority Bill Gunston, 'these engines are generally regarded as the inspiration – some have said the prototypes – of the many thousands of engines made for the German and Austro-Hungarian aircraft during the war, by Mercedes, Benz, BMW, Hiero and others. From the start the modest rpm and general stress levels resulted in good reliability.' 'In some respects,' seconded engine expert Fred Starr, 'the 1912 Austro Daimler water-cooled "Six" can be regarded as a forerunner of all in-line types subsequently built.'

'At the outbreak of the First World War,' said aviation historian Herschel Smith, 'the world's most efficient and reliable aircraft engine was the 120-hp Austro Daimler. This water-cooled in-line six, the work of Dr. Ferdinand Porsche, was the prototype of all the Central Powers wartime engines and of many built elsewhere.' 'It was an outstanding six-in-line,' added engineer Kenneth A. Hurst. 'Between 1911 and 1914 it was one of the best-selling engines and it set a standard for aero-engine design for years to come.'

Porsche's six soon showed its mettle in flights to high altitudes. In 1913 the height of 16,436ft was reached in a Lohner aircraft with two passengers. At Leipzig in the following year the world's altitude record was unofficially surpassed by Oelrich in a flight to 24,800ft in an all-metal DFW biplane – still with the nominal 120bhp engine. This was clearly an exceptional engine, on the scene none too soon for European tensions foreshadowed a conflict in which military aviation would be born.

All the European powers struggled to get to grips with their aviation responsibilities at the outbreak of war in August 1914, none more than Austria-Hungary. Where, for example, would its air resources best be mustered? Envisioned was fighting in the widely dispersed and disparate fronts of the Balkans, Italy and Russia. Deciding on the right equipment and its timely deployment was no easy task.

Among the many dual-monarchy aircraft powered by Austro Daimler's early sixes was Aviatik's B series of scouting and training biplanes, built at Vienna as an offshoot of the design's German originator.

---

* Some engines for particular installations had single to double-throat downdraft carburettors.

Exemplifying Austro Daimler's aviation involvement is the sight of the fuselage of an Etrich aircraft being towed through the works at Wiener Neustadt in 1910. Ferdinand Porsche stood in the foreground with a case of drawings.

Nor was the dual monarchy's structure entirely up to the job. The testing and sourcing of engines and airframes was assigned to the Fliegerarsenal, the authority that had done the job from the start, under the civilian Ministry of War, to be formalised in that role in March 1915. Described as having 'an extremely complex bureaucratic structure', the 'Flars' was a power unto itself. Fortunately for Ferdinand Porsche, he and his allies at Lohner had made their bones with the Flars at an early stage.

Change came to the Austrian Motor Aircraft Company with the arrival of Skoda as an active partner of Austro Daimler at Wiener Neustadt. This led to a restructuring of the company in 1915 as the Austrian Aircraft Factory AG with ÖFFAG as its acronym.* Both Karel von Skoda and Ferdinand Porsche had shares in the firm, which was managed by Karl Ockermüller. It became a producer of indigenous designs by Theo Weichmann and Josef Mickl, as well as the successful German Albatros fighters, under licence, with Austro Daimler engines.

An elegant design by Ernst Heinkel, the Hansa-Brandenburg B 1 was produced at Fischamend under licence with Austro Daimler power. It was the dual monarchy's most popular training biplane.

Hailed on its introduction in 1911 as the finest aero engine of its class, Austro Daimler's 120hp six was soon widely adopted. It powered numerous early Lohner types as well as aircraft made by the Austrian Aviatik Works at Vienna, where Karl Illner was in initial charge. Among its productions with the AD six were three bomber training planes with twin engines as well as designs by Julius von Berg. Budapest's Hungarian Lloyd Aircraft Factory used the 120hp six in its second prototype of the summer of 1914, designed by technical director Tibor Melczer.

Fischamend, the site of dirigible design and construction, also took on a military role with its designation as a proving ground for the Army's first Etrich Taubes and other advances. In May 1914 it achieved the first radio transmission in Europe between an aircraft and the ground. In 1913 Fischamend became a repairer and eventually producer of aeroplanes, including Lohner and

* The Österreichische Flugzeugfabrik AG.

In 1912 Porsche, with bowler, conversed with Etrich works director Karl Illner (in flight suit) during trials of one of his engines in a Lohner-built biplane. The aircraft was named Pfeil or 'Arrow' after its swept-back wings.

A 1911 Austro Daimler advertisement celebrates not only successes on land but also the triumphs of Aëro-Daimler, including overland flights and records for altitude including that with a passenger.

Brandenburg designs. The latter included many with Austro Daimler's new six.

Thus those aeroplane makers who were active before the war were helped into the air by Austro Daimler's first six. Soon a domestic rival was on the scene, based in Vienna at a firm founded by pioneer Alex Warchalowski. This was Otto Hieronimus, four years younger than Porsche and his most creative home-country competition on land and in the air. His 'Hiero' engines roughly paralleled those of Austro Daimler. Neither firm ventured into the light, air-cooled rotary engines popular in France, Germany and Britain because Austria-Hungary lacked access to the castor-based oils that were thought essential for reliable operation of the earliest types.

At the head of the queue for Ferdinand Porsche's new six was friend and aeronautical pioneer Igo Etrich. One of his Taubes was among the first aircraft to have the 120hp engine. With exceptional audacity Etrich prepared this plane for the *Daily Mail* Circuit of Britain Air Race, starting from Brooklands on 22 July and covering the country counter-clockwise, including Edinburgh and Glasgow in the north.

Uniquely among all the entrants having a crew of two, Lieutenants H. Bier and C. Banfield, the Taube impressed by virtue of its substantial size, elegance and the distinctively throaty roar of its Austro Daimler six. Although it retired from the race early, owing to damage from a hasty descent beyond Hendon with radiator problems, the Taube drew kudos. *Flight* considered that it 'is universally ranked among the most successful and most scientifically designed of aircraft. Its appearance in England was a revelation to our English constructors, and its influence will doubtless have an effect on current designs.'

One person who paid close

attention during the *Daily Mail* race was Samuel Franklin Cody, an American who made the first flight in the United Kingdom with his self-built aircraft in 1908.* Cody was on his third aeroplane, already much rebuilt, in August 1912 when the British military staged trials of potential designs for the Royal Flying Corps on Salisbury Plain.

In search of more power for a biplane that rivals called 'an obsolete monstrosity', Cody obtained a six-cylinder Austro Daimler in April 1911. In the trials, said Bill Gunston, 'its completely reliable 120 hp carried all before it' and won for the American the £5,000 first prize.† The magnitude of that award may be appreciated in relation to the price of the six, which was £850. Gunston was not the only person to attribute Cody's success to his use of a superior engine. Cody testified that he had flown a complete season with the engine with no overhaul.

At Vienna's flying meeting at the beginning of July 1912 an AD six took Lieutenant Blaschke and a colleague to an altitude of 13,976ft, well above the official record held – naturally – by a Frenchman. When the Austro Daimler engines were shown at London's Olympia in February 1913 they drew appreciation, which was not lessened when Karl Illner climbed to 16,436ft on 18 June in a Lohner Pfeilflieger.

Headed by Francis M. Luther, Austro Daimler's London agency at 112 Great Portland Street happily took orders for more such engines. In its developed form Tommy Sopwith's Bat Boat, Britain's first practical amphibian aircraft, had Austro Daimler power. His tractor biplane was already powered by an AD six in early September 1913 when Mr and Mrs Winston Churchill arrived at the Admiralty flying ground at Hamble near Southampton.

Lieutenant Spenser Grey took the Bat Boat's controls for a flight of about a quarter-hour, carrying the First Sea Lord and a colleague. Wearing a cap and mackintosh, his wife of five years Clementine then took her turn for another flight, almost as long, while her husband paced fretfully. 'It was beautiful!' she smiled on her return to the ground. Porsche's handiwork had been relied upon for two important aeronautical debuts.

Another user of the engine was Britain's Royal Aircraft Factory. The six powered its FE 6, a pusher-propeller prototype. Another of its products, the RE 5 biplane, began life with AD sixes. In the spring of 1914 Norman Spratt flew one to 18,900ft according to his barograph, thought a record for the British Isles.

War was actually under way when Noel Pemberton-Billing disposed two Austro Daimler sixes ahead of the pusher propellers of his Pemberton-Billing PB 29E. This was an extraordinary quadruplane – four wings stacked from top to bottom – created to loiter in the British skies to intercept German airships. It carried a searchlight and machine guns to mow down the dirigibles being used for

Seen from its exhaust side, Beardmore's version of the Austro Daimler 120bhp six followed the proven original closely. In 1914 it launched Glasgow's Beardmore into the aero-engine business in time for the First World War.

* Cody didn't discourage those who thought he was related to the famous 'Buffalo Bill' Cody, but in fact his name was originally Cowdery. He changed it in honour of his boyhood hero.

† Some reports say £10,000. The result was much to the credit of Porsche's engine. However, Cody's 'weird biplane' was in fact outperformed in the trials by a government-built aircraft, not eligible to compete, piloted by Geoffrey de Havilland. His was the design that ultimately prevailed.

bombing missions over England. It had not been long under test at Chingford before a crash ended its career.

War between the Entente and Central Powers did not deny British aeroplane makers their supplies of Porsche-designed sixes. On the wings of an order from HM Government for two dozen engines to power RE 5 bombers, at the end of October 1913 Glasgow's Beardmore and Company announced that it had snared exclusive rights to build and sell the 120bhp engines.* Its board ratified the agreement on 17 November. While other foreign makers of aero engines had spurned British enquiries, Wiener Neustadt had responded positively to this welcome source of licensing income.

With the support of subcontractors, the actual manufacture of the engines took place in Dumfries at Arrol-Johnson, also a property of industrialist William Beardmore. Otto Stahl and a cadre of Wiener Neustadt engineers journeyed to Britain's north to help the Scots master the production techniques. 'At Arrol-Johnson in Dumfries,' said Stahl, 'I set up the production of the first engines, which were entered in a competition of the English military authorities and thus had to be made in England. As far as I know the engines for this company were the only ones produced in England before the start of the war.'

Just before the conflict's outbreak, in May 1914 the Royal Aircraft Factory began an elaborate test of aero engines in the grand Naval and Military Aeroplane Engine Competition – the trials to which Otto Stahl referred. Twenty-six engines arrived but only half of them proceeded to the actual proving of power, economy and efficiency. Of these, eight survived a 6-hour eliminating trial to be assessed in the final rating.

Announced on 15 October, the findings resulted in an award of £5,000 to Britain's Green Engine Company as the winner for the performance of its 100bhp six. The other finishers also received awards, including £200 to the Beardmore Austro-Daimler Engine Company. In spite of this positive finding, Gustavus Green's engines found relatively little wartime use, their designs being quickly outpaced by the advances of others.

With the onset of war the company name was changed to Beardmore Aero-Engines Ltd. Caught on the hop in Dumfries, Beardmore's borrowed Austrian engineers were interned. Barred from access to Porsche's later designs, the company had to shift for itself, although one of Beardmore's directors, Thomas Charles Willis Pullinger, fancied himself an engine designer. Following detail changes the company's first major step was to enlarge the Austro Daimler's bore to bring its capacity to 16.6 litres. Output rose to 160bhp and ultimately to 185bhp with twin carburettors feeding both variants.

Beardmore-built sixes found a home in one of Britain's workhorse combat and reconnaissance craft, the pusher-propeller FE 2b biplane. Both 120- and 160hp versions powered the almost 2,000 produced. To meet the demand Arrol-Johnson installed engine production at its Speedwell plant at Coatbridge and at Manchester's Crossley Motors.† A tally of availability as of 30 November 1918 showed 1,902 of the sixes to be on hand, net of those lost in combat.

Thanks to its many installations, wrote Beardmore historian Kenneth Hurst, 'it is estimated that the 160hp and 185hp engines carried more bombs over Germany and German-occupied territories than any other aerial power unit.' Both the RE 5 and the RE 7 made contributions to this achievement, which perhaps was not entirely to the liking of the leaders of the Central Powers.

---

* Information published by Beardmore stated that the engine 'in 1911 had gained nine world records through speed, distance and time trials and, by 1912, had gained four awards for altitude, including one for vertical speed. During 1913 its passenger-carrying abilities were proven when awards for duration and height were gained for carrying an average of five passengers.'

† By the end of the war the 'Beardmore' name had been phased out in favour of 'Galloway', the marque the company would use for a range of post-war Fiat-style cars.

CHAPTER 4

LAND TRAINS AND MORTAR MOVERS

Effective though they were at shifting mortars and their ancillaries on well-prepared road surfaces, the Austro Daimler tugs were less suited to the poor roads in Austria-Hungary's southern provinces of Bosnia-Herzegovina. In the First World War these were perilously exposed to Serbia, an ally of the Entente Powers of France, Russia, Italy and Britain. Roads were sketchy, tortuous and interrupted by bridges of low capacity. Nor were rail networks well developed in the region.

It was the brainstorm of an army major, Ottokar Landwehr von Pragenau, to provide transport to the region by means of multi-wheeled road trains. Landwehr turned to Porsche and Austro Daimler to realise his dream.* Their proposal of 17 December 1908 to take on the task for a fee of 37,000 crowns was approved.

Major Landwehr laid down strict criteria for his train. Its laden loading per axle was not to exceed 3½ tons. It had to negotiate sharp turns. Both power and braking should allow it to climb and descend grades of up to 23 per cent. It should be able to reverse direction, backing up in its own wheel tracks. Its wheels were to be adaptable to allow it to run on both road and rail. And the train had to achieve all this in a simple and efficient manner.

Partnered by another engineer, Karl Säckward, Ferdinand Porsche applied his electrical experience and ingenuity to Landwehr's demanding criteria.† They powered their road train with a tractor unit carrying a 20.3-litre six-cylinder engine composed of three pairs of two cylinders, developing 120bhp at 1,000rpm. It drove forward to a huge direct-current dynamo that generated a current of 235 amperes at 300 volts.

Using control methods well proven in Porsche's Mixtes, current was provided not only to the tractor's rear wheels but also through cables to motors at the wheels of every alternate trailer in the ten that followed. Instead of Porsche's famous wheel motors the engineers used chassis-mounted motors, paired in aluminium housings, driving the wheels through internally toothed ring gears whose outer surfaces served as brake drums.

Braking proved a tough nut to crack. At first Porsche tried a cable solution, backed up with electric short-circuit braking through the motors. He finally settled on pneumatic braking powered by hoses between the trailers. Another problem was the behaviour of the powered trailers, which shimmied in a zig-zagging, 'dancing' manner. Fitting hydraulic dampers on both sides of the couplings cured this.

Each coupling was controlled by gearing that kept the trailers exactly on track. The last trailer had a steering gear and controls with which a driver could pilot the whole train in reverse. If a bridge or other road hazard was too weak to support the motor unit when attached to the trailers, as was often the case in

---

* Appropriately enough, Landwehr's surname could be translated as 'country defence' or even 'national defence'. He was later promoted to the rank of General.

† Porsche was not the first to be given this task by Landwehr. An Ing. Müller in Berlin had essayed a design in 1905 that had no issue.

Based on trials with the earlier A-Train Porsche and Austro Daimler developed the multi-trailer B-Train, ten of which went into service during the war to deliver supplies to the army's huge mortars and howitzers.

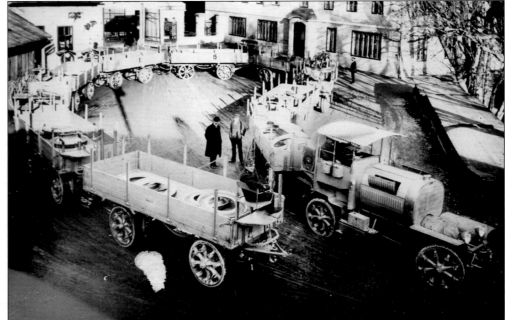

The ingenious steering mechanism developed by Porsche and his team allowed a B-Train to turn forwards or backwards in its tracks as it demonstrated spectacularly in the Austro Daimler works courtyard.

From the B-Train's generator car, with its 120bhp six-cylinder engine, electrical power for propulsion was delivered by cable to alternate trailers among the ten allocated to each train.

Bosnia-Herzegovina, the two could be separated by a length of cable and the trailers powered independently.

The first 'Elektrotrain System Daimler-Landwehr' – also known as the *A-Zug* or A-Train – was complete in 1910, when a second was ordered.* Several years of development passed before such trains were service-ready. 'Porsche had to overcome many setbacks,' said one account, 'before every trailer would follow in the tracks of the generator vehicle. But it would not have been Ferdinand Porsche had he not applied the time, patience and perfection needed to hand over a vehicle that functioned flawlessly according to his standards.'

The final version, in service from March 1913, could carry 30 tons in five two-axle trailers or 20 tons in ten single-axle trailers. Its running cost was rated as one-quarter that of an equivalent horse-drawn load train. Ten in all of the final design, known as *B-Züge* or B-Trains, were in use by 1914. Many carried the equipment that supported the batteries of Skoda's new and formidable 305mm mortars.† They were ready in good time to assist in sieges of defences in Belgium and France.

Eager to view his equipment in action, late in 1914 Ferdinand Porsche toured the western battlefields in the company of Karel von Skoda and a director of the

Starting in 1913, the military activities of the B-Trains generated valuable service revenue for Austro Daimler. This Wiener Neustadt team is overhauling the generator car of a B-Train.

The B-Train could be converted for rail operation, allowing it to take advantage of a rail infrastructure which, however, was not a strength of the dual monarchy. One was on a passenger-hauling mission in 1914.

* In spite of its obvious military potential, this early work on such road trains was not kept especially secret. An article in Vienna's *AAZ* of 2 October 1910 on a visit to the Wiener Neustadt works described the A-Train in considerable detail.

† After the war, as reparations Italy received a number of B-Trains complete with mortars. Making use of spare parts supplied by Wiener Neustadt as late as 1925, these remained in service for some years.

Skoda Works, Moritz Paul. They convened in Berlin and drove west in the Austro Daimler with which Porsche had finished the 1914 Alpine Trial with a clean sheet. 'Understandably I was interested in seeing our motorised batteries on the scene of their activity,' the engineer told the *AAZ*. 'Even though they are dismantled into their constituent parts (barrel, carriage and bedplate), the respective components of the 30.5-centimetre mortars are the highest weights which, given normal roads and bridges, can be transported over large distances at higher speeds without difficulty.'

During their journey the party was warmly received by Germany's Kaiser. Porsche learned that although his road trains were meant to be followed by troops on foot they were too fast for that, so the soldiers had taken to climbing aboard: 'In order to have the operating crews always present and in place they are allowed to sit on the front truck and the gun trailers. It makes no difference at all to the stalwart engine that it now also has to pull an operating crew of more than a hundred people!'

Visiting Brussels, Ghent, Bruges, Lierre and Antwerp they saw the damage done by the mortars, as also at the fortresses of Givet, Namur, Boechem, Waelhem and Kessel. With bridges down over the Schelde, their Alpenwagen was floated across on a barge. After seeing one of the batteries in action they returned through Liège to Cologne, where Porsche 'used the train to return to Wiener Neustadt, whither pressing projects called me'.

The destruction they witnessed was wrought both by the Austro-Hungarian weapons and by the German Krupp mortars of 420mm calibre. 'During my sojourn in France,' Ferdinand Porsche related, 'a chief engineer from the Krupp company complimented me on the towing trucks and asked if we could not build such vehicles for 42 centimetre howitzers. He had observed the performance of our mortar trucks with amazement and, recognising the importance of this rapid transport, wished to have appropriate towing trucks also for the "42"s. After subsequently viewing the "42" I understood the remarks of Krupp's chief engineer. Even with traction engines the "42"s are moved forward on roads with little speed.'

For Karel von Skoda's part, his sight of the massive Krupp weapons inspired him to think bigger. Demand from the field was for even more powerful artillery, especially to attack Italy's mountain fortresses. Moved by what he witnessed and learned, von Skoda decided to offer his dual monarchy a siege weapon that would rival the Krupp mortar.

Again Skoda's corporate technical director Oswald Dirmoser oversaw work on the new machine. Directly concerned was his brother Richard, chief engineer in charge of the design office for heavy guns. At a meeting of the three men in Pilsen on 25 April 1915, wrote Michal Prášil, Skoda 'authorised Richard Dirmoser to work on the project of a heavy gun which would fire in the higher group of angles, with both a maximum practical calibre and a range of minimally 15 kilometres [9.3 miles] and that would be minimally as mobile as the 30.5 mortar. Richard Dirmoser was granted a month of leave so that he could work on this highly secret mortar at home. He finished the design during his leave.'*

The result was the design of a 380mm howitzer, this term describing a gun with a relatively greater range and muzzle velocity than a mortar, albeit still capable of high elevations. On 16 May Karel von Skoda was able to present the

---

* 'This is the official story of the gestation of the new gun,' remarked Michal Prášil, 'but it is much more probable that Richard Dirmoser did not work on the project alone. Considering the specialisation of the various armaments departments it was impossible for one person, however gifted, to design, calculate and master all the details pertaining to the design, construction and mode of transportation of such a gun. Similarities between the new gun and the 30.5cm mortar, as well as the history of co-operation with Ferdinand Porsche, allow us to presume that co-operation between a small circle of the "initiated" took place.'

The first of two such awesome 380mm Skoda howitzers, Barbara *was given her initial tests in January 1916. The complete M 16 assembly, with barrel, cradle and base plates, scaled 90 tons.*

project to the dual monarchy's Minister of War, General Alexander von Krobatin. With no little daring he said that he would shoulder the cost of 1,119,000 crowns to produce such a gun if the army would agree to buy it if its tests proved positive. Von Krobatin agreed, asking for not one but two such weapons and insisting on the same secrecy that had shrouded work on the 305mm mortar.

The barrel of the first of the new guns was ready for test-firing on 21 January 1916. Officially the M 16, the first of its kind was dubbed *Barbara* after St Barbara, patron saint of artillerymen. With barrels almost 22ft long, the two 380mm howitzers – the second named *Gudrun* – were capable of firing a missile weighing 1,630lb and carrying 140lb of explosive a distance of 9 miles. Both were destined for duty on the Italian front. By the end of the war the army would have ten in total, with two more being readied at the works.

Earlier Skoda had developed an even bigger howitzer of 420mm bore, the

Gudrun was the second Skoda M 16 howitzer to be completed. The total produced during the war came to ten of these 380mm guns. They could only be transported by dismantling them into their constituent parts.

Two C-Train generator cars posed for the camera in 1915. With engines developing 150bhp at 1,200rpm, they supplied 300 volts to motors in their own rear wheels and to a motor for each of their trailer's wheels.

M 14, for coastal defence at Pola. A barrel nearly 37ft long ejected a missile weighing 2,940lb, including a 230lb charge, to targets as much as 13 miles distant. First tested in 1913, the M 14 was so massive that no thought had been given to making it mobile. However, the success of the M 16, and the means devised to make it mobile, led to a transportable version of this huge weapon as well.

The task of shifting these great howitzers was immense. These were truly colossal weapons. The barrel of the 380mm howitzer weighed 22.8 tons and that of its bigger 420mm brother 28.7 tons. All-up weight of the smaller of the two was a daunting 90 tons, with the larger scaling 143 tons in firing position. Geography was the enemy as well, with some of the vantage points for firing more than a mile high in the Italian Alps.

Porsche and the Dirmosers put their heads together to find the best way to transport these mammoth weapons. Although the system that they devised was known as the *C-Zug* or C-Train, this was misleading. It resembled the B-Train only in having a tractor or generator unit. Instead of powering a series of trailers, each C-Train generator car had only a single trailer in its wake. Five such combinations in all were needed to move a howitzer: one for the barrel, one for the carriage, two for the bedding (half each) and one for the munitions. By war's end the Central Powers had five mobile 420mm howitzers, each requiring five generator cars. In all, 32 C-Train vehicles were made.

Itself weighing 9¼ tons, the C-Train tractor unit or 'generator car' was powered by the same 20.3-litre six-cylinder engine as the B-Train, now uprated to 150bhp at 1,200rpm. A gearbox behind its multiple-disc clutch doubled this speed to 2,400rpm at the input to the dynamo. Operating voltage was still 300 with an output of 450 amperes, able to rise to 600 for short periods. With a louvred ovoid-

Here towing the barrel of a 420mm howitzer in 1917, a complete Austro Daimler C-Train needed five tugs, each pulling a separate component of the complete gun: its barrel, chamber/pivot, two-part foundation and munitions.

section shroud for its engine behind a massive radiator, the C-Train generator car was itself a tremendous motor vehicle. Shifting 65 per cent of its weight to its electrically driven rear wheels allowed its front wheels – still carrying 6,400lb – to be manually steerable.

Cables carried electric power to the trailers. These had eight wheels, each of which was powered by its own electric motor, driving its wheel through a pinion and internally toothed gear. They were grouped into front and rear four-wheeled bogies, both of which were steerable for manoeuvrability. The chassis of three of the four gun-carrying trailers were interchangeable with that of the munitions trailer, so the latter could be brought into use if one of the others were damaged.

'The driving torque and speed of the train,' said Michal Prášil, 'were regulated by changing the rpm of the petrol engine with a hand lever, by changing the excitation current [in the dynamo] and by connecting the electric motors parallel or in series.' In series they developed more torque for difficult terrain while in parallel they attained higher speeds. 'The main electrical cable with four cores, 70mm in diameter, ran through the whole train.'

Wheels of both tractors and trailers were fashioned to run on rails, with solid Dunlop rubber-tyred road wheels attached to them. Transition from road to rail and back again took only 4 hours, thanks to clever design. The trailers could also be integrated with normal rail freight. Speed using the C-Train tractors was limited to 22mph on rail and reached 8.7mph on a prepared road like an Autobahn with one trailer, or 7.5mph with two trailers, which it could handle under such conditions. With a single trailer it could cope with a 26 per cent grade, with two a 20 per cent slope.

Fuel demand was understandably heavy. For a trip of 100 kilometres in mildly hilly country, the fuel used would be 300 litres or 66 gallons, slightly better than 1 mile per gallon. On more difficult roads and grades consumption could be twice as high. Over the same distance the machine would need 31lb of engine oil and 13lb of vacuum oil. Servicing the complex machinery was demanding and time-consuming. Wheel bearings were to be lubricated every 100 kilometres and fuel filters emptied every 2 or 3 hours.

The knotty problem of braking was solved by adapting a pneumatic system already in wide use on trains in the Empire's Czech and Moravian regions. An engine-driven pump produced a vacuum which was piped to the rear wheels of the tractor unit and all the trailer wheels, to rub iron brake shoes on the rims of the railway wheels.

Austro Daimler began the C-Train's first tests at Wiener Neustadt on 25 January 1916. Trials commencing at Skoda in March verified that the laden tractor and trailer could climb a 26 per cent grade at a speed of 0.8mph. Its performance, said Alfred Neubauer of the Austro Daimler test department, 'was a half-century ahead of its time. With the Skoda cannon we drove through the most difficult landscape. Once Director Porsche sent us to the Seeberg Saddle. That was a twisting, narrow, steep mountain road. Twenty-one per cent grade! I can remember that very well, although we ourselves didn't believe it at first.'

'The gasoline-electrical trains were extremely modern and efficient machines,' summed up Michal Prášil, historian of the Skoda armaments. 'Their performance aroused admiration wherever they appeared. All their single parts were designed in such a way as to remain operational over even the harshest terrain and even under combat conditions. This was achieved by doubling certain mechanisms, by extremely robust construction and by the possibility of eliminating defective or damaged parts without affecting the working of the others.'

At 6:00 a.m. on 15 May 1916 the first two 380mm howitzers, *Barbara* and *Gudrun*, opened fire. Their C-Trains had taken them high in the Italian Alps to support an assault against the Italians, who had belatedly joined the fray on the side of the Entente Powers. Richard Dirmoser oversaw the siting of the huge guns, which was not the work of a moment. Emplaced at Costalta, 4,577ft high, on her first day *Barbara* set a record with 57 shells fired at Fort Punta Corbin, 8¼ miles away. Their attacks strongly supported Austria-Hungary's advance in the Tyrolean

Heavily laden with mortar components and supplies, a C-Train module made its way up a mountain road on the Italian front in 1916. Porsche's C-Train concept offered exceptional flexibility.

Next to a laden C-Train trailer in a wintry 1917 Ferdinand Porsche – second from right – discussed his system's applications with Austrian and Bulgarian military personnel.

Alps, led by none other than the General von Krobatin who had supported Karel von Skoda's 380mm adventure.

Soldiers in the high peaks were tremendously encouraged to have these immense guns on their side. 'For us mountain troops,' said one of them, 'it was important to know that these heavy weapons could be lifted to such heights by Porsche's tractor engines and that thereby he placed the Panzer-smashing "big brother" at the side of fighters from the ranks of the Kaiser's hunters and the German Alpine Corps. That was his contribution to the mountain weaponry, and thus we too commemorate it.'

With Russia collapsing on all fronts, in October 1917 the Central Powers forces combined to overwhelm those of Italy in battles centred on Caporetto. Their supply lines were too stretched, however, to permit them to deliver a decisive blow. In May 1918, in support of another southern offensive, C-Trains manoeuvred two of the 380mm howitzers to the greatest heights ever negotiated by such immense guns. To reach a firing point near Monte Erio, at an altitude of 4,530ft, their elements had to be transported over a pass at an elevation of 5,250ft. After completing their mission they were retrieved over the same route.

Late in the war a 240mm Skoda cannon was developed and also taken into positions by the C-Trains. More were commissioned to cope with the increase in the number of weapons. 'In July 1916,' wrote Michal Prášil, 'the army ordered another 48 generator cars and 42 trailers. Another order was placed at the beginning of August 1917. Fifty-five complete trains were to be delivered in the first half of the year 1918.' In all, he continued, 'Austro-Daimler produced at least 87 trains.' The success of Porsche's solution to this daunting task was self-evident.

The peacetime fates of the C-Trains were manifold. Some were used for howitzer transport by the forces of Italy and Romania. To carry the 240mm cannon the Czech arm bought 31 generator cars and assorted trailers to match from Austro Daimler through their home-country supplier, Skoda. In 1922

Towing a mixed load in its powered trailer, an Austro Daimler C-Train gave a hint of its colossal dimensions. Not without reason mountain troops saw the guns and their electrified deliverers as their 'Panzer-smashing big brothers'.

Fighting in mountainous terrain required howitzers to fire from high altitudes. In 1916 this C-Train's units brought a weapon's components up astonishingly steep grades to a new position near Folgaria in the Italian Dolomites.

Austro Daimler promoted the availability of its newly named 'Gigant' load-shifting system, showing a generator car and trailer coping with a 10-ton transformer and a vast boiler. The Czech forces loaned out their units for similar civilian tasks.

In Czech military service during the 1930s the units were updated with new pistons and spark plugs for their engines. Boži systems replaced the original Hardy vacuum brakes. From Matador in Bratislava pneumatic tyres were sourced to replace the solid-rubber Dunlops. Not surprisingly these greatly improved the ride, once the correct diameter was ascertained.

When in 1934 Czechoslovakia decided to begin fortifying its borders, the heavy cannon were moved to fixed emplacements, carried there with their equipment by the C-Trains. Special trailers for the trains were built so that they could transport the heavy turrets and cupolas for the fixed cannon. So weighty were these that two or even three generator cars were needed to shift their cargoes. But most of this immense effort was vitiated by the Munich Pact of 30 September 1938 which gave Germany the northern Sudetenland and *de facto* control of the rest of Czechoslovakia.

No sooner had this occurred than wide interest was expressed in obtaining the Czech mortars, cannon and their C-Trains. France and Britain were among those thought likely to bid, but soon Germany stepped forward with negotiations led by foreign minister Joachim von Ribbentrop. A contract signed in Berlin on 11 February 1939 found Germany paying 5.6 million Reichsmarks for a substantial suite of cannonry including 17 of the 305mm mortars and 32 C-Trains.

After training of the German batteries by Skoda cadres the C-Trains were used to bring the heavy weapons to bear on targets in the Somme and near Liège during the invasion of May 1940. The rapid movements and sodden countryside of the invasion of the Soviet Union in June 1941 meant that the generator cars struggled, needing help from huge Hanomag half-tracks. By the time of the siege of Leningrad, however, their cargoes were in place and firing.

Although Skoda was content to maintain its cannon for the Wehrmacht, it abjured similar work on the C-Trains. 'It asked that the repairs should instead be entrusted to an expert manufacturer,' said historian Prášil, 'as Skoda had never produced them and did not have either their drawings, tools or spare parts.' Having been transformed into an aircraft producer, Austro Daimler no longer existed, while the trains' designer and his team had other Wehrmacht responsibilities. In June 1942 the C-Trains were ordered shipped to a works in Paris that had been assigned their overhaul. Delivered there were 12 generator cars and 16 trailers. Five and three respectively were added in October.

In July 1943 nine generator cars and eight trailers were on their way to Leningrad to support the continued siege of that city. By early 1944 all but three generator cars and matching trailers from this expedition, repatriated to Skoda in Pilsen, had been captured by the Soviets. At the end of the war, after confiscation by the Czechs of the Wehrmacht's materiel, 23 generator cars and 9 trailers, most in need of repair, were accounted for at or near Pilsen by December 1945. After discussions and negotiations, the remains of these complex 'marvels of military technology' – as they were called in their lifetime – were knocked down to scrap merchants. None is on display in Czech museums.* That the C-Trains worked so effectively over so many years for three entirely different operators is stunning tribute to their designers and builders. Also evident is the immense respect that they engendered in their users, on whom they imposed a rigorous and complex maintenance regime.

In May 1916 Italy's Fort Punta Corbin, high in the southern Dolomites, came under attack from the Skoda howitzer Barbara, *lifted into position by a C-Train. This was the result of her assault with fifty-seven shells.*

* Some elements of the trailers may be seen, with a 420mm Skoda howitzer, in Romania's National Military Museum. Possibly C-Train constituents may be in Russian or post-USSR depositories.

Many Austro Daimler passenger cars served as military transport in the First World War. One was pictured in government service in 1914 in the Kingdom of Galicia, since 1772 part of Austria-Hungary.

Ferdinand Porsche's team also proved themselves capable of creating less elaborate prime movers for the defence of their realm. In fact many Austro Daimler passenger cars were requisitioned or commandeered for use on Austria-Hungary's fronts, suitably equipped. Crewing them were enthusiastic and patriotic members of the dual monarchy's Volunteer Vehicle Corps, headed by reserve officers who perceived the potential of the motor vehicle in military service.

Late in the war Austro Daimler developed a two-wheeled tractor whose mission was replacement of increasingly scarce horses. Primarily designed to tow light artillery across country, it was powered by an air-cooled engine of 1.7 litres developing 14½bhp at 1,400rpm. This had an F-head valve arrangement with side valves for the exhausts and pushrod-operated overhead valves for the inlets. The latter were in separate cages at the top of each cylinder which, when removed, allowed the exhaust valves to be drawn out. A squirrel-cage blower at the end of the engine's crankshaft provided air to cool the finned cylinders. The top and bottom halves of the four's aluminium crankcase extended rearward to enclose clutch and gearbox as well, making the whole assembly both rigid and compact.

Seen in a 1915 test on the Austro-Daimler grounds, this was the initial experimental version of Porsche's Pferd or 'Horse', a two-wheeled tractor that could tow an artillery piece or other field equipment.

Steering was effected at a pivot between the tractor and the cargo being towed in a design patented by Otto Köhler.

Appropriately nicknamed *Pferd* or 'Horse', this was an uncommonly versatile and manoeuvrable prime mover that achieved 'quite remarkable performance over the countryside', said Otto Stahl. These Horses also hauled field kitchens and other equipment in campaigns in Galacia and Italy.

In designing the Horse, Porsche faced an impossible challenge. Its huge steel wheels could either be smooth, in which case they would lack traction on loose surfaces, or fitted with bold cleats, which would make the Horse all but unusable on hard surfaces. He needed a simple means, controllable by the driver, of enjoying both worlds. Assigning the task to his drawing office, Porsche was presented with impractical or unworkable or complex solutions. To motivate his struggling team he offered a 500-crown prize to the man who could crack the puzzle.

So desperate was Porsche for a valid solution that all his staff were drawn in, including a raw recruit named Karl Rabe. An only son from a family in Pottendorf, north-east of Wiener Neustadt, Rabe had joined Austro Daimler in the summer

In its developed form the versatile Austro Daimler KP I Pferd was steered at the coupling between the power unit and its cargo, as Ferdinand Porsche personally tested and demonstrated. It also had many post-war applications.

of 1913 at the age of 17. He was not yet 20 when, after a few days, he laid a possible mechanism before chief engineer Otto Köhler. Not daring to trust his own eyes, Köhler called Porsche out of a meeting. The office fell silent as the Austro Daimler chief scanned the drawing. Time seemed to pass slowly as he regarded it for a quarter-hour, then more. Finally with a wave Porsche cried, 'Pay him out! Pay out the Rabe!'

In one of the first of the many technical contributions he would make to the Porsche legend, Karl Rabe had found a way to use centrifugal force to control the cleats, which protruded through slots in the wheel rims. His solution even provided for several stages of extension of the cleats, which were linked to mechanisms at the wheel hubs. Thanks in part to Rabe's ingenious solution, the Horse came to have many post-war applications, among which farming was only the most obvious.

To meet the need for a light air-cooled engine to power small railcars or switching engines Ferdinand Porsche used two of the Horse's four separate cylinders to make a parallel twin of 855cc. It developed up to 6bhp at 1,400rpm. Assuming that it was likely to be used by unskilled personnel, he equipped it with a simple planetary transmission whose lever moved it clutchlessly through neutral to first, top and reverse. This would also be a post-war Austro Daimler offering.

An even smaller two-cylinder air-cooled engine was built in large numbers during the war as a replacement for the horses that were in short supply, like its larger sister making itself useful in rail yards. With a capacity of 510cc in its two side-valved cylinders, it produced 3hp. Cooled by a pair of three-bladed fans, it too was supplied with an integral transmission.

Austro Daimler's rail-equipment shops did not draw the line at small fry. They produced heavier machinery as well. In 1918, as project P844, they built a double-ended switching locomotive. Powered by a big six-cylinder engine like those in Austro Daimler's road tugs, it had all-electric drive. A dynamo at the centre of its chassis powered a geared motor at each of its eight wheels. Braking was pneumatic, with a hand brake at each of its two posts for a controller.

Open to the elements under a parasol roof, this light locomotive was a pioneering effort by Porsche and his team that drew heavily on his knowledge in the electrical field.* A number were made, including at least one that also transmitted electrical power to motors at the wheels of attached carriages in the manner of the B-Train.

For both Ferdinand Porsche and Austria-Hungary 1916 was a year of mixed fortunes. Reverses struck the Central Powers in the war, Russia retaking Galicia from Austrian troops and France and Britain attacking over the Somme, exploiting for the first time a new secret weapon built under the code name 'tank'.† The Kaiser sacked his chief of the General Staff and replaced him with grizzled veteran Paul von Hindenburg.

At Wiener Neustadt employment was climbing from 1,000 to the 6,500 it would reach by the end of the war. Many of the added workforce were Russian prisoners of war, housed in barracks in the factory grounds. Both these rough buildings and the permanent structures, including the glass skylights in the main halls, were badly damaged in the summer of 1916 when a tornado swept through.

---

* Although electric drive from engine to wheels was to become standard in internal-combustion locomotives, this was a very early example. Contemporaneously in America General Electric was building similar light locomotives, but was unable to master the complexities of controlling both the engine and the drive system. We can be confident that this did not trouble Ferdinand Porsche.

† If there is an irony in the application of Porsche's resources in the war it was the failure of the Central Powers to commission him to design and build tanks. The French fielded more than 3,000 light tanks and 800 heavy tanks and the British 5,000 in all, while the German General Staff thought little of the new invention and commissioned only a handful of a clumsy 32-ton tank, the LK II. After the war this project was sold to Sweden. Paul Daimler's armoured-car initiative at Wiener Neustadt had not been followed up.

Ferdinand Porsche received the Officer's Cross of the Emperor Franz Josef Order in a ceremony on 6 July 1916 in the courtyard of his works at Wiener Neustadt. It was signal recognition of his robust leadership of Austro Daimler.

Porsche had been away. On his return his eyes welled with tears when he saw the scope of the devastation. Fortunately the cost in human lives was low.

A happier moment for Ferdinand Porsche that summer of 1916 came on 6 July. In an august gathering of the civil and military in the courtyard of his factory he was presented with the Officer's Cross of the Emperor Franz Josef Order with the Band of Wartime Service 'for excellent performance in the military-technical field'.* With his remarkable transport systems Porsche had made a significant contribution to the fighting capability of his country and its allies. That they were ultimately obliged to sue for peace was more a reflection of a shortage of resources, both at home and in the field, than any inadequacy of their fighting equipment.

Said a contemporary report about his award, 'This recognition of Director Porsche was received in all circles with great satisfaction. Not only does he rank among experts as an authority, but also people at large have learned to value his contributions in the Daimler motor works. Thus there was great joy in all quarters when his work was granted the highest possible recognition. The Daimler Motoren AG can be proud of its director.'

The presentation of the award was followed by a parade of Austro Daimler's wartime production. Overhead the guests heard the droning roar of a circling aeroplane powered by an Austro Daimler engine.

* On 28 May the Emperor had approved the award to Porsche. After ruling the Austro-Hungarian Empire since 1848, he died on 21 November of the same year.

CHAPTER 5

ADVANCING AVIATION POWER

At the war's outbreak Austria-Hungary's Air Service disposed of nine flying companies with 10 balloons, 39 serviceable aircraft and 85 pilots. Airframes were made by Lohner and at Fischamend while Hiero and Austro Daimler made engines. Many German imports had to fill the gap while the dual monarchy's producers geared up to provide suitable equipment.

Ferdinand Porsche accepted two challenges on behalf of Austro Daimler. One was acceleration of manufacture while the other was response to demands from his military flyers for still more power. As a stopgap Porsche produced an uprated version of his six-pushrod engine by enlarging its bore to raise capacity to 15.7 litres. Starting at 150bhp at 1,300rpm, it was ultimately credited with up to 165bhp at 1,350rpm. Like its British-made counterpart, this engine was in urgent demand at the outset of war.

In parallel Porsche began work on a completely new in-line six. Starting from a base in which he had confidence, it had the same stroke as its antecedent and, at first, a bore smaller than the one in its enlarged predecessor to give 15.0 litres. Porsche's aim was higher reliable power through better breathing efficiency and slightly raised crankshaft speed. The key, as a few other aero-engine producers had hinted, including Mercedes in 1912, was an overhead camshaft to gain greater valve-opening duration and overlap reliably.

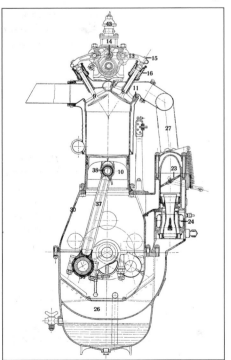

Features of the 15.0-litre 160 PS Austro Daimler six are illustrated, including its valve gear with roller-tipped rocker arms and Porsche's patented drawing of induction air through the crankcase to prevent carburettor icing.

A superbly integrated design, Porsche's 160 PS six of 1915 had shaft-and-bevel drive to its overhead camshaft at the crank's drive end. A single twin-throat updraft carburettor replaced the 120 PS engine's twin instruments.

Porsche's first overhead-cam aero engine, a completely new design, was ready for production in the second half of 1915.* Vee-inclined valves were opened by rocker arms from a single overhead camshaft, driven by a vertical shaft and bevel gears at the six's propeller end. Rollers at the rocker tips followed the cam lobes. Two valves per cylinder were closed by coil springs. Instead of iron, the new engine had cylinders of cast steel.

Pistons were aluminium initially with cast steel being used later. For compatibility with the cylinder bores this required coating the pistons with white metal or aluminium. Porsche relinquished the *désaxe* cylinder in favour of conventional alignment. Gone was the faithful Friedmann oiler, supplanted by recirculating lubrication with a reciprocating pump below the bevel drive. A duct cast into the sump delivered the engine's air, simultaneously cooling the lubricant and warming the air reaching the carburettor to ward off icing. This happy solution was awarded a patent.

Twin Bosch magnetos were at the front of the overhead-cam six, driven from the vertical shaft at one and a half times engine speed. A special hand-cranked magneto for starting was in unit with the carburettor-side magneto. The water pump was at the rear. Induction was now through a single updraft carburettor, simplifying adjustment, on the left flank. It consisted of two carburettors in a single housing, retaining ring-shaped floats to give consistent fuelling.

Introduced in 1915, Austro Daimler's 160hp six was a mainstay of the dual monarchy's aeronautic arm. Ferdinand Porsche climbed on the cart transporting one to a test bench to check its overhead-cam valve gear.

The new six initially matched its predecessor by developing 160bhp at 1,300rpm. More elaborate in its valve gear and more rugged in its crankcase, the new engine's weight of 640lb was 90lb higher than its simpler antecedent. This was a robust foundation for future development, which by 1916 persuaded its output to 185bhp at 1,400rpm. As the Dm 185 this powered countless types of the dual monarchy's aircraft.

At the request of the Imperial Navy, which aimed to build up a fleet of flying boats, in 1915 Ferdinand Porsche and his team used their six's cylinders to create their first-ever V-12 engine to provide adequate power. Displacement was doubled

Designed by future Porsche colleague Josef Mickl, Austria-Hungary's Seaplane G8 was powered by three 160hp Austro Daimler sixes. It was well suited to operations on the Adriatic.

* Although it made flights in the Möwe, as described in chapter 3, the overhead-cam Prince Heinrich four was not designed as an aero engine.

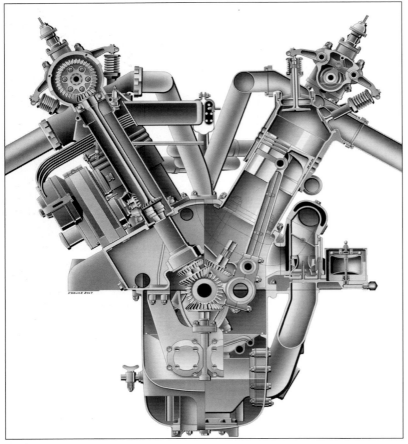

Components of Porsche's 160 PS six of 15.0 litres were combined in a 60-degree vee to produce a 30.1-litre V-12 for the Royal Austrian Navy. Wolfgang Franke's drawing shows its master-and-link-rod bottom-end design.

Derived as it was from the 160 PS six, the Austro Daimler V-12 had two valves per cylinder and shaft-driven overhead camshafts. With inlet piping in its centre, its fresh air/fuel charge was warmed by the crankcase.

thereby to 30.0 litres in a handsome 1,025lb engine. Although officially rated at 300bhp at 1,250rpm, later raised to 345bhp at 1,350rpm, its output could be as high as 382bhp at 1,500rpm.

Initially fork-and-blade connecting rods were used, to accommodate the facing cylinders on the same crankpin. These were unkind to their bearings so the bottom end was changed to master rods on the right side and link rods on the left. Induction pipes in the central vee drew mixture from two double-throat carburettors on the left side of the engine in the passages through the crankcase that cooled the engine oil and gently warmed the fresh mixture.

In a pusher configuration the V-12 powered numerous Type W 13 Brandenburg flying boats for use from the Navy's bases at Trieste and Pola. With production only stuttering into gear, however, numerous W 13s sat awaiting their engines. An agreement between the forces entitled the Navy to one-eighth of Austro Daimler's production. With this far from meeting their needs, however, the V-12 was licensed to Munich's Karl Rapp to be built exclusively for marine aircraft. Rapp too struggled to get their manufacture up to speed.

The apotheosis of Austro Daimler's airborne sixes was attained in 1917 with the introduction of a new cylinder design for the 15.0-litre six with the major enhancement of four valves per cylinder. Each rocker arm was forked at its end to open paired valves. Ram air was channelled from front to back of the sump to provide oil cooling. With aluminium pistons and a 5.0:1 compression ratio, this six initially delivered 210bhp at 1,400rpm.

Like all other production aero engines from Wiener Neustadt, the new 24-valve engine passed a 100-hour full-power test. Its final iteration saw an enlarged

With Porsche's powerful twelve in a pusher position, Hansa-Brandenburg's W. 13 was capable of 100mph and a range of 500 miles. In all the Austro-Hungarian Navy received 128 of these biplanes from three different producers.

Another seaplane design by Josef Mickl was the Type R, the first of which was pictured at the ÖFFAG awaiting its V-12 engine. Its top speed of 108mph was promising but production was curtailed by the armistice.

Valve cooling and durability were improved by the adoption of four valves per cylinder for the 16.2-litre six introduced in 1917. Rated at 225bhp at 1,500rpm, its twin magnetos were driven from the king shaft at the front.

By 1917, when this line-up of 225 PS engines was pictured at Wiener Neustadt, the versatile Austro Daimler six was also produced in Munich by Karl Rapp under licence for use in a number of Central Powers aircraft.

As shown in a Lyndon Jones cutaway, Austro Daimler's 24-valve six of 1917 had four-bolt connecting-rod big ends and forked rocker arms to operate its duplicated valves. Channels of induction and lubrication are depicted.

bore that brought capacity to 16.2 litres. In spite of a reduction in compression to 4.8:1, it gave an increase in output to 225bhp at 1,400rpm. Its weight was 705lb.

The last Austro Daimler aero engine to enter volume production during the war, this 24-valve six saw service in a wide range of Central Powers aircraft. To supplement Wiener Neustadt's capacity it was also made in Munich under licence by Karl Rapp, whose company was a direct predecessor of BMW.

Ferdinand Porsche may well have had a chance to introduce this, his newest aero engine, to the most renowned fighting ace of the Central Powers. In June 1917 the Red Baron himself, Manfred von Richthofen, set off on a good-will tour of Austria-Hungary's aviation facilities. His first stop was Vienna, where he was based for a few days until he had to return to Germany for the funeral of a comrade. Had he not met Porsche on his whirlwind rounds it would have been a surprising omission.

One application of the latest six was in an Aviatik-Berg scout plane, an example of which was downed on the Italian front in April 1918. In good condition, both aeroplane and engine were referred to the Royal Aircraft Establishment (RAE) for analysis. The engine, a 24-valve six with its original 15.0-litre swept volume, was subjected to meticulous investigation and testing, the detailed findings taking up 17 pages in three issues of *Flight*. The engine, which seemed to have had only a few hours' service, weighed 729lb by the RAE's standards and delivered its placarded power on the button with 200bhp at 1,400rpm. Output was 186bhp at 1,300rpm, 212 at 1,500 and 222 at 1,600. Ten and a half hours of operation included an hour at

1,400rpm and 202bhp plus speeds as high as 1,700rpm, at which its running was found 'very steady'.

'Both in its general layout and in most of its details of construction,' the RAE opined, 'this engine undoubtedly possesses more originality in design than the majority of enemy engines up to the present time. The design of the lubrication and oil-cooling systems has evidently been carefully considered, as have also the carburettor and induction systems.

'The 200 h.p. Austro Daimler engine would be very suitable for installation in any of the small enemy scout machines which use the 180 h.p. Mercedes engines,' the RAE report concluded. As if fearing it had gone too far, the RAE added that, 'speaking generally, the new Austro Daimler engines need hardly be considered as a serious effort to compete with any of the Allied aero engines of the same rating and proportions.'

The meticulous detail on the Austro Daimler engine thus made available in Britain lends credibility to a story that emerged after the war when Ferdinand Porsche attended the first major motor shows in Paris and London. While in Britain he met with a Rolls-Royce technical director. He was later proud to recall a conversation that the Englishman related. During the war a decision had to be made at Rolls-Royce about a technical solution, he was told. 'We'll do it the way Porsche does,' said the company's engineering leader, Henry Royce, 'because what he does is always correct.'

Near the end of the war Porsche had several other engine projects on the stocks. In 1918 he was working on a larger-bore version of the 24-valve six that displaced 18.6 litres. Significantly lightened as well, this saw as much as 280bhp at 1,400rpm in tests but was never released for production. Neither was a high-compression version designed to be throttled at sea level so that it could produce its best power in the thinner air of high altitude.

By the time Porsche's 24-valve six was introduced, many other sixes were in the field from manufacturers on both sides of the conflict. But as the experts testified in chapter 3, Porsche had been there early. Extrapolating a four to a six was no easy matter, as many pioneers had discovered. The torsional vibrations of its longer crankshaft defeated many. But Porsche and his team had shown that it could be done, at first with their 120hp six and later with developed versions. This was a signal contribution to the evolution of the aviation engine and the high-performance aeroplane.

The last 18 months of the war found Ferdinand Porsche and his colleagues actively at work on all-new engines of even higher power. They built and ran a single-cylinder test engine which as a six would have displaced a massive 38 litres. Chief tester Otto Zadnik reported that it gave the best specific torque that Porsche had yet achieved in an aero engine. It would have approached 600bhp at 1,400rpm in a full engine.

Equipped to allow trials of different sparking-plug positions, the single-cylinder engine had two inlet valves and three exhaust valves for better cooling of the latter. Finding the triple exhausts unreliable,

When the war ended Porsche was working on this 27.5-litre six. After single-cylinder trials of five-valve heads he reverted to four valves and sophisticated inlet manifolding to deliver 350bhp at 1,400rpm. Only a pre-series was produced.

Ferdinand Porsche's novel W-9 was under development at the end of hostilities, planned as the precursor of engines with multiples of its nine-cylinder 'trident' module. This 10.9-litre Type RBI developed 300bhp at 3,000rpm.

An end view of the 'broad-arrow' RBI showed its central master connecting rod, 40-degree angles between cylinder banks and cylinder and valve-gear components designed by Porsche to be capable of high-volume production.

---

* Porsche obtained a patent on this valve gear. It included an important refinement. For this to work, both valves had to be in position. If one were stuck open – a not-unknown occurrence – the balance beam would not open its neighbour. Porsche built limit stops into the stirrup that would catch the balance bar and hold it in place if this happened.

however, Porsche reverted to a four-valve head for a six based on these tests. With dimensions that gave 27.5 litres it produced as much as 360bhp at 1,400rpm. This was made in a small pre-series that took to the air in a few airframes, including some from the ÖFFAG.

Also interrupted by 1918's armistice was Porsche's work on another aero engine, his most advanced and ambitious. Called the RBI, it was a W-9. Its three banks of triple cylinders radiated from the crankshaft centreline with 40 degrees between them. Each trio of cylinders was fabricated of steel as a single block. This was achieved by welding together sets of pre-formed components in a process, patented by Porsche, that would allow high-volume production at low cost. Although the first such engine was a nine-cylinder, Porsche saw it as a concept that could easily be scaled up to multiples of nine-cylinder modules.

Each cylinder bank had a single shaft-driven overhead cam. This operated four vertical valves per cylinder in a unique manner. Instead of being paired at the sides of each cylinder, the pairing was across the cylinder. Above both valves Porsche placed a rocking beam with a central pivot. At this pivot was a roller contacting the cam lobe. Holding the pivot were the prongs of a stirrup that was hinged to the cylinder head. When the cam lobe pressed the roller and pivot down, the rocking beam opened both valves at the same time, guided by the stirrup.*

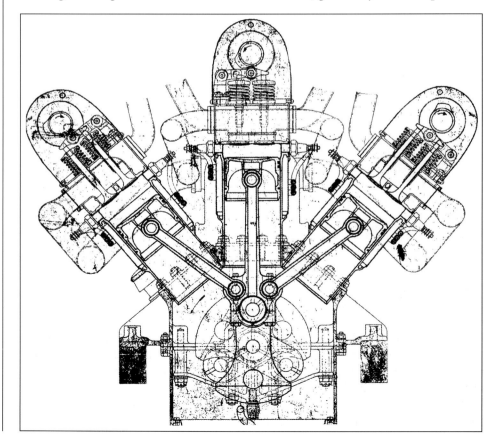

Ferdinand Porsche solved the knotty problem of the triple-bank nine's bottom end by giving master connecting rods to the central cylinder bank and link rods to the two outer banks. Massive four-bolt caps held the main bearings of its counterbalanced crankshaft. Balancing the W-9's bottom end was important, for this 10.9-litre engine was designed to produce power through speed, not capacity. It was to run at 3,000rpm, twice the usual rate, with its output geared down to 1,500rpm to suit available propellers. Compact and light, the RBI was producing an exceptional 300bhp when the armistice was declared.

Yet another experimental aero engine, dated to 1912–13, was the most enigmatic of this era. Its four cylinders had exceptionally long-stroke proportions giving a 10.0-litre capacity. This was even more swept volume than Porsche's smaller 90hp six of 1911. Air-cooled with very fine finning, its cylinders had Porsche's single-rod control of overhead valves.

In an extremely unusual layout, the four's cylinders were horizontally disposed as two 20-degree vees, opposite each other, on a common crankshaft. Ancillaries were typical of the period with a Friedmann oiler, twin magnetos and a single central carburettor feeding both cylinder pairs through long pipes. No information has come down to us about the performance of this engine or the reason for its construction. It has all the earmarks of a unit built by Wiener Neustadt to the unusual requirements of a particular customer.*

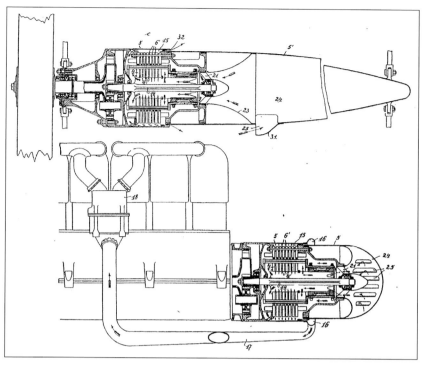

Ferdinand Porsche patented a Mixte drive for aviation use. Both its engine-driven generator and its propeller motors were to be internally cooled. His Mixte drive promised unparalleled flexibility of propeller positioning.

Ferdinand Porsche's restless creativity attained an apogee in three more aviation concepts worked out before and during the war. One, for which a patent was applied in November 1913, placed two engines side-by-side to drive a single propeller. The connection was through a differential gear, two of whose input bevels were driven by cone clutches, at the respective engine outputs, that could act both as clutches and brakes.

The arrangement, said Porsche, 'renders it possible either to use two engines simultaneously, each working at less than full power, so that they are not overstrained and can work permanently, or in case of emergency to use one engine, which it is true would result in reduced speed, but the work would be only slightly reduced compared with the work of the two engines'. In the event of one engine failure, the inventor continued, the pilot 'can support the aeroplane at a height by the work of the other engine, at least until he has found a suitable landing place or until he has corrected the fault of the first engine'.

In another patent, applied for in June 1916, Porsche applied his Mixte concept to aircraft. It would use the same basic high-speed rotating components for both

---

* Rewriters of history, eager to trace the history of the Volkswagen back to 1913, have described this as a 'boxer' engine, i.e. as a horizontal four like that of the VW. This it is emphatically not. It is a peculiar form of radial engine about whose bottom-end design we can only conjecture. So alien is it to the mainstream of Porsche's work that it is difficult to credit its configuration to him.

generator and motors. This would allow a single gasoline engine, powering the dynamo, to drive several propellers placed wherever the airframe designer desired. Porsche worked out, and patented, a way to use both the engine's inlet suction and the available airflow to cool the electrical machinery. It was an astonishing leap of an idea from one medium to another.

The third aeronautical concept traced its origins to 1908 when, as project P451, Austro Daimler first explored the idea of an electrically powered tethered helicopter as an artillery observation platform. Balloons and kites had been used for this purpose, each having obvious disadvantages. Why not, thought Porsche, use electric power to support a platform? A drawing of 1908 showed a wheeled platform underneath an electric motor driving counter-rotating blades. Power was to be fed to the motor by a cable from the ground to lift a unit weighing 400lb to an altitude of 330ft.*

The idea simmered until the outbreak of hostilities. Then, in 1915, the army's Major Stephan Petróczy von Petrócz brought it up again. Petróczy, commandant of the training battalion at Wiener Neustadt, and Porsche decided to tackle the challenge using a petrol engine. Austro Daimler undertook to supply an engine producing 200hp that weighed no more than 550lb. This was within shouting distance of a lightened version of Porsche's first overhead-camshaft six. The project found neither backing nor priority over Wiener Neustadt's other tasks at that time.

Petróczy was not easily discouraged. He revived his idea in the last months of the war at the Fischamend research and development base, where Richard Knoller was in charge. There Knoller transformed the vast airship hangar into an experimental station, complete with a dedicated wind tunnel for propeller experiments. Their work included the dissection and analysis of enemy aircraft and perfection of the synchronisation of machine gun with propeller invented by Anthony Fokker. Methods for sealing bullet holes in fuel tanks were developed.

Carried forward by Major Stephan Petróczy von Petrócz at Fischamend, the idea of a powered observation platform was addressed by Porsche but without success. A final effort produced unmanned flights powered by Le Rhône rotary engines.

---

* At the request of Porsche's nephew and secretary Ghislaine Kaes, in 1968 a Porsche engineer assessed this design. He estimated that with 1908 technology its motor would produce no more than 5hp. 'Based on this power,' he said, 'and on the basis of its own weight and that of the necessary cable, it is practically impossible to lift it to the desired altitude, quite apart from the size and position of the rotors and the speed developed by the motor.'

Electricity was back on the agenda for the powered observation platform, thanks to Ferdinand Porsche's commitment to produce a motor able to deliver 300hp at a speed of 6,000rpm, geared down to 1,000 to drive the rotors. It was to weigh only 550lb as part of a complete apparatus which, with observer, defensive machine gun and power cable, would scale 2,650lb. While Austro Daimler would build the motor, the ÖFFAG would produce the airframe and rotors.

Even with his ideas for cooling it, such a motor proved beyond Porsche's capabilities. Overheating kept him from maintaining its rated speed and power. Problems with poor rotor performance also hampered the project's progress. With Knoller too busy with other work at Fischamend to assist, to the rescue came a Budapest-born scientist who in 1912 had founded an aerodynamics institute at Aachen. Theodor 'Theo' von Kármán, already on active duty, was reassigned to work at Fischamend in August 1915.

There, said his biographer Michael H. Gorn, von Kármán 'began a lifelong affiliation with military aviation. The air corps may have been pitifully weak in flying machines and equipment, but it lavished facilities and staff on the young scientist.' Helped by their wind tunnel, he and his colleagues succeeded in vastly improving rotor efficiency. After model trials a full-scale vehicle was built using three Le Rhône rotary engines. Tethered unmanned flights to more than 150ft were made before engine failure caused a crash that ended the project.

This crossing of paths of von Kármán and Porsche was historic. While Porsche remained loyal to his homeland, almost to a fault, Theo von Kármán would move to California in 1930 and become an American citizen in 1936. Six years younger than Porsche, von Kármán would build a brilliant career as a pioneer aerodynamicist and innovator of the jet and rocket age.

Ferdinand Porsche did not luxuriate in von Kármán's freedom. With the capitulation of Austria-Hungary in 1918, he was in sole charge of an enterprise on which thousands in the surrounding countryside depended for their livelihoods. During the war his company had made a major contribution to Austria-Hungary's total production of some 4,000 aero engines to be fitted in approximately 5,000 airframes. Mortifyingly, this was only one-tenth of Germany's production of both categories. Even Italy, which entered the war a year later, managed to produce 20,000 airframes and 38,000 engines.

Now, after the armistice, Austria's condition was increasingly grave. The armed hijacking of a barge on the Danube gave a good indication of the state matters had reached in Vienna in 1918. Laden with grain in the Ukraine and bound for Germany, the barge floated west on the great river through Austria-Hungary. So desperate was an Austrian general for foodstuffs for his troops and their families that he commanded it to halt, his weapons at the ready, in Vienna's harbour. The German army's quartermaster general, Erich Ludendorff, was so infuriated by the incident that he threatened to wage war on Austria.

Fortunately for Austria, Germany's war leaders Von Hindenburg and Ludendorff had greater preoccupations in 1918. Their big summer offensive against France had failed. No good news reached the Central Powers from their other fronts. 'With every successive war year our sustenance was worse, people more and more dissatisfied,' recalled racing-driver-to-be Manfred von Brauchitsch, a 13-year-old in 1913. 'There were strikes in the munitions factories; when the inadequate rations were handed out there were often riots. People could and would no longer live under these conditions.'

What was bad in Germany was even worse in Austria-Hungary. The general who commandeered the grain barge on the Danube was none other than Ottokar Landwehr von Pragenau, the erstwhile major whose need for transport was met by Porsche's B-Train. In 1917 Landwehr was placed in charge of the dual monarchy's internal food supplies, which suffered from divided authorities and failures in the region's hastily built and still-exiguous rail network.

As the war continued, Austria-Hungary's effort was increasingly subordinated to Germany's military aims. It was decidedly secondary in equipment and supplies. Conditions on the home front were dire. Although predominantly agrarian, the empire was not self-sufficient in foodstuffs. Lack of men and equipment meant that by 1917 its output of oats had fallen to only 29 per cent of its 1913 level, of rye 43 per cent and of wheat 47 per cent. The empire's nutritional mainstay was Hungary, which had supplied 85 per cent of Austria's cattle and wheat. Preferring to sell its surpluses directly to Germany, Hungary spited its partner in the dual monarchy by cutting its shipments to a trickle in 1914.

Most remote from agricultural areas, big cities were hardest hit by these shortages. Those who were able followed the example of Porsche, who erected a farm building near his home. With deliveries of milk to Vienna plunging by 69 per cent, he could keep a cow to have his own supply. Louise and Ferry found the shed ideal for their ponies and raucous play out of earshot of their elders. Although they suffered disproportionately, the captured Russian workers at Wiener Neustadt had the best rations that Porsche could cadge for them.

With another frigid winter looming, the patchwork quilt of nations that was Austria-Hungary began to unravel. The disintegration was internal; the Entente Powers had not sought the empire's collapse. The Habsburgs' creation had had its day. New, more nationalistic ideologies were in the ascendant. The dual monarchy's constituents began declaring independence, Czechoslovakia on 28 October. Revolutions in Vienna and Budapest on 31 October marked the beginning of the end. On 3 November Austria-Hungary agreed an armistice with the Entente Powers and on the 11th Emperor Karl I relinquished his office. The former empires became republics, exiling the Habsburgs.*

Among the agreements defining the region's new borders was the Treaty of Saint-Germain, which established the new Austria. It was a shadow of the old. Peripheral domains were stripped away to leave a German-speaking rump nation – 'six impoverished mountain cantons' – with a population of only seven million, one-tenth the size of Austria-Hungary. Major industrial areas in the north and west were ceded to Czechoslovakia, including Porsche's birthplace. With a minuscule home market, Austria's remaining industries would struggle to establish export markets in a Europe whose heightened nationalism strongly favoured indigenous products.

This was the daunting prospect that faced Ferdinand Porsche and Austro Daimler. The engineer/manager could be said to have had a good war. His aero engines and unique transport systems had earned him world recognition. On 20 June 1917 Vienna's Royal Technical Institute had named him an honorary Doctor of

---

* Hungary experienced a communist revolution, only to have pro-Habsburg Romania intervene and re-establish a monarchy. Karl I attempted to take its throne in 1921 but failed, dying in exile in Madeira the following year.

Rump Austria's sad state after the Armistice and the Treaty of Saint-Germain is exemplified by the loading of Porsche's precious aero engines into a railcar for scrapping. Austro Daimler would soon lose its managing director.

Technical Science 'as the intellectual leader of a large domestic undertaking that has rendered the highest service to the furtherance of automobilism and aviation technology'. Henceforth he would be addressed formally as Dr h.c. Ferdinand Porsche.

Another who had a good war was Trieste-born Camillo Castiglioni. From a minnow of moderate size in the vast sea of Austria-Hungary, Castiglioni had become a shark in the back-yard pool that was the new Austria. Only 37 years old at the time of the armistice, the canny Castiglioni had taken positions in more than 80 Austrian enterprises through his business and banking connections. In the latter sphere his main vehicle was the General Deposit Bank.* After the war that bank controlled Austro Daimler. Reciprocally, the Wiener Neustadt company was one of the jewels of Castiglioni's holdings. And the banker knew that Ferdinand Porsche had made it so.

Austria's dramatic shrinkage meant that neither Camillo Castiglioni nor Ferdinand Porsche remained citizens of that nation. According to their birthplaces the financier reverted to Italian citizenship and the engineer to Czechoslovakian. This was advantageous to them, for it meant that none of the constraints on travel suffered by the Germans and Austrians applied. Of immense importance and value to both were their travels abroad, vital to the development of desperately needed exports. Austro-Daimler had ended the war with 500 officials and 6,500 staff

---

* The Allgemeinen Depositenbank.

working flat out to fill urgent orders. Now it had to start again almost from scratch.

Ferdinand Porsche had a gift for long-lasting relationships. Otto Köhler was still heading his engineering, while Otto Grünewald, whom he had known from his Egger days, was leading his workforce. His nephew recalled his character: 'The "Ferdl", as his workers called him – either reverentially or anxiously – was quick-tempered, irascible and forgiving all in one. He could pop up unexpectedly. In an instant he was everywhere and he saw everything. People feared that. If something had gone wrong, or otherwise a bit awry, he would grab the derby from his head, throw it on the ground and trample it. If people prattled on without saying anything – stealing his time – he called them "steam blatherers" and left them standing. Nevertheless they all honoured him, indeed loved him. Why was that? I don't know.'

After the war several other long relationships began. Karl Rabe was already established as a member of Porsche's team. Twenty-seven years old in 1918, Bohemia-born Alfred Neubauer, an artillery officer, had been assigned as liaison between the army and Austro-Daimler. With the military's collapse it was natural enough for him to look for work at Wiener Neustadt. Standing out with his 'sonorous voice' and 'straight, stiff bearing', Neubauer was placed in charge of the department in which finished cars were tested and made ready for sale. Others in the department were Otto Kaes, Fritz Kuhn and Georg Auer.

Another who entered orbit around Planet Porsche during the war was Josef Mickl. From Austria's Carinthia, the trained engineer had been with Taube builder Edmund Rumpler during the war and then at Fischamend, creating seaplane designs for the ÖFFAG and others. Mickl would apply his mathematical skills to a wide range of projects, including aerodynamics. His first assignment at Wiener Neustadt in 1918 was to create a mathematical model of the stresses generated in an engine's crankshaft.

From 1919 onward Ferdinand Porsche moved heaven and earth to provide employment for the more than 6,000 souls whose families were completely dependent on the continuation of Wiener Neustadt's activities – without the benefit of orders from an empire fighting a major war. His was the total responsibility, as general director, for all aspects of the company's activities.

Porsche kept not only Austro Daimler but also the former aircraft factory, the ÖFFAG, busy with contracts both at home and abroad. With anything to do with aeroplanes prohibited by the Allies, the ÖFFAG – another Castiglioni holding – turned to the production of vehicle bodies. Two of Porsche's wartime creations, the two-wheeled 'Horse' and the 'Goliath', with its four giant steel wheels, continued in the Austro Daimler post-war catalogue. Also remaining in production were light railcars for inspection and servicing, powered by Porsche's air-cooled engines. The horse-replacing parallel twin was offered in powers up to 6bhp.

When he had extra capacity, Porsche used it to develop new initiatives for the future. Meritorious though they were, these initiatives were also costly. They faced a strong headwind in spiralling inflation in both Austria and Germany. So desperate was the financial situation that Porsche discovered that the entire value of his life insurance would barely suffice to buy a pair of shoes.

Meanwhile Camillo Castiglioni was just as active in fields that interested him. Lingering Skoda participation in Austro Daimler finally faded at the end of 1920, when Castiglioni began share exchanges to form 'common interests' with two other Austrian enterprises, Fiat's licensee in Vienna and Graz's Puch, a maker of cars and motorcycles. Under Porsche's guidance this led to money-saving sharing of activities in purchasing, sales and design.

In the straitened Austrian economy, tensions between the two men grew towards the end of 1922. The tragic death of one of his drivers in a race meeting at Monza became a stick with which to beat the engineer. Although Porsche could and did show that the car's crash was not the fault of his design, his board still blamed him for the driver's death. Other issues arose when the research and development budget for 1923 was being prepared. Ferdinand Porsche's first submissions were rejected on the grounds of excessive cost at a time of poor business conditions, although a subtext would have been an attempt by the board to rein in what it saw as Porsche's excessive zeal for product improvement.

Finally, at a board meeting in Vienna in February 1923, Castiglioni laid down conditions he must have been confident that Porsche could not wear. Within two weeks, he said, it was the general director's task to slash the workforce by 2,000, fully one-third of the staff.* The engineer was furious, outraged. This would cut the heart from his already struggling company. Hurling imprecations at the board members – some say even hurling a candlestick – a choleric Porsche shouted, 'I won't go along with that! And if you don't like it, you can carry on making your garbage on your own – without me!'† and stormed from the boardroom, slamming the door.

It took the engineer three-quarters of an hour to drive back to Wiener Neustadt. Arriving at the factory gates he found his path blocked by a commercial official. He was told that he was barred from the works grounds. Not one to give up easily, Porsche arrived at work the next morning as usual. Soon a two-man delegation from the board confronted him. 'Herr Porsche,' one said, 'we have decided to accept your resignation of yesterday.' Exploiting the engineer's trip-hammer temper, Castiglioni had achieved his objective. Word swept through the factory, where the reaction was astonishment, incredulity. To an entire generation of Austrians, Ferdinand Porsche *was* Austro Daimler.

Engineering at Austro Daimler was taken over by Porsche acolyte Karl Rabe at the tender age of 27. Early in 1922 Rabe had been granted his degree in engineering. This earned him a warm letter of congratulation from Porsche, who was 'extraordinarily pleased' by the news and sent his 'heartiest congratulations'. Porsche and Rabe were only at the beginning of a technical and personal association of epic proportions.

---

* Some sources suggest that this drastic action was timed and planned by Castiglioni to exploit – or provoke – a slump in the Amsterdam stock market. Subsequently he did shut the plant for a number of weeks and cut the staff substantially.

† The expression that Porsche was said to have used as 'garbage' was Saubagasch. 'This was not unusual,' said his grandson Ernst Piëch. 'Bagasch is already "lowest of the low" and Saubagasch is even worse!'

CHAPTER 6

MERCEDES MAKES MILITARY

Another chief engineer was falling foul of his board. Paul Daimler, whom Porsche had succeeded at Wiener Neustadt, was at odds with his colleagues at Stuttgart's Daimler Motoren Gesellschaft. He had committed some of his company's models to Knight-patent sleeve-valve engines, which were not turning out to be the answer to everyone's dreams. Daimler had been a board member at the DMG until 1920, but then was dropped from their ranks. When in 1922 the company rejected his plans for an eight-cylinder engine, Paul Daimler looked elsewhere. In July of that year he accepted the post of chief engineer at Horch.*

The DMG had been casting sidelong glances at Ferdinand Porsche for some time. At 1921's Brussels Salon, Daimler's Richard Lang was admiring the display chassis of Austro Daimler's impressive new six-cylinder AD 617 when production chief Otto Stahl, a fellow Swabian, ambled over. After some pleasantries Lang, an engineer in his own right, asked Stahl, 'Do you think Porsche could be moved to change to Untertürkheim?' Stahl reported the remark to Porsche, who at the time still hoped to realise his dreams in Austria.

Porsche knew that conditions in Germany were not all that much better than in Austria. Alfred Neubauer described Germany's economic plight: 'Stocks fell while prices and taxes rose, and almost daily another zero was added to the figures on our banknotes. You paid a million for a tram ride and a billion for a loaf of bread. Inflation roared through the exhausted nation. Sheer existence was obliterated, the wealthy melted like snow in sunshine and shrewd speculators profited from widespread destitution.'†

German inflation soared relentlessly from 1918. From its value of 4.2 to the dollar before the war the mark fell to 49 at the end of 1919, 73 at the end of 1920 and 190 to the dollar at the close of 1921. Following a severe depression the first chaotic collapse of Germany's currency occurred in 1922, which ended with a dollar worth 7,353 marks. The year 1923 experienced the demoralising absurdities related by Neubauer. Relative stability only came at the end of 1923 when the respected Hjalmar Schacht was named to head the Reichsbank. In 1924 Schacht restored the mark to its former value of 4.2 to the dollar.

These were the turbulent economic conditions in which the Daimler-Motoren-Gesellschaft had to find its way after the war. It was led by an authentic veteran, Ernst Berge, as general director. Born in 1868, Berge was a businessman in Hamburg before joining Daimler in 1902 as joint commercial director with Eduard Fischer at its restructured Österreichische Daimler-Motoren-Gesellschaft at Wiener Neustadt. Berge came to Untertürkheim in 1905 and became a management-board member in 1907.

---

* At Horch, Paul Daimler realised his dream of a powerful straight eight. His engines of 4.0, 4.5 and 5.0 litres for the Zwickau company were regarded as Germany's finest in-line eights.

† 'That is when the seeds of hate were planted that would bear their malevolent fruit many years later,' Neubauer added. That this would happen was not inevitable, but it is a matter of historical fact that it did occur.

Ernst Berge was named the DMG's general director during the company's dramatic wartime expansion. The DMG and Berge were under the watchful eye of their supervisory board, chaired since 1910 by an enterprising banker, Alfred von Kaulla. Born in 1852, von Kaulla had proven his entrepreneurial skills at Stuttgart's Württemberg Vereinsbank, which had close and successful links with the Deutsche Bank. Called 'an enterprising, bright, robust man', Alfred von Kaulla had been a member of the DMG board since 1902. As supervisory-board chairman he took a close interest in all the automaker's affairs.

In mid-1922 the DMG found itself without a chief engineer. Richard Lang was doing his best to streamline manufacturing at Untertürkheim, but its cupboard of future products was all but bare. With von Kaulla retiring from the supervisory board, Ernst Berge asked a neutral third party, Paul Eberspächer of the eponymous components firm, to sift the candidates and make a recommendation. Eberspächer, familiar with the impasse facing the company, argued that only a man of the topmost category could provide the products it needed. Ferdinand Porsche would be costly, he told the supervisory board, but worth it.

Ernst Berge remembered Porsche, the youthful prodigy, from his spell in Vienna 20 years earlier. The DMG decided to engage him. Porsche's first official board-meeting presence would be on 4 May 1923. Nevertheless that February Ferdinand Porsche was already ensconced in Stuttgart's Hotel Marquardt and hard at work. He soon arranged for other members of his Austro Daimler team to join him, including his trusted design engineer, Otto Köhler, and the head of his vehicle running-in department, Alfred Neubauer.

The Bohemian engineer 'was welcomed with anything but open arms' at Untertürkheim, wrote Porsche's biographer Peter Müller. Resentment over a man parachuted in at such a high level was only to be expected from senior engineers, especially if they were puffed-up 'steam blatherers' of the ilk that Porsche heartily despised. His no-nonsense style of direct and personal involvement in the drawing offices, shop floor and test stands was in startling and unnerving contrast to that of Paul Daimler, who preferred to oversee his domain from on high.

Like Austria-Hungary but to a far lesser extent, Germany suffered territorial losses after the First World War. When Porsche moved to Stuttgart's Daimler in 1923 his new nation's economy was in chaos.

Daimler Motoren Gesellschaft (DMG) general director Ernst Berge took the initiative in bringing Porsche to Stuttgart. Porsche lost a supporter when Berge left in 1925 to head sales for a German truck company.

Allowed by the Allies to produce armoured vehicles for border defence, the DMG presented its DZVR in 1919. First for that purpose and later for internal policing, the DZVR continued well into Porsche's Daimler-Benz years.

---

* Porsche was asked to sign Stuttgart's golden book of high honours. He was also awarded an honorary doctorate by Stuttgart's Technical Institute, granted 'in recognition of his outstanding merit in the field of motor car design and particularly as designer of the winning car in the 1924 Targa Florio'. Another honour for Porsche came in January 1926, when Italy's King Vittorio Emanuele III added his name to his list of Cavalieri, a knight of the realm. This smacked of the recognition he had received during his years in the dual monarchy.

Brusqueness and irascibility were integral to Ferdinand Porsche's portfolio. His style as he passed through his drawing offices from board to board was abrupt. If he disliked what he saw he would wave a hand and shout, 'That's shit! That's no good! We don't have to go any further with that! Change it! Think of something!' Praise was not forthcoming, but if Porsche liked what he saw he would say, 'That's good, that's the direction we have to go! But I'd like to see this detail done differently! Perhaps like this,' taking the designer's pencil and sketching an alternative with a few deft strokes. Porsche was a facile technical artist whose freehand drawings elegantly depicted the most complex machinery.

Porsche soon learned how narrow-minded Stuttgart's Swabians could be. He had his new business cards printed up with 'Dr. h.c. F. Porsche', as before, only to be told that this was unacceptable. A Viennese honorary doctorate was all very well in Austria, he was told by Württemberg's Interior Ministry, but fell short of Germany's standards. The cards would have to be reprinted. 'Porsche responded with work and performance,' wrote a nephew about this petty episode.

There was no shortage of work. In fact Porsche found himself in the midst of final preparations for the most ambitious racing effort ever undertaken by Daimler, its entry of three supercharged 2.0-litre four-cylinder Mercedes to race 500 miles at Indianapolis in America in May 1923. Although this was not successful, Porsche drew upon the materiel left him by Paul Daimler to build a revised car that won Italy's renowned Targa Florio in 1924. This achieved the recognition that Porsche needed in his new environment. In 1929 he added an honorary Dipl.-Ing. degree from the Württemberg Technical Institute, restoring the status he had earlier been denied.*

Wearing his passenger-car hat Ferdinand Porsche faced a turbulent decade. From 1924 into 1926 a courtship between Daimler and Benz led to marriage in the latter year, creating the Daimler-Benz Aktiengesellschaft (DBAG). To Porsche fell the task of co-ordinating the engineering of Benz at Mannheim with that of Daimler at Stuttgart-Untertürkheim, two proud organisations that were hitherto bitter rivals. He led the design of new models to top the Mercedes-Benz range, using the driver-engaged supercharging that Paul Daimler had introduced, while launching work on smaller cars to broaden the company's sales base.

Benz and Daimler had been stalwart producers of aero engines and other materiel for the Kaiser's army. As well, in co-operation with Krupp a counterpart to Porsche's Austro Daimler tugs had been produced by Daimler at Stuttgart. This model, the DZ I, had huge artillery wheels with steel rims at first and later hard-rubber tyres, doubled up at the rear. Four-wheel drive was on the lines established by Paul Daimler in Austria in 1904. Used for various purposes including gun-carrying, some with nose winches, the DZ I continued in production post-war, as of 1924 at Daimler's Berlin-Marienfelde plant.

A vast version of this vehicle, its huge wheels requiring a stepladder for the driver, was the DZ III, made near the end of the war as a counterpart to the Austro Daimler M 17. Although required by the Versailles terms to be dismantled, the DZ III was quietly reassembled in 1926 during Porsche's regime. It found no further use, however.

In 1919, before the Allies had made up their minds about Germany's fate, the DZ I chassis was recast as the DZVR to serve as the basis of armoured cars for border defence, especially in the east. Still using a 12.0-litre four developing 100bhp, as developed it had equal-sized front and rear artillery wheels on a 147.6-inch wheelbase. By the spring of 1920 output of armoured versions of the DZVR had reached 94. Some Krupp-Daimler tug chassis had been rebodied as armoured scouts.

With the promulgation of the terms of the Treaty of Versailles, signed on 28 June 1919, Germany's emasculated Reichswehr was forbidden the use of any kind of armoured vehicle whatsoever. The language was plain: 'The manufacture and the importation into Germany of armoured cars, tanks and all similar constructions suitable for use in war are also prohibited.' This unequivocally included the Daimler-based armoured cars.

With the DZVR-chassised vehicles on the brink of extinction, a loophole was found. A supplementary treaty granted up to 150 armoured vehicles to Germany's internal forces to be used solely for the purposes of the *Schutzpolizei*, the defensive police. Each vehicle could carry two machine guns. Providentially a role had been found for vehicles that now carried the 'Schupo' designation.

A commission representing the German states placed orders with Benz, Daimler and the Ehrhardt-Automobil AG for more such vehicles to bring the numbers up to the permitted maximum. Deliveries extended through 1924 into 1925, well into Porsche's stewardship of DMG engineering.

In fact the Reichswehr succeeded in retaining a three-figure fleet of armoured vehicles by classing them as protective transport for troops. All were built on the Daimler DZVR chassis, their development continuing through 1928 when some were used for experiments with radio communication. Capable of 31mph, these vehicles could cover up to 180 miles per day.

With chassis like the DZVR getting on in years, the Reichswehr had to take what steps it could to infuse its equipment with new blood. One of the lessons of the Great War had been its demonstration of the usefulness of vehicular mobility. Now, with its army limited to 100,000 men, providing mobility for Germany's troops was all the more important. This was the firm opinion of the chief of army command, the decisive and far-seeing General Hans von Seeckt. He envisioned a future in which his forces would prevail thanks to combined arms operations of artillery, infantry, armour and air power that would concentrate superior firepower on specific objectives. To achieve this he needed a more modern style of horsepower.

Not until 1923 did von Seeckt issue parameters providing for the inclusion of passenger cars in the manoeuvres of his downsized army. One class of cars would be small vehicles for reconnaissance and column use with open bodies, capable of some 45mph. Existing passenger cars were equipped with suitable coachwork for operations in the field. An early adopter was the Coburg firm of Karosseriefabrik Nikolaus Trutz, which pioneered seating in individual buckets or *Kübeln* that held occupants in place over difficult terrain. Cars so bodied were known as *Kübelsitzwagen*, soon shortened to *Kübelwagen*.

General von Seeckt also foresaw the need for larger vehicles with six-passenger capacity in open or closed bodies, able to attain up to 60mph. These, it was said, would 'make the middle and upper staffs mobile and allow them frequent, direct exchanges of ideas and a feeling of closeness to the troops'. The Mercedes-Benz offering, based on a design initiated by Porsche, was the Stuttgart 200 and later the more powerful 260, with six-cylinder side-valve engines. Others active in this category were Adler, Stoewer and Wanderer.

Driven through only two wheels, such vehicles were considered to be suited solely to decent roads, rare enough in combat zones.* From its experience in the field the post-war Reichswehr decided that the best solution to passenger-carrying agility would lie with a six-wheeled configuration in which four or even six wheels were driven. A key proponent of this layout was Major Werner Kempf, who closely followed its applications by vehicle makers.

As an element of its fresh initiatives in 1924 the army decided to commission prototypes of such vehicles from three automobile producers: Daimler, Horch and North Germany's Selve. Within the Army decisions about new vehicles were made by the *Heereswaffenamt* (HWA) or Army Weapons Office. Rather than disclose its role in the matter, however, the HWA arranged for the Reich Transport Ministry to issue the purchase orders. The vehicles were to be six-wheeled transports in the ¾-ton load category, driven through at least four of their wheels.

For this mission the Porsche-led Daimler engineers created their Type G 1, known internally as the W 103 in the new designation system for vehicle projects created during the merger with Benz. With long straight frame rails, gently kicked up over the rear-wheel pairs, its construction was more lorry-like than automobile-style. A 3.1-litre side-valve six from the DMG inventory, producing 50bhp at 3,000rpm, drove through a five-speed transmission and a final-drive ratio of 6.30:1 that gave a top speed of 37mph.

While the forward axle of the rear pair was braced to the frame by radius rods, the rearward axle received guidance from a torque tube anchored in the forward axle's housing. Pivoted to the frame at its centre, a single semi-elliptic spring at

---

* As will be related later, Porsches father and son would struggle to gain military acceptance of their two-wheel-drive wartime version of the Volkswagen.

each side joined to an axle at each of its ends. Wheels were steel discs, doubled at both rear axles. At the extreme rear was a 26.4-gallon fuel tank, below the frame. Above it was a cable winch, driven by a long shaft from a power take-off at the gearbox.

Completed in 1926, the purposeful G 1 carried its driver high, behind a vertical windscreen, and its passengers even higher in an open tonneau with ample space for six. At 5,300lb it was the lightest of the three prototypes in competition. Five each of the sample vehicles were built by Daimler-Benz and Horch, the latter a 65hp design by Paul Daimler. Selve's entry also drove its front wheels, using a patented design by Voran, a producer of front-driven cars.

Making improvements, the DBAG built another G 1 in 1927 and a final one, with Porsche's inputs, in 1928. Nevertheless the prize for the best design went to Horch, which benefited from an order for 60 vehicles put to use by the Reichswehr. The latter, however, ultimately proved less than enthusiastic about this class of vehicle. The six-wheelers turned out to be complex and costly, by their lights, without compensating benefits.

In its report of 16 June 1929 on these vehicles the HWA did not discriminate among the three producers, saying that 'in the interest of openness the class is presented in general without identification of the producing companies. The names of the makers are not mentioned because in addition to Daimler, Horch and Selve more and more companies are taking up this challenge.' One was Krupp, for which a more rugged six-wheeled vehicle was designed by Ludwig Baersch, known as the 'Krupp-Boxer' after its 3.3-litre flat-opposed four. Introduced in 1934, some 7,000 were built to carry troops and tow artillery.

Ferdinand Porsche's final effort on a six-wheeler for the DBAG surfaced in the spring of 1929 as the G 3. Powered by a 3.5-litre six of 60bhp, it had hydraulic brakes instead of its predecessor's mechanical system. Wheelbase to the first axle was much longer, 118.1 inches against the G 1's 93.0 inches. The frame's side

To meet a 1924 request from the Army for a versatile six-wheeled communications vehicle the DMG created its G 1, in competition with Horch and Selve. Only prototypes were made, the Army being uncertain about its needs.

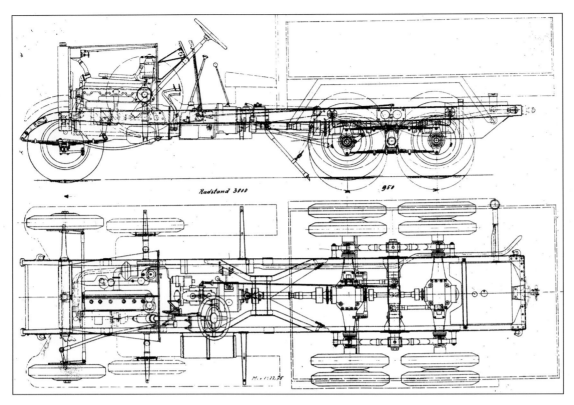

Before leaving Daimler-Benz Porsche oversaw work on another six-wheeled military vehicle, the G 3. Pivoted and paired semi-elliptic leaf springs gave its rear driving axles great flexibility for rough surfaces.

Bodied for field work and camouflaged, the Mercedes-Benz G 3 looked suitably militaristic when it made its debut in 1929. Although some were made for various purposes, this was still not what an indecisive Army was looking for.

Porsche's work on the G 1 and G 3 set the stage for the G 4, to which Otto Köhler made design contributions. Its introduction in 1934 was just in time for use by senior Wehrmacht figures in their tours of conquered territories.

members were dead straight, now holding two centre-pivoted semi-elliptic springs at each side. Final drives were worm-type with the worms at the tops of their differentials. With wheels that were still doubled up, the rear axles could flex over a 25-degree lateral range to provide traction over rough terrain.

Although the G 3 failed to make a breakthrough, it set the stage for the G 4 of 1934. Key elements of its design were contributed by erstwhile Porsche colleague Otto Köhler. Powered initially by an unsupercharged version of the Mercedes-Benz 5.0-litre straight eight, the G 4 had large-section single tyres at all six wheels. Its four-speed ZF-Aphon gearbox worked in series with a two-speed auxiliary box that gave direct drive plus a 3.06:1 reduction for cross-country and parade work. The G 4 saw a lot of the latter, for on its lengthened 122.0-inch wheelbase – 37.4 inches between axles – it was an imposing machine, part luxury car and part war-commander carriage, that satisfied the solipsism of Nazi nabobs.

Classed internally as the W 31 by Daimler-Benz and as a 'Heavy Command Car' by the Wehrmacht, the G 4 was built in an initial series of 11 units through to 1936. With a 5.3-litre eight 16 more were produced through to 1938. The final tranche, of which 30 were made, were laid down with 5.4-litre power in 1939. G 4s were prized transport for Adolf Hitler and others for such ceremonies as the occupation of Czechoslovakia and Austria as well as tours of the front lines.

This final evolution of the six-wheeled line met the criterion of 'a feeling of closeness to the troops' set for the Army's motorised vehicles by General von Seeckt. After creating a Reichswehr that was in the best possible condition allowed by the Versailles constraints, well organised and staffed by future combat leaders, von Seeckt left his post in the autumn of 1926.* His successor was General of the Infantry Wilhelm Heye, who held the office until his retirement in October 1930.

Although Germany's military were still bound by the stipulations of the Versailles Treaty, a sense of liberation attended the cessation of inspections by the Military Inter-Allied Commission of Control in February 1927. In the immediate aftermath of this event the Reichswehr's HWA issued specifications for several new vehicles in which Daimler-Benz and Ferdinand Porsche were asked to take a direct interest. Supporting this activity was Generalmajor Alfred von Vollard Bockelberg, who headed the Inspection of Motorised Troops from October 1926.

One requirement was for an advanced and versatile reconnaissance vehicle. The machine had to reach at least 40mph both forwards and backwards, change from one mode to the other in only 10 seconds, be as quiet as possible, cope with 33 per cent grades and 5ft ditches without assistance and be undeterred by water 3ft deep. Although an initial requirement for full amphibious capability was later withdrawn, builders undertook to meet it anyway.

Here was a fascinating challenge for the DBAG's Porsche-led engineering team. Their company was chosen for the task in competition with Büssing-NAG and Magirus, each required to build two prototypes. On 22 March 1927 Daimler-Benz signed its contract, which set delivery targets for the two vehicles at 31 March and 15 April 1928. It provided fixed-price compensation of RM 250,000, payable one-third on signing, one-third after completion of the main design activity, one-sixth on delivery of the first prototype and the final one-sixth on acceptance of both vehicles.

---

* Von Seeckt was cashiered by his civilian chief, Otto Gessler, who thought him too political, ambitious and demanding. He sought in Heye a more biddable personality.

In 1927 Germany's Reichswehr issued a requirement for an agile and amphibious eight-wheeled scouting vehicle. At the DBAG this became the MTW 1, of which a wooden model was carved to support its candidature.

Commissioned to build two MTW 1 prototypes, Porsche arranged for a storage-tank maker to build their hulls of half-inch steel. Vertical slots showed the amount of travel provided for powered wheels guided by their swing axles.

---

* This was an acronym for *Mannschaftstransportwagen* or Crew Transport Vehicle, as bland a designation as the Army could have desired for its secret project. Another specific designation for the Daimler-Benz prototypes was DB-ARW.

Acceptance required completion of two 310-mile test sessions, as close together as possible, each covered at an average speed of at least 19.9mph. Costs of its personnel and fuel for these would be borne by the DBAG, which effectively provided a six-month warranty covering the trials. Cessation of inspections or no, the company was 'obliged to take all possible measures to keep this contract secret'.

In response to this request Ferdinand Porsche oversaw work on a concept that added materially to his reputation as an engineer both innovative and ingenious. Opaquely designated the MTW 1,* it was of advanced full-monocoque construction with drive to all of its independently sprung wheels, whose steering system was sophisticated. Its shape minimised its profile, making it ideal for scouting missions. Although a small turret for a 37mm cannon was part of its design, there is no evidence that one was fitted to the prototypes.

Sloping and rounded contours gave the MTW 1 a shape both appealing and functional, able to deflect machine-gun fire. Produced by Martini & Hüneke Maschinenbau AG, whose stock in trade was underground storage tanks, its steel hull was at one and the same time body and frame. Its external skin was just over ½-inch thick, providing a modicum of protection from light weapons. Although welding was clearly the desired fabrication means – and was used by rival Magirus – the DBAG prototype tubs were assembled with rivets and screws. Resulting weight was 5,900lb.

The power source for the MTW 1 was a new Daimler truck engine, the 7.8-litre Type M 36 developing 100bhp at 2,000rpm. An in-line six, it had two iron blocks of three cylinders each on an aluminium crankcase. Removable cylinder

Although a ZF Aphon transmission was ultimately fitted, the MTW 1's original gearbox was an in-house design. Placed next to the driver, its control levers and hand brake were immediately accessible.

Powering the DBAG eight-wheeler was a dual-ignition six of 7.8 litres delivering 100bhp. With two iron cylinder blocks on an aluminium crankcase, the side-valve six was an in-house rival for the Maybach engines the Army often specified.

heads revealed side valves and carried two spark plugs per cylinder for reliability in this application, one set driven by coil and distributor and the other by a magneto. Gearing from the tail of the crankshaft drove a string of accessories along the right side of the crankcase. Exhaust was piped upwards to an external silencer.

Always a challenge in this class of vehicle, cooling was solved with a pair of radiators in the bulkhead between the first and second wheels on the left. Belts and shafts from the nose of the crankshaft drove a cooling fan for these and another on the right-hand side. Initially using the DBAG's own gearbox, the MTW 1 later had a ZF-Aphon unit that offered four speeds both forwards and in reverse. Also providing a low 'crawler' ratio, the gearbox carried a robust external-contracting brake.

Wheels rose and fell on wide-based swing axles carried by a central member that extended the full length of the MTW 1. Vertically hinged at its centre, this member thus divided the undercarriage into front and rear four-wheeled bogies. Because the steering system changed the angle between the front and rear halves, this gave sufficient turning angle to the two wheel pairs closest to the vehicle's centre. Full steering action and outboard universal joints were only needed by the wheel pairs at the extremes, controlled by ingenious linkages that Daimler-Benz patented.

Extending from the tub to the hub carriers, quarter-elliptic leaves gave springing for each wheel. For each of the centre pairs of wheels this took the form of a semi-elliptic spring attached to the tub at its centre. Brake drums showed signs of racing expertise with deep cooling fins around their periphery. Drive to all eight wheels was by a central shaft to the final drives, universal-jointed at its centre to allow the two bogies to steer.* Movement of the wheels gave the steering required in water, for which a propeller at the rear was engine-driven.

---

* References show two types of final drive, one using low-placed worm gears and the other bevel gears. The latter has the advantage that these gears rather than a universal joint could allow the necessary central pivoting of the member carrying the front and rear drives and suspensions.

Cascades of straight-cut gears packed the MTW 1's transmission, which lowered the drive from the engine to fore-and-aft shafts. Noise from these gears prepared the way for ZF's new Aphon, which had much quieter helical gears.

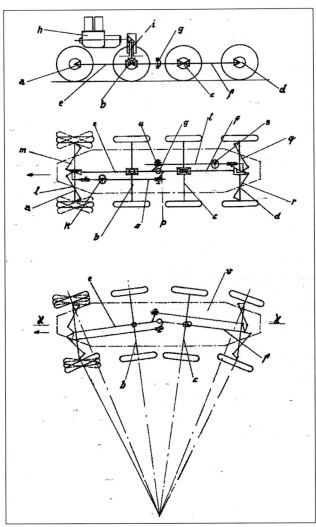

Daimler-Benz's patent for the MTW 1's design showed the spinal member that was hinged at its centre to create front and rear four-wheeled bogeys that both steered and drove. Front and rear wheel pairs received additional steering angle.

A schematic showed the internal layout of the Daimler-Benz MTW 1, including the location of its controls for driving in reverse. Remotely driven fans cooled the radiators and engine bay.

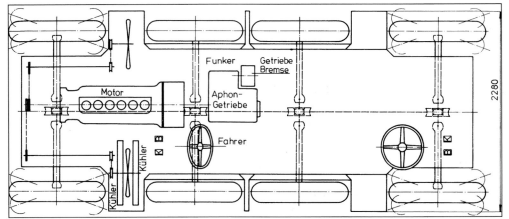

The first MTW 1 under construction showed the leaf springs, each acting as a quarter-elliptic to control its respective swing axle, and the insertion in all available apertures of cork to enhance buoyancy.

Forward steering was on the left and that rearward on the right of the MTW 1's chassis. Rear-motion gear, steering wheel and levers, was made removable, purportedly to disguise the vehicle's intentions from the Control Commission – although what other purpose the MTW 1 could have would be hard to divine. For marine buoyancy the wheelhouses and external skin, extending out and up to a flat top, carried coatings of cork bricks.

Created in Stuttgart, the radical MTW 1's components were sent to the DBAG's Berlin-Marienfelde works to be married to their hull. Serious assessment did not commence until 1929, a year in arrears, and was on the brink of completion in 1930, with trials with troops scheduled for the late autumn. Hopes were high on both sides for sums of RM 250,000 and RM 275,000 in successive years to be available for further development.

Although the eight-wheeled Magirus design showed promise, with its aluminium wheels and Maybach engine, Porsche's creation for the DBAG seemed to have an edge with its 40mph speed. The Büssing-NAG entry was out of the running early, its total of ten wheels not taking well to being braked to provide its steering. The DBAG and Magirus saw their prototypes accepted by the Reichswehr by the beginning of 1931. A good sign was selection of the MTW 1 for further trials at the secret Kama proving grounds in Russia in 1932.

With the economy in sharp reverse, however, no further development could be

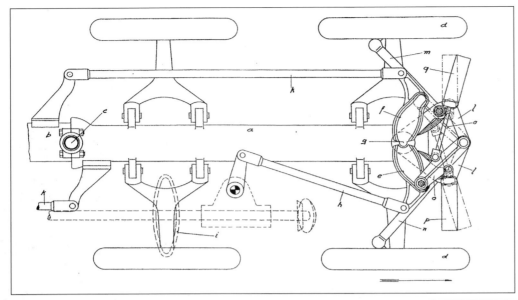

The DBAG patented the ingenious linkage that controlled both the movement of the MTW 1's front and rear bogies and the further turning of the most remote wheel pairs. Wide bases carried the swing axles guiding the wheels.

Carrying out its initial trials in 1929 bereft of its turret, Porsche's MTW 1 was as sleek as a seal and obviously able to cope with a variety of conditions. Its speed of 40mph met the Reichswehr's requirements.

Although far less attractive with its carapace of cork, the MTW 1 lost none of its traction over difficult ground. Turrets were not thought necessary for trials extending into 1931, when the DBAG design was accepted.

funded so these promising projects languished. A straw in the wind had been a meeting on 18 March 1930 in which a representative of the HWA said that 'for the time being the manufacture of these vehicles is out of the question, since vehicles of this type and size are too costly for the current state of the Reich's resources'.

Nevertheless it was obvious that six-wheeled auto-derived vehicles of the ilk of the Mercedes-Benz G 1 would not be up to the demands of reconnaissance in modern warfare. 'The results so far with the multi-wheeled vehicles promised favourable prospects,' judged military historian Walter J. Spielberger. 'They were the only vehicles that combined good off-road performance with high speeds on roads, two requirements that absolutely have to be fulfilled by a strategic reconnaissance vehicle. Thus the viewpoint of January 1931 that the development of multi-wheeled vehicles could only occur over the long term no longer applied.'

Ironically the winner of a fresh competition was first-round loser Büssing-NAG, based in Leipzig's Wahren district. On traditional lines, albeit with all-wheel independent suspension and steering, it fashioned a much larger armoured scout car that came into use in the late 1930s. An angular monster, it was bereft of aquatic aspirations. Proving highly versatile, 1,235 of this version were made. Of a successor with more power and an integral body-frame, introduced during the war, some 2,300 were ultimately produced.

Although with a minimum of freeboard, the MTW 1 was a good swimmer in both directions. Plans for further trials in Russia in 1932 were dropped although the underlying concept had acknowledged potential.

CHAPTER 7

ARMY TANKS AND NAVY TWELVES

Daimler-Benz accepted another Reich contract in parallel with its work on eight-wheeled vehicles. It was signed on 26 March 1927, only four days after its agreement to build a multi-wheeled scout car. This was a massively more ambitious project: design and construction of a new battlefield tank. Although originally mooted as early as 1925, to be addressed by Krupp and Rheinmetall as well as Daimler, the project took time to mature. The tanks were given the cover name *Grosstraktor* or 'Big Tractor'.*

Stuttgart's DMG cut its teeth on tanks with the A7V of 1917, a design by Joseph Vollmer powered by two 100hp fours. With 18-strong crews, several took part in history's first tank-to-tank battle in April 1918.

For Daimler this was not virgin territory. Having witnessed the impact of the first British tanks in 1916, Germany decided to have her own. This was the responsibility of the General War Department's Section 7 for Transport and Communications, abbreviated as A7V. In November 1916 it commissioned Joseph Vollmer to commence design of a tank using Holt caterpillar tracks, samples of which had been left in Austria after the American company's aborted production efforts there.

Having created the first cars made in Gaggenau by Orient Express in 1895, Vollmer was no babe in the woods of vehicle design. After serving Berlin's NAG as chief designer he set up his own engineering consultancy in 1906 with his colleague Ernst Neuberg, specialising in truck designs. Thus he was available to assist A7V in a project which received only moderate support from a military that still thought its troops could cope with the new-fangled British weapons on their own. Commissioned a captain, he set to work to build the best tank he could with the material at hand.

Vollmer worked in partnership with the Daimler Motoren Gesellschaft to produce the 'A7V' as the vehicle was known. Powering a pair of its Holt caterpillar tracks was a brace of Daimler engines, four-cylinder sleeve-valve units that produced a total of 200hp. A single aero engine would have been a more logical choice, but demand for these for aviation was intense. The Vollmer design was a two-storey concept that resembled nothing so much as a rolling steel-sheathed pillbox with its single cannon and six machine guns. Scaling 35 tons, it could saunter at up to 10mph.

After viewing a wooden mock-up in January 1917, the A7V ordered a hundred copies of its namesake. In October of that year the first was delivered. The model's participation in the first tank-versus-tank encounter in history occurred on 24 April 1918 when three A7Vs met a similar number of British Mark IV tanks after a fleet of 13 of the German machines had taken Villers-Brettoneux. An

---

* Work was also launched on samples of a Leichttraktor or 'Light Tractor' in the range of 10 to 12 tons.

inconclusive contact left heavy damage on both sides in what was the only significant wartime encounter of its kind.

In all Daimler completed 20 A7V tanks before the armistice. Although not without merit, their inability to vault trenches was a disadvantage, as was their 50-mile range and their high crew demands – up to 18 men. 'If the clumsy model contrasted unfavourably with the usual perfection of German technology,' wrote William G. Dooly, Jr, 'it reflected Germany's flair for the bizarre in weapons.' So did the initial work that Vollmer and Daimler had completed on a 150-ton monster tank.

Such was subsequent progress in tank design that Daimler's experience with Vollmer's creation left little of value to Porsche and his team. However, he had knowledgeable support from Otto Köhler and Benz stalwart Max Wagner. Confronting them were detailed requirements including a weight of 16 tons, maximum speed of 25mph and minimum of 1.9, ability to traverse an 8ft trench and full amphibian capability with a water speed of 2½mph. Thankfully for the latter requirement only light armouring was specified.

Although a top-mounted turret was provided, housing a short-barrelled 75mm cannon, visually the Daimler-Benz GT I* tank resembled its British antecedents with tracks completely surrounding its side sponsons to provide the ultimate in trench bridging and obstacle mounting. All three entrants were identical

Among the Army's requirements for their 'Big Tractor' tanks of 1927 was a rear-mounted turret. This excrescence, as seen on the DBAG's GT I, was soon excised by the Reichswehr.

Propelling the GT I was a decade-old aero engine, the 31.2-litre Type DIV b, modified to suit its new role. In spite of its 260bhp, it would prove to be a weak link in the design of the Daimler-Benz tank.

Starting the main engine of the GT I through a train of gears at its flywheel end was a 10hp DKW twin, placed low at the right. Hydraulic pumps and other accessories were driven from this aggregate.

---

* Some references quote 'G.T. 1' and others 'G.T. I.' A preponderance of uses causes the author to lean towards the latter, expressed for simplicity without stops.

The aluminium housings of the GT I's transmission were loosely assembled for trial purposes. Hydraulic cylinders on the smaller case in the foreground shifted the two-speed gearbox giving the tank a choice of overall ratios.

The GT I transmission revealed its two-speed gearing and, behind it, the planetary transmission that gave three forward ratios and reverse. It was controlled hydraulically from the steering wheel, a distinctive Porsche touch.

in this respect as well as mounting a small rear turret and gun, an idea which was soon abandoned. All three also used in-line six-cylinder aero engines, a BMW for Krupp and Rheinmetall and for the DBAG entry a 31.2-litre unit from Daimler's inventory. Type DIV b, this was rated at 260bhp at 1,450rpm. Complete with a 10hp DKW two-stroke engine serving as a starter motor, the power aggregate weighed 1,320lb.

Aiming to make life as easy as possible for the GT I's driver, the DBAG experts provided him with gear-ratio selection by a lever at the hub of the steering wheel. This sent electrical signals to a controller that manipulated hydraulic pressure to make the shifts. Both innovations were straight from the Porsche playbook. While the use of electrical commands was derived from the engineer's profound experience in this realm, the wheel-hub shift control was a feature first used on his 1907 design for Emil Jellinek's short-lived Maja, in which it controlled a patented Diamant transmission.

From the GT I's engine a train of three gears took its torque to the transmission. This was planetary with four controlling brakes, its output driving through a conventional two-speed gearbox which was also hydraulically shifted. It gave a total of six speeds forwards and two in reverse, which the driver selected by moving the lever to the appropriate quadrant position. A foot pedal freed the multi-disc clutch.

Drive through a large drum footbrake, also hydraulic, was rearward through bevel gears and a cross shaft to a gearbox at each side. Within each a planetary gear train was controlled by a brake which reacted under hydraulic control, through a worm drive, to change relative track speed for steering.

Propulsion of the GT I to meet the Reichswehr's amphibious requirement was by two propellers which were turned to direct the craft. Its drive combined with its steering to create a complex assembly.

Manual brakes were also included in each box, from which a gear reduction turned the drive sprocket at the rear of the relevant track.

From the left-hand gearbox a drive powered two three-bladed propellers at the rear of the GT I, engaged by means of a lever. Aquatic steering was by changes in the angle of the propellers. Within the GT I a six-bladed fan cooled radiators on both sides, a design detail shared with the eight-wheeled MTV 1. Housing machinery and crew was a hull weighing 9,245lb, fabricated as specified of mild steel 0.55 inch thick.

Supporting the GT I on its tracks were three four-wheeled bogies on each side, plus a fourth on the slope at the front. Each wheel had a hard-rubber Continental tyre. The bogies consisted of pairs of wheels on a walking beam, two of the latter joined at their centres by the ends of a pair of semi-elliptic leaf springs. In turn, the pivoted centre of the leaf-spring assembly was guided by a trailing link from the hull. Next to the pivot, the centre of the spring grouping abutted against a pair of hydraulic energy absorbers that transmitted impact forces to the hull. Thusly combined these elements gave flexibility which, said the DBAG, 'should allow greater off-road agility to be achieved'.

Using paired leaf springs with hard-rubber rollers on balance bars, Porsche's design for the GT I's track bogeys had features in common with Britain's near-contemporary 6-ton Vickers tank.

A year after the contract's signing, Daimler-Benz submitted an invoice to the HWA for work that had been required on its GT Is over and above the agreed emolument. The added cost came to RM97,130 for design and manufacture. At that time the plan was to have the prototypes ready in the summer of 1928. Testing was to follow in 1929 and 1930 with production starting in 1931. Each vehicle was to cost the Army some RM150,000.

With components made in Untertürkheim the GT Is were assembled at the same Rheinmetall facility in Unterlüss, between Celle and Uelzen, that was being used by that company to ready its competitive tanks. This was effectively an Army facility because in 1925 the German Reich acquired a majority stake in Rheinmetall, which in 1921 had been allowed to restart small-scale production of medium-calibre weapons. To maintain secrecy no vehicle trials were permitted in Germany. Only firing tests of the cannon were sanctioned.

Trials of the *Grosstraktor* prototypes would take place in Russia. 'In what seemed an improbable concurrence of interests,' wrote historian Alfred M. Beck, after the war 'Soviet Russia and Germany began a cautious approach to each other driven by economic necessity and their common status as pariahs in international circles of the time. A Russo-German commerce treaty was signed in May 1921 and a secret economic agency had already begun the process of reopening Russian markets for German capital investment.' After the signing of their Rapallo Treaty in 1922, Beck added, 'a season of open political and economic co-operation began between the two European outcasts'. Confirming this was the Berlin Friendship Treaty of April 1926.

Under the resourceful General von Seeckt, in the military sphere this co-operation took the form of three major experimental stations in Russia. They were a gas and chemical school near Saratov, an aviation establishment at Lipetsk and an armoured-warfare centre near Kazan, capital of the Tartar Republic on the banks of the Volga.* Cleverly the creators of the armoured centre gave it the code name 'Kama' after Kazan and its first commandant, Lieutenant Colonel Malbrandt, only to discover that their cover was blown by the Kama River, flowing nearby as a major tributary of the Volga.

Although technically not breaching the terms of the Versailles Treaty because no German weapons were made there, the *Panzertruppenschule Kama* was a vital link in the reestablishment of a German armoured capability. Also used by the Russians, the centre would train almost 150 German tank commanders before its closure to the Reich in 1933. Establishing and maintaining it was a moral wrench for von Seeckt, who was violently anti-Bolshevik, but he was a pragmatic manager who recognised and reacted to the imperatives of the day.

In mid-June 1929 the first Daimler-Benz GT I was packed in a case and delivered to Stettin for its sea voyage to Leningrad, where it was transferred to a train. Arriving at Kazan station, it was trucked to the Kama grounds 4 miles distant. Testing began in July. Its turret was removed and set up in a jig for sighting of its cannon. Before road trials its drive train was subjected to static tests. Arriving later, the second GT I prototype followed suit.

The lack of testing at their production sites soon showed itself in problems with the tanks from all three makers, not least a deep dunking for Rheinmetall's sample during its amphibian exercise. Although the DBAG entry's travails were centred on its adventurous hydraulically controlled transmission, the GT I's adapted aero engine contributed faults of its own. Equipment and working conditions at Kama were inadequate to make the design changes required. The immense distance from home factories was a major handicap.

Although the Krupp and Rheinmetall offerings managed 39 and 83 miles respectively in 1929, the GT I's running was negligible. Total mileage for the Daimler-Benz 'Big Tractors' at Kama was 16 miles in 1930 and 19 miles in 1931.

Among the tanks assembled in August 1935 at Munster for a review by General Werner von Blomberg were the two Daimler-Benz GTIs plus prototypes from Krupp and Rheinmetall. A Mercedes-Benz G 4 was their transport.

'It's incomprehensible,' Walter Spielberger later reflected, 'why Daimler-Benz didn't dig in to deal vigorously with the evident design faults.' In fact a major deterrent was the absence of the tank's moving spirit, for a bitter disagreement with his board led to Ferdinand Porsche's departure from the DBAG at the end of 1928. For neither side was the amelioration of these tanks a priority.

None of the competing tanks reached production at this early clandestine stage of Germany's rearmament. They contributed, however, to an immensely improved understanding of the issues of speed, power to weight, design and drive of tracks, types of steering, relationship of chassis to firing ability and development of radio communication, knowledge that could not have been gathered in any other way before 1933. They also demonstrated conclusively that the idea of an amphibian tank was a non-starter.

Not until 1933 were the sample tanks returned to Germany, those of Daimler-Benz to the Berlin-Marienfelde plant. There they were put in condition to be mustered for special occasions such as an August 1935 visit by General von

---

* Movements to and from these establishments were conducted in secrecy, so much so that the corpses of any fatalities in Russia were returned to Germany in crates marked 'machine parts'.

Blomberg to a display of Germany's Panzer power at Münster. Both GT Is ended their careers as gate guardians at military bases. One stood watch at Wünsdorf south of Berlin, home of the Fifth Panzer Regiment. The other took up its defence of the First Panzer Regiment at Erfurt.

A footnote to this undertaking was Daimler-Benz's role in another HWA initiative, the creation of prototypes of a *Leichttraktor* or 'Light Tractor'. Commercial director Carl Schippert, who had signed the *Grosstraktor* contract, told the DBAG management board on 22 December 1927 that 'a new Panzer business' was in the wind, the building of smaller prototype tanks that should weigh some 8 tons. The board said it was in agreement as long as supervisory board chairman Emil Georg von Stauss approved.

Formal letters requesting participation went out on 16 June 1928 to Krupp, Rheinmetall, Daimler-Benz and separately to Ferdinand Porsche. A hint of hesitancy on the DBAG's part was the provision that each should build two prototypes, but if the Stuttgart company did not participate the remaining two companies should make three apiece. In fact Daimler-Benz chose not to build Light Tractors, saying that it would work instead on a Panzer supply vehicle. It was in the frame in any case because the other makers chose its M 36 six-cylinder engine for their vehicles.

With engines a special forte of the DBAG, even under the Versailles constraints it could not neglect the field of aviation in which it had performed magnificently during the First World War. It was still free to produce small powerplants for light planes, an opportunity that Paul Daimler had not overlooked and that Ferdinand Porsche seized with both hands.

Let down by problems with their engines and transmissions, the two GTIs ended their careers as gate guardians. For Ferdinand Porsche, however, they were a first essential taste of the demands of Panzer design.

Pushrods from a central camshaft opened four valves per cylinder of the F7502 through siamesed rocker arms. Development of the useful Versailles-permitted engine continued under Porsche's aegis.

Inherited from Paul Daimler by Porsche was a horizontal twin-cylinder aero engine, the Type F7502, producing 20bhp from 105lb. For optimum cooling its exhaust ports faced forward into the wind and prop blast.

In 1925 Daimler's Sindelfingen Works introduced this Klemm-designed parasol monoplane, the L 21. Equipped with two F7502 twins, a sister L 21 to this aeroplane won its 40bhp class in 1925's Round-Germany Competition.

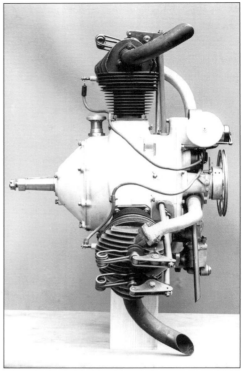

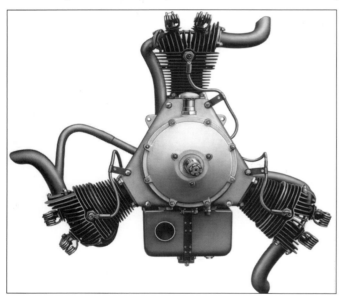

Seeing an opportunity, Porsche's team designed their F1, a three-cylinder 1.5-litre aero engine giving a maximum of 34bhp from its 126lb. Hemispheric combustion chambers were fed by two vee-inclined valves.

Pushrods and rocker arms opened the valves of the F1, which had one spark plug per cylinder. Its opportunities were spoiled by the split between Daimler and aircraft designer Hanns Klemm, who set up his own company in 1926.

Offered into the Porsche era was the F7502, displacing 884cc in two air-cooled cylinders that opposed each other. Cylinders made of steel were bolted to the crankcase and screwed into their iron cylinder heads, each with a complement of four valves. Opening them were pushrods and rocker arms from a gear-driven camshaft positioned above the roller-bearing crankshaft. Its output of 20bhp at 3,000rpm was geared down to a more appropriate 1,000rpm by a planetary reduction gear. Complete with its Bosch magneto sparking one plug per cylinder the F7502's weight was 105lb.

Under Porsche's direction the F7502 was joined by a radial three-cylinder engine, dubbed the F1. Although this had only two valves per cylinder, they were vee-angled for hemispherical combustion chambers in aluminium heads and closed by light and efficient hairpin valve springs. The F1's vertical cylinder housed its master connecting rod, to which the other two were joined by link rods. Roller bearings were liberally fitted and oiled by a pressure pump from an external reservoir.

With 1.5 litres the air-cooled F1 was rated at a peak of 34bhp and a continuous 30bhp at 2,800rpm. Thanks to its integral planetary reduction gear this gave a propeller speed of 933rpm. The attractive F1 triple weighed 126lb. Starting it by hand from the pilot's seat was possible, Mercedes-Benz said. Continuing into

1928, the F1's development struggled when suitable customers were found to be thin on the ground.

Among applications for these engines were aircraft designed by Hanns Klemm for an arm of Daimler that began producing light aeroplanes from 1919. In the mid-1920s the construction of Daimler aircraft was suspended until pending corporate arrangements were settled. This occurred during 1926 with the founding of the Klemm Leichtflugzeugbau GmbH in the Stuttgart suburb of Böblingen. Hanns Klemm conducted his own company with considerable success.

Maintaining his contacts with the relevant ministry, Ferdinand Porsche told his board on 26 March 1926 that the government was preparing its thoughts on future engine requirements. In the meantime he and his colleagues began flexing their aviation muscles more robustly that year with orders for production of one of their First World War favourites.

Both during and after the war many training planes were powered by the 120hp D II six, a reliable overhead-cam engine internally known as the F12556. Trainers built not only in Germany but also in Holland, Norway, Denmark, Sweden and even Russia depended on this engine, supplies of parts for which were dwindling. A solution was found in a contract with a French distributor of aeronautical components, Marabini-Aviation of Paris.

Although Marabini would brand the D IIa engines as its own, stating that they were produced under licence, they were actually made in Untertürkheim. The first to be built completed a 100-hour proof test in December 1926. Now rated at 130bhp at 1,450rpm, it could rev higher to deliver 145bhp. Equipped with a self-starter, this 450lb engine allowed a new generation of aviators to discover the merits of aero engines made by Daimler-Benz.

Feelers to Germany's Army aviators for the development of more modern engines had borne little fruit for the DBAG, Porsche concluding that his erstwhile colleague Camillo Castiglioni had insinuated his BMW into pole position with the relevant authorities. It was thus all the more rewarding when the Navy, instead, came knocking.

Leading the Navy's delegation on the morning of 24 November 1926 was Ministerialrat Wilhelm Laudahn. Scant weeks younger than Porsche, Laudahn was an engineer who had introduced diesel power and directed other advanced technical projects for Germany's Navy. Accompanied by two colleagues, Laudahn met with Wilhelm Kissel and Richard Lang as well as Porsche at 10 o'clock.

As soon as possible, said the visitors, they would like to see a supercharged aero engine developing some 800hp in air- and water-cooled alternatives. This led to a lively and detailed discussion between the two technicians about the implications of this request. Kissel asked whether this was likely to develop into a regular business for Daimler-Benz and also whether the Navy was prepared to provide financial support for the engines' creation.

Asked what they had in mind, the DBAG men withdrew to prepare a response. Returning, they avowed that such a programme would cost in the order of 400,000 to half a million marks and that the Navy could see an engine by October 1927. The naval mission said that that sum could be forthcoming at least by April of the coming year, depending on reaching the necessary agreements. All concurred

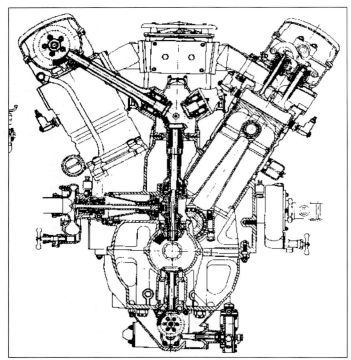

Alfred Berger and Hans Nibel supported Porsche in the design of the F2, a 53.9-litre V-12 commissioned by the Navy in 1926. A single camshaft above each bank was driven by shafts and bevel gears from the tail of the crankshaft.

that discussions should be verbal to the extent possible rather than written to maintain the utmost in confidentiality.

Here, for Ferdinand Porsche, was a happy reminder of the way the Austro-Hungarian Navy had commissioned a V-12 engine from Austro Daimler during the First World War. As in that undertaking the aim was to have a suitable power unit for the Navy's future large flying boats. The horsepower target was right at the cutting edge for contemporary aero engines. While Renault, Lorraine-Dietrich and Farman produced engines of 700bhp, only Fiat exceeded that – and the Navy target – with its 900bhp A25.

For this project Porsche drew on his own experience and that of Benz men Hans Nibel and Alfred Berger, architects of Mannheim's First World War aero engines. All had built V-12 engines, the configuration that they chose for this assignment. A key issue would be capacity. How big did an 800hp engine have to be? Leading recent designs were developing 16–17bhp per litre without supercharging. That suggested a swept volume of between 47 and 50 litres. Any risk of failure was skirted by settling on a capacity of 53.9 litres for the base engine with supercharging as a bonus. Each of its cylinders displaced 4½ litres, as much as a luxury car of the era.

Far bigger than his 30-litre twelve of the First World War, this would be the largest engine that Porsche had yet built. Like that engine, the F2, as it was dubbed, would be a 60-degree V-12 with each bank having a single overhead camshaft. An array of shafts and bevels at the engine's anti-drive end turned the camshafts, twin water pumps along the flanks, ignition apparatus, dry-sump lubrication pumps and, indirectly, two Junkers fuel pumps. Four vertical valves per cylinder were opened by individual roller fingers for the exhausts and a single forked finger for the inlets. For security the coil valve springs were duplicated.

Roller bearings were used throughout the bottom end, which had master connecting rods and link rods to cope with the big-end bearings of the facing cylinders. Individual fabricated-steel cylinders with integral heads were mounted on a split magnesium crankcase. Its underside was enclosed by a multi-bolted pan that could be removed for inspection and/or attention while the engine was mounted in its aeroplane. A dynamo in the central vee topped up an electrical system that included a Bendix-Eclipse inertial starter and permitted coil ignition for one set of spark plugs. The other set in the F2's dual ignition was magneto-sparked.

The tail of the crankshaft drove an integral supercharger, centrifugal with an impeller just over 14 inches in diameter. Stepping up its speed was a set of planetary gears. These had their own internal housing which also carried the unit's ring gear. When supercharging was not needed the housing and its planetary gears idled. At the pilot's command a band brake anchored the housing, putting the gearing and blower into action.

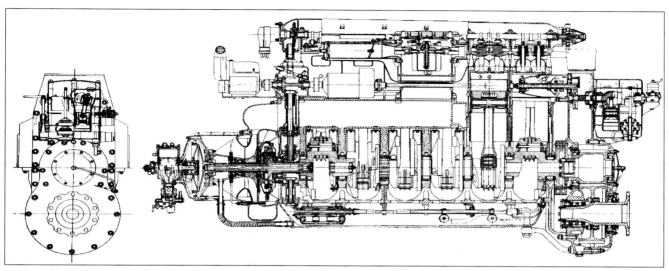

A longitudinal section of the F2 concept depicts the planetary drive of the supercharger that ran free until engaged by a restraint on its outer housing. On this marine BF2 version a large clutch engaged a pneumatic starter.

An inertia starter protruded from the blower housing of the F2 V-12, one of whose pair of water pumps is visible. The engine pictured carries a reduction gear almost halving the crankshaft speed for the propeller's benefit.

Pressurised or not, induction air was fed to four carburettors within the vee of the Mercedes-Benz F2 V-12. Its continuous sea-level output was 800bhp at 1,580rpm from a weight of 2,160lb with its reduction gear.

This modus operandi was an ingenious echo of the Daimler-Benz automobile policy, which gave drivers of its supercharged cars the option of whether or not to use the blower. When engaged in the F2 the charger delivered pressure air to a block of four horizontal Daimler-Pallas carburettors and through manifolding to the inlet valves. The blower was sized, said the DBAG, to maintain sea-level pressure up to 4,000 metres or more than 13,000ft. The V-12 was to be offered either with direct drive or a 1.985:1 reduction gear.

In both April and July 1927 Ferdinand Porsche briefed his board on the engine design's 'favourable progress'. At its meeting of 22 December the board decided to cease all designing and testing of aero engines save for the F2 project. Since Klemm needed the F7502 twin for his aircraft, it concluded, he could be offered the licensing of its production. This did not eventuate, Klemm switching instead to engines from Hirth, Argus, BMW and even a DKW two-stroke for a *Volksflugzeug*.*

On 10 May 1928 the board learned from Ferdinand Porsche 'that the new large aero engine is assembled and will be on the test bench next week. There is great interest in this engine, and the Dornier company has asked us for a meeting so that consideration can eventually be given to the manner of its installation.' Closely allied with the Navy and a world leader in seaplanes and advanced aircraft, Dornier was a logical candidate to put this engine to use.

Developed through 1929 and into 1930, the F2 delivered 800bhp at 1,580rpm at sea level, its rating for continuous use, and 880bhp at 1,600rpm as take-off power. Maximum supercharged output was 1,030bhp at 1,700rpm. It core weight of 1,875lb rose to 2,160lb with all accessories including the reduction gear.† Its output figures ranked the F2 among the most powerful aero engines of its time, along with the Fiat A25 of 950bhp and Sunbeam's massive Sikh III airship V-12, rated at 1,000bhp. Weighing well over 2lb per horsepower, however, save in its boosted form, the big Daimler-Benz V-12 was behind the curve of contemporaries achieving 1.6–1.7hp per lb.

As early as October 1928, with Ferdinand Porsche still in residence, Daimler-Benz released sales information on the F2 in German and English. An engine displayed at London's Olympia in mid-1929 had reduction gears. *Flight* noted that it was unique among the supercharged engines on show in having a blower that boosted into the carburettor(s) instead of drawing from it as did the others. This reflected the DBAG's passenger-car practice.

Dornier's interest notwithstanding, the two prototypes of the F2 remained the only such units produced. Neither reaching installation in an aeroplane, they were the subject of testing through 1932. As in the GT I tank project, Porsche's departure at the end of 1928 seemed to deflate the company's interest in this aeronautical project, which had obvious military implications. It did, however, have significant issue, not least the air-cooled version that the Navy had also requested.

Unlike other air-cooled aero engines, which exploited the medium through which they flew, the F3 – as it was designated – carried its own internal cooling. Its architecture was the same as that of the F2 as regarded crankcase, reciprocating parts and valve gear, with its cylinder bore reduced from 165 to 155mm to make room for the fine finning that surrounded each individual cylinder. Separate ports were given to all 24 exhaust valves.

---

* At a board meeting in May 1928 Porsche fought a rear-guard action on behalf of his air-cooled F1 triple, saying that it had been well received as being ideal for light training aircraft and thus could deserve the DBAG's funding of further development. After extensive discussion the board decided that this could only continue if, in the absence of other customers, Klemm were to engage in a substantial contract for the F1 at a profitable price, secured by an advance payment. This did not take place.

† Weight figures are from data collated from DBAG internal documents. Weights quoted publicly at the time were more optimistic at 1,650lb and 1,800lb respectively.

A shaft down the F3's central vee drove six sirocco blowers, one between each pair of the V-12's cylinders. Lifting two of its internal baffles revealed the finely finned cylinders. Its heads as well as cylinders were fashioned individually.

The entire interior of the F3's central vee was given over to its cooling system. Powered by a shaft running the length of the engine, six sirocco blowers rotated at high speed, one placed between each of the facing cylinder pairs. Drawing air from above, the blowers pumped cooling air through carefully baffled passages that funnelled the main outlet past the crucial exhaust valves and ports.

Displaced from the central vee by the cooling system, induction arrangements were elsewhere in the F3. Equipped with two inlets and two outlets, the supercharger was continuously engaged. Induction manifolding was a log alongside the inner surface of each cylinder head, receiving mixture from both ends to give the best possible distribution. Pipes along the sides of the crankcase carried pressure air to four updraft carburettors, two at each end of the V-12, which fed the manifolds. The dual spark plugs had to be accessed from the F3's central vee.

Here was an astonishing engine by any standard, another remarkable work from the pens of Porsche, Berger and Nibel. With its integral cooling it could well have suited as a power unit for tanks, in which its all-up weight of 2,410lb would have been tolerable. The output of the only F3 built was ascertained as 600bhp at 1,450rpm, 640 at 1,500 and 700 at 1,600. It was and remained a fascinating and doubtless costly engineering exercise.

A version of the F2 with greater sales potential was the BF2, a marine version of the 53.9-litre V-12. In all respects its features echoed those of the F2, with its reduction gear inverted so that it suited a low-placed drive to a propeller. With its cooling system still preferring fresh water, preferably distilled, it required a heat exchanger to interface with seawater. Fuel was to be either leaded petrol or a half-and-half blend of petrol and benzol to produce a supercharged output of a steady 800bhp at 1,600rpm or 950 at 1,700, sustainable for half an hour at a stretch.

Here was a significant return from its investment for the *Kriegsmarine*. In the post-Porsche era three of the BF2 versions of his design powered Germany's first

Finning surrounded the individual exhaust ports of the F3, the air-cooled version of the Navy's V-12. With blowers occupying its vee, the F3's induction system was fed from both ends through four updraft carburettors.

Twin inlets and twin outlets were features of the F3's supercharger. With the smaller capacity of 47.6 litres the air-cooled engine developed a maximum of 700bhp at 1,600rpm for a weight of 2,410lb. Only one was made.

For the BF2 marine version of Porsche's Mercedes-Benz V-12 the reduction gear was inverted, the better to mate with a ship's propeller shaft. Two Junkers fuel pumps are driven from the sump.

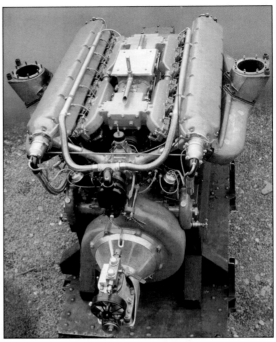

While outlet manifolds carried the exhaust to the deck above the marine V-12, the BF2's inlet passages remained deep in the vee feeding Pallas horizontal carburettors. Its cooling water released heat to the sea through heat exchangers.

Differing from its sisters in its twin overhead camshafts on each bank, the OF2 version of the V-12 was oil-fuelled. Naturally inducted, it was developed under Fritz Nallinger to produce continuous power at 1,750rpm of 750bhp.

In this configuration the BF2 V-12 had triple ignition with two plugs on the outside and single plugs in the vee. Its pneumatic starter motor had its own cooling fan. Maximum power was 950bhp at 1,700rpm.

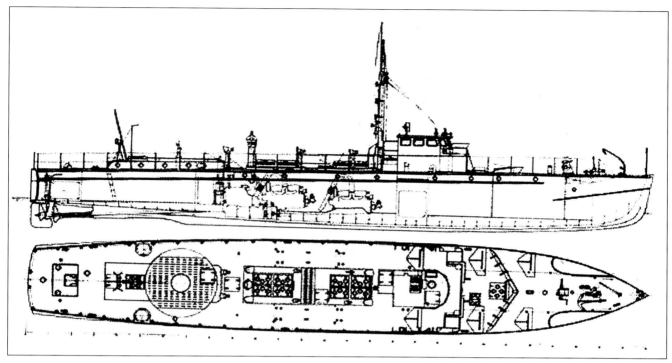

modern *Schnellboot*, a fast attack boat. Commissioned from Bremen's Lürssen in August 1930, this displaced 51.6 tons and was mahogany-planked on an aluminium frame. Ultimately known as the S-1, first of the *Schnellboot* line, it was powered by three BF2 V-12s. While the centre and port engines had conventional rotation, the starboard BF2 had reverse rotation to minimise torque effects.

Fitted with two torpedo tubes, a 20mm anti-aircraft weapon and a machine gun, three boats were launched in this initial series. All had BF2 power that delivered a sustained speed of 34 knots and good manoeuvrability. This was an important foot in the door for Daimler-Benz, which competed with MAN for the supply of engines for subsequent S-boat generations. With petrol engines rejected on the grounds of their flammability, diesels were preferred. The DBAG eventually prevailed with its marine diesels.

Those seagoing diesels from Daimler-Benz traced their origins to yet another version of Ferdinand Porsche's F2 V-12. First run on 14 November 1930, this was the OF2, the 'O' standing for *Ölmotor* or oil engine. Created with aviation in mind, at a time when a number of companies were developing airborne diesels, it had completely new valve gear with twin overhead camshafts on each bank, still shaft-driven. At the centre of its vee were two six-cylinder Bosch injection pumps in line, delivering fuel oil to the pre-chamber at the centre of the four valves atop each cylinder.*

With a compression ratio of 16.0:1, no supercharger was fitted to the OF2, which in its initial version carried propeller reduction gears. Called from Benz to Stuttgart after Porsche's departure, as Hans Nibel's right-hand man Fritz Nallinger took charge of testing, including work on high-speed diesels. Bringing this new kind of engine to life, Nallinger defined its continuous power delivery as 720bhp at 1,720rpm. For five-minute periods, he concluded, 750bhp at 1,750rpm could

Their construction starting in 1930, each of the five first ships in the Schnellboot *line was powered by three BF2 engines. Creation of these S-boats, launched by Porsche and the* Kriegsmarine, *initiated the most successful line of their kind.*

---

* Early in the 1920s both Benz and Daimler had developed diesel engines, which meant that at the time they forged an alliance in 1924 a choice had to be made between their conflicting technologies. This choice was Ferdinand Porsche's responsibility. In 1925 he resolved the conflict between Untertürkheim and Mannheim in favour of the latter's pre-chamber technology as developed by Prosper l'Orange. This promised simpler and lighter construction that would be less costly to produce.

be delivered, with 800bhp at 1,790rpm available for one minute. Including a small two-cylinder engine for starting, the OF2's weight was 2,365lb.

Test results with this experimental engine led directly to the design of the 1933 Daimler-Benz LOF 6, soon renamed the DB 602. This was an 88.5-litre diesel designed for airship propulsion. Its cylinder dimensions of 175 x 230mm were scaled up from the OF2's 165 x 210mm for 16 rather than 12 cylinders at a vee angle of 50 rather than 60 degrees. This tighter angle and the use of pushrod valve gear meant that its four Bosch oil-injection pumps were placed laterally at the front instead of in the central vee, where space was limited.

Developing a maximum 1,320bhp for its weight of 4,360lb, the DB 602 powered two Zeppelins, the LZ 129 *Hindenburg* and its sister the LZ 130 *Graf Zeppelin*. In these it was partnered by a 50bhp four-cylinder diesel to drive generators and accessories. Compressed air was used not only for starting but also to shift its single camshaft to reverse the engine's rotation for manoeuvring. Accessible on board to the Zeppelin's engineers for repairs, the DB 602's external design was handsomely executed. In the LZ 129 the engines drove pusher propellers while in the LZ 130 they gave tractor propulsion.

Their Zeppelin assignment was a coup for Daimler-Benz, for the airships had long been Maybach-powered. Appropriately for such a starring role, the V-16s were handsomely designed and finished. Under its MB 502 designation for the *Kriegsmarine* the same Daimler-Benz diesel V-16 powered four of the 1933 series of S-boats, each of which had three engines. The MB 502 also served as the starting point for the V-20 MB 501 diesel, whose 2,000hp output gave it a range of military applications including later and faster S-boats. All were direct descendants of Porsche's V-12s.

Although supporting a massive workload in the automotive products of the merged company, Ferdinand Porsche made profound contributions on land, sea and air to the efforts of Daimler-Benz to find a military role in the difficult conditions of the financially strapped Weimar Republic. He had accepted the technical helm of Daimler and then Daimler-Benz at a time of phenomenal turbulence in the company and its host nation.

At the end of 1928 Porsche was disappointed by a board that said it was reluctant to go to the banks for new-model funding at a time when some of its new models were being criticised for various problems that Porsche was confident could be fixed. On 15 November he mustered a defence of his engineering colleagues who, he said, in the recent past had developed 14 new types, not including experimental engines. They were overloaded, he said, by too many tasks. Two years earlier he had gone to bat for them in search of higher pay for some of his key people, only to be turned down by his board.

In addition to his military work, Porsche could point to his production-car progress. From only 30 per cent in 1926, Daimler-Benz's utilisation of its productive capacity rose to 60 per cent in 1927 and remained at 58 per cent in 1928. Production of cars was 7,908 in 1927, 6,275 in 1928 and 10,014 in 1929 with models from Porsche's drawing boards. It was a not-inconsiderable record. Like Austro Daimler, Daimler-Benz would live for years on models designed by Ferdinand Porsche.

CHAPTER 8

PORSCHE AND HITLER MEET

In December 1928 Ferdinand Porsche was shown the door by Germany's – perhaps the world's – most prestigious vehicle producer. While compensation was forthcoming from Daimler-Benz, for his contract did not expire until 23 March 1929, Porsche's expensive lifestyle did not admit of inactivity. Since his Lohner days he had always been well rewarded and he did not propose to lower his standards.

Porsche's financial requirements did not accord with some of the offers he now received. At the age of 53, with a glittering career behind him, he could be choosy. He was unconvinced by prospects at Skoda, which extended an invitation. Fortunately a valid option was available. This was Austria's Steyr-Werke AG. It was well known to Porsche, of course, for the 1916 decision of weapons-maker Steyr to commence car production gave him and Austro Daimler unwelcome competition in the drastically shrunken post-war Austrian automobile market.

Realising that it needed stronger technical leadership, the Austrian firm made Ferdinand Porsche an attractive offer. He took up his new post at Steyr-Werke at the beginning of 1929. Karl Jenschke, who had been with Steyr since 1922,

At the family villa in Stuttgart Ferdinand Porsche (left) *and son Ferry posed with colleagues and a Type 30 Steyr designed during Porsche's 1929–30 tour of duty with that Austrian maker of cars and weaponry.*

became a top deputy to Porsche. Other technical talents that Porsche discovered at Steyr included Walter Boxan, Josef Kales and Karl Fröhlich.

The resonance of Porsche's return to Austria was tremendous. 'Our great countryman, who first earned his fame here, is coming home again,' hailed Vienna's *AAZ*. 'When this news was spread by the daily press it caused justifiable excitement in all automotive circles, for Dr. Porsche is one of the few automobile designers who have known how to create an international reputation through their achievements. Knowing that a man with the engineering brilliance of Dr. Porsche is again in Austria must fill the Austrian automotive world with pride, and it was a great coup on the part of the Steyr-Werke AG to secure his employment.'

With his customary vigour Ferdinand Porsche plunged into the problems that he discovered at Steyr. He found a capable company – almost too capable, with its in-house manufacture of all components and its assembly-line production. But its volumes were too low to allow it to compete economically in continental markets. While sprucing up existing products, Porsche initiated work on two new models: a large luxury car that could command the prices Steyr would have to charge and a new smaller model, the 2.1-litre Type XXX, which could have volume aspirations.

Porsche would have success with one and disappointment with the other. His Type XXX, also known as the Type 30, wrote historians Jerry Sloniger and Olaf von Fersen, 'set the engine pattern for every further Steyr produced up to the outbreak of the Second World War – to say nothing of establishing a standard for high-performance long-lived production engines throughout Germany and Austria for at least the same time span.' They added that it 'could be considered an ancestor of all the non-racing high-performance engines of exceptional durability which appeared in German-speaking countries through the thirties'.

With his luxury car, the 'Austria', Porsche gifted Steyr a masterwork. That, at any rate, was the verdict of visitors to the Paris Salon near the end of 1929. It was a beautiful machine with a superb swing-axle chassis and straight-eight engine. With 24 October 1929 going down in history as 'Black Thursday' on New York's stock market, however, its timing could not have been worse. Caught up in the financial chaos was the Allgemeinen Österreichischen Bodenkreditanstalt, the august bank on which the Steyr-Werke was heavily dependent. Porsche's magnificent Austria would be stillborn.

The exigencies of 1929 found the Steyr-Werke's board agreeing a co-operative production and marketing policy with Austria's other major auto maker, Austro Daimler-Puch, a 1928 amalgamation of Porsche's old Wiener Neustadt company with Graz-based car and motorcycle producer Puch. Behind this merger, and the alliance with Steyr, was the Kreditanstalt, a financial vehicle of none other than Ferdinand Porsche's erstwhile ally and now nemesis, Camillo Castiglioni.

With a shrunken Austria struggling to support its car makers, it was evident to Porsche that the 'co-operation' would soon intensify its grip on Steyr.* And Castiglioni's ways were not those of Ferdinand Porsche. He was informed by the Steyr directors that in view of the appalling business conditions his costly three-year contract was being wound up after only one year.† Behind this lay the negative views of Camillo Castiglioni, who considered Porsche a ruinously innovative engineer.

---

* In 1934 the Steyr-Werke merged with Austro Daimler-Puch and in 1935 the company was reformed as Steyr-Daimler-Puch AG. Passenger-car production was halted at Wiener Neustadt and concentrated at Steyr.

† In his book with Günther Molter, Ferry Porsche said that the contract was wound up at Porsche's initiative, but in his earlier book with John Bentley Ferry said that the break was at the behest of the Austro Daimler-Puch group. I find this the more credible testimony.

Porsche officially left the Steyr-Werke on 23 April 1930. He remained in Austria to hatch plans for the future. He weighed his alternatives with care. His next move could, should, be his last. But to join another company? He had already achieved far more as both technical director and managing director than most engineers in their lifetimes. Placing himself at the beck and call of another board of financial executives was intensely unappealing to this man of independent thought and action.

'At my age I shouldn't take a step backwards!' he told nephew and secretary Ghislaine Kaes, saying that some firms were keen for him to lead engineering but reluctant to give him a management role. As Porsche summed up to his son, 'It makes no sense for me to keep going from one company after another.' Ferry put it differently: 'My father found that when he signed a contract with a firm, they could live another ten years on his designs, but he couldn't!'

From these musings arose the idea of setting up an independent design office for automotive engineering. With the help of a small team, Porsche could contract with companies to design and build their vehicles. He could concentrate on the engineering he enjoyed while suffering a minimum of the company politics that he detested. 'My father was a man who knew exactly what he wanted to do and how he wanted to go about it,' Ferry recalled, 'and it irked him at times beyond endurance to have to defer to opinions which he knew instinctively were not right or were intended to deflect him from his goal.'

The idea of an independent engineering office had its precedents. We have already seen the example of Joseph Vollmer and his German Automobile Design Company. Karl Slevogt, a year younger than Porsche, gave up corporate life in 1927 to set up his own consultancy. Almost four years older, Berlin-based Edmund Rumpler had made a good fist of an independent life as a car and aeroplane designer. Rumpler's example in particular confirmed for Porsche the importance of obtaining and exploiting royalty-earning patents, an art in which the Austrian had already proved proficient.

From the beginning of 1930 Ferdinand Porsche's thoughts were concentrated on the plan to establish his own company. He had a promise of financial support from Adolf Rosenberger, a successful Mercedes-Benz racing driver and wealthy dealer in iron and steel. The two men knew and respected each other from racing adventures in the 1920s. 'Rosenberger', said Porsche biographer Peter Müller, 'was one of the few people whom Ferdinand Porsche would allow to contradict him.'

Another who would assist was Porsche's son-in-law Anton Piëch, who joined as legal adviser while maintaining his Vienna law practice. To help realise his design projects Porsche turned to a lieutenant of old, Karl Rabe. With prospects for Austria's indigenous industry anything but promising, the inventive Rabe was able to accept the invitation of the 20-years-older Porsche with an easy mind.

Wealthy Alfred Rosenberger was an enthusiastic and successful racer of Porsche's designs for Mercedes-Benz in 1928–1930. In the latter year he agreed to back Porsche's establishment of an independent design office.

Pictured in later years, Vienna solicitor Anton Piëch was not only Porsche's son-in-law but also a valuable aide in Porsche's new company. He successfully unwound the engineer's contract with former employer Daimler-Benz.

Such was his respect for Porsche that Karl Rabe left a leading position at Austro Daimler to join the engineer's new company. His was the key role of leading the technical team to create in metal the large-scale concepts that Porsche visualised.

In Porsche's new design team mathematician Josef Mickl (left) was the purest theoretician. A former designer of seaplanes at Fischamend, Mickl was in demand for his expertise in the dynamics of metals, fluids and gases.

From Steyr, followed by a stint at Daimler-Benz, Erwin Komenda joined the Porsche team as body designer and engineer. He manifested phenomenal creativity in his assignments for the team well into the 1950s.

Others attracted to Porsche's new studio were Steyr's Josef Kales, Josef Zahradnick and Karl Fröhlich. After post-graduate tutoring by Steyr's Walter Boxan, Ferry Porsche joined the team, as did Porsche's nephew Ghislaine Kaes, who had been exposed to all the departments at Steyr. Franz Xaver Reimspiess, 'friendly and relaxed', came from Austro Daimler after the closing of Wiener Neustadt, as did Otto Zadnik, while theoretician Josef Mickl was a veteran of Porsche's airship days before the First World War. Steyr and Daimler-Benz had been the training grounds for body designer-engineer Erwin Komenda, youngest of the new team after Ghislaine and Ferry.

Germany was Porsche's choice as a base. It offered both customers and markets. A resident of Pforzheim, Adolf Rosenberger urged Porsche to settle there, but the engineer preferred nearby Stuttgart. In addition to having a villa there, Porsche knew the city and its capabilities well from his Daimler years. Renowned local firms like Bosch, Mahle, Reutter and Hirth stood ready to help build his prototypes. Rosenberger found quarters for the engineers on the second floor at number 24 Kronenstrasse in a modern building known as the Ulrichsbau, a convenient block from the main train station. They moved in on Monday, 1 December 1930.

News of their arrival broke on Christmas Day, when a newspaper report said that Ferdinand Porsche, 'probably the most popular man in the field of automotive design', had set up a 'neutral design bureau' so that he could be 'active as a guest designer, soon working for this respected firm or that one, soon too for two or three houses at the same time'. On 25 April 1931 the new company officially registered to trade as the Dr. Ing. h.c. F. Porsche GmbH.

Porsche's reputation was recognised in his industry with such depth and breadth that he had every reason to hope for commissions for his new engineering consultancy. However, it would take more than a few newspaper reports to lure the customers that Ferdinand Porsche needed. The nation and its auto industry were mired in the depths of a crippling depression. In 1931, wrote Ferry, 'there were months when I drew no salary at all for the simple reason that we had no funds to back us up'. He admitted that 'we probably had far more technical skill and enthusiasm than business acumen. We did not always figure out our contract costs and fees so as to have a reasonable margin of profit left over.'

Porsche's design office in Stuttgart's Ulrichsbau was a forest of drafting boards, overseen by the Führer. Extra floors were leased to cope with the Auto Union and Volkswagen projects respectively.

Early design contracts came from small-fry Wanderer, which soon became a member of the Auto Union group with Horch, DKW and Audi. Rear-engined small-car prototypes were built for tiddlers NSU and Zündapp. In these projects the Porsche team bid low on the engineering in order to benefit from royalties on production – production that failed to eventuate.

Early in 1932 a delegation of three German-speaking Russian technicians arrived for a visit to the Ulrichsbau. This was promising, thought the Porsche men, surely indicative of a commission to design the tractors that Russia desperately needed. But the visitors evinced little interest in the tractor schemes they were shown. They had much more ambitious intentions.

'We are at the eve of great changes in our country,' they said, 'the change from an agrarian to an industrial country.' Stalin, they added, said that Russia always lost her battles because she was behind the curve technically. The great nation's first five-year plan, a Stalin creation in 1928, was nearing its end. 'We are fifty or

The Porsche team's first project was the design of a small car for Wanderer in Siegmar, here with Ferry Porsche at its wheel. Introduced in 1933, the six-cylinder models remained in Wanderer's product range until 1938.

a hundred years behind the advanced countries,' said the dictator. 'We must make good this distance in ten years. Either we do it, or they will crush us.' With the second five-year plan in preparation, Ferdinand Porsche was on Russia's radar – to use an anachronistic expression.

After appropriate palaver the Soviet delegates disclosed their intention. It was to invite Porsche to tour the Russian industrial centres over a two-week period, including the train journey from Berlin to Moscow and return, so that he could see for himself the progress made since the revolution. Curious about many aspects of engineering in the east, not least the wider-gauge railways, Porsche agreed. Three weeks later he began his all-expenses-paid tour.

Accompanied by an interpreter, Ferdinand Porsche was first taken to the Central Drawing Office where vehicle engineering was concentrated. At Stalingrad he toured a large tractor factory. Other trips were to Kiev, Kursk, Odessa and Nizhny Novgorod to view factories making aircraft, turbines, tanks and trucks. 'The extent of the technical impressions captivated him,' wrote Richard von Frankenberg. 'Naturally they had made him promise not to divulge what he had seen and it was obvious that he would keep his promise. He was even taken as far as the secret industrial centres beyond the Urals.'

On his return to Moscow Porsche suffered a leg injury that kept him bedridden for a fortnight. He had much to mull over, for the Russians had offered him the post of czar of all automotive design and development throughout the huge nation. He and his entire family would, they said, have living conditions on a par with the most luxurious the nation had to offer, provided free of charge as part of his contract. The engineer thanked his interlocutors and said he would let them know after he returned home.

Contemplating the extraordinary offer, Porsche found it tempting. What a fascinating challenge! He already knew of the potential of Russia, where his tank prototypes had been tested soon after his departure from Daimler-Benz. 'But at my time of life,' he told his son, 'I'm too old to begin a new career. Especially one which would involve working for others in a strange land whose people I don't know and whose language I know even less. It's one thing to translate an ordinary conversation,' he said of the use of interpreters, 'but quite another when you try to put over some highly abstract technical ideas. It just could not work.'

Ultimately the language problem was the major barrier to Ferdinand Porsche's acceptance of the Soviet invitation. The Russians relied instead on the agreement with Ford that they had signed in May 1929 as part of the first five-year plan. It required the American company to build a mass-production facility at Nizhny Novgorod, gifting to the Russians their obsolete Model A and a Model AA light truck. Both were swiftly adapted to military demands.*

On Porsche's part, ironically, he did indeed find himself beginning a new career at the age of 57, for two decades of the most demanding effort lay ahead of him, much of it in entirely new disciplines. As to language, here too he had to adapt. Ahead lay Germany's Third Reich, which had its own oft-impenetrable array of acronyms, evasions and euphemisms.

The Nazi Party's leader first met Ferdinand Porsche when the engineer was in his pomp as the technical chief of Daimler-Benz, which had been officially created a fortnight before their encounter. It occurred in Berlin on Sunday, 11 July 1926.†

---

* How much Porsche was aware of Stalin's Great Purge, which saw the deaths or imprisonment of many prominent people from 1936 to 1939, we do not know. However, this was the fate of some Ford personnel who stayed on in the Soviet Union after the factory was completed.

† Ghislaine Kaes maintained that the first meeting occurred on 12 September 1926 at a race meeting at the Solitude circuit on the periphery of Stuttgart. Hitler was said to have been impressed by the big supercharged Mercedes-Benz driven by Otto Merz to win the main race of the day. The author prefers his version, which takes into account Hitler's previous awareness of Porsche's fame and career.

Adolf Hitler was drawn to crowds, and the city had seldom seen crowds like the more than 200,000 who swarmed to the Avus, the fast toll road from Berlin towards Potsdam. The German Grand Prix had just finished and officials and local bigwigs were mobbing the Mercedes-Benz that had bested 43 other entrants over 243 miles of racing and its plucky driver, young Berliner Rudolf Caracciola.

Rain pelted on the celebrants, among whom was Porsche, the winning car's designer. Pushing through the crowd to congratulate him was Adolf Hitler, wearing his trademark belted raincoat. It was important to him to seize this opportunity to greet Porsche, for Hitler was a fellow Austrian who had long admired his countryman's work and achievements. A quick handshake and a few words were all that the moment allowed. Porsche would quickly forget them. But he had just met the man whose passions, desires and demands would dominate the last two decades of his life.

Born in 1889, politician Adolf Hitler was 14 years younger than Porsche. He was 37 when the two men met in Berlin. He had only been out of prison for a year and a half. His path to power would not be smooth. But in less than seven years this enthusiast for cars and technology would take totalitarian command of the Third Reich he envisioned for Germany.

From the autumn of 1907 until the spring of 1913 Adolf Hitler lived in Vienna, the city in which Ferdinand Porsche and Austro Daimler were constantly in the news in those dramatic years of the engineer's career. Hitler's reading limned Porsche's newsworthy achievements: his launch of the Maja, his remarkable Mixtes, his Prince Heinrich campaigns and triumphs, his Alpine successes and his first aviation engines, for one of Hitler's ambitions at the time was the design of an aeroplane. In fact it is thought that he was an enthusiastic viewer of some of the pioneering Porsche-powered flights at Wiener Neustadt's aerodrome.

When Adolf Hitler left Vienna – whose louche lifestyle he loathed – for Munich in Bavaria he carried with him a positive recollection of Ferdinand Porsche's drive and creativity. Nor can the possibility be excluded that during Hitler's service in Germany's army in 1914–18 he was aware of the air and ground machines that Porsche produced for the use of the Central Powers.

While the Depression proved problematic for Ferdinand Porsche's new business, it created an environment in which Adolf Hitler's National Socialist Party flourished. In 1932, with 230 seats in the Reichstag, Hitler's NSDAP – 'Nazi' for short – was Germany's largest political party. Although its representation fell to 196 in November 1932, Hitler's momentum was such that he was named chancellor of Germany at the end of January 1933 with a coalition-party cabinet. It took Adolf Hitler a scant few months to seize full control, strip the Reichstag of its powers and ban all political parties save his own.

The new chancellor had not lost his well-known enthusiasm for cars and motor sports. Before seizing power he had disclosed his ideas for the future in private meetings

At the 1935 Berlin Show Josef Werlin (left) *and Adolf Hitler were shown around a special exhibition of classic Mercedes-Benz models by Christian Lautenschlager, who famously won the 1908 and 1914 French Grands Prix for Mercedes.*

At a 1937 racing-car display the Führer shook hands with driver Hermann Lang, Manfred von Brauchitsch in the background. Behind Hitler was Adolf Hühnlein, chief of the National Socialist Motoring Corps.

Hitler and Porsche greeted each other warmly at a Berlin auto salon. 'The situation was in some respects as though my father were also Hitler's father,' said Ferry Porsche of the nature of their relationship.

After Mercedes and Benz merged in 1926, Alfred Rosenberger raced a mid-engined Benz that had first competed in 1923. His enthusiasm for its design contributed to a similar configuration for the 1934 Auto Union.

with Jakob Werlin, the Daimler-Benz agent in Munich, and well-connected Mercedes racer Manfred von Brauchitsch. He also spoke with famed and popular racer Hans Stuck, who had raced Austro Daimlers and knew Porsche well.

These were reasons enough for the cash-strapped Porsche engineers to concur with the drafting and sending to Hitler of a telegram signed by Porsche complimenting the newly minted chancellor on his encouragement of motor sports in his speech at the opening of the Berlin Show in February 1933: 'As the creator of many renowned designs in the realm of the German and Austrian motor and aviation field and as a co-combatant towards the present success for more than thirty years, I congratulate Your Excellency on the profound opening speech for the German Automobile Exhibition.'

Hitler's arrival on the scene meant the prompt departure of the Jewish Adolf Rosenberger. In January 1933 he decamped to France, where he represented Porsche's patent rights, helped by the loan of one of the rear-engined prototypes designed for NSU. Later he emigrated to California, where he was well-known in auto circles as 'Alan Roberts'. Porsche's business affairs were now managed by Baron Hans von Veyder-Malberg, a wealthy enthusiast and one-time Austro Daimler racer who acquired Rosenberger's 15 per cent shareholding. Veyder-Malberg would prove to be an excellent ally for Ferdinand Porsche.

Not long after the Berlin Show Porsche and Hitler met properly for the first time. The topic was the new racing car, the P-Wagen, for Auto Union. Its initiator was Auto Union's Baron Detlof von Oertzen, who hoped to persuade the Reich to provide funding for new racing cars not only to Daimler-Benz, which had the inside track, but to his brash new company as well. But no one would let him see Hitler, who was immensely busy in February 1933 with the tasks of the first weeks of his administration.

'Then I went to see his deputy, Rudolf Hess,' von Oertzen related. 'He and I were pilots of yore; we knew each other from the Great War. I asked Hess to get us an appointment with Hitler. Hess then arranged it for the beginning of March.'

With Porsche at Berlin's AVUS track in 1934 was Willy Walb. A designer, builder, tester and racer of Benz competition cars, including its mid-engined models, Walb was manager of Auto Union's new team.

Although the Auto Union racing cars were built at the Zwickau Works of Horch, a member of the AU group, Ferdinand Porsche and his team closely monitored their assembly. Horch's skills were well up to the demanding job.

The appointment was set for Friday, 10 March 1933. Von Oertzen: 'To this meeting I took Dr. Porsche and the racing driver Hans Stuck, who unlike myself was personally acquainted with Hitler.' Porsche and Stuck compared notes the previous evening in the latter's flat in Berlin-Charlottenburg. 'Under his arm he had a thick portfolio of drawings,' Stuck recalled of Porsche. 'He didn't yet know Hitler and asked "what sort of fellow" he would be.'

To von Oertzen's opening overtures at the Chancellery, Hitler gave not a hint of acquiescence. The Baron persisted, saying he owed it to Auto Union's 10,000 employees to press his case for support. Turning away from the emissary, Hitler addressed Porsche, who opened his portfolio on the glossy surface of the massive conference table. To the engineer's surprise Hitler reminded Porsche that they had met in Berlin in 1926. For 20 minutes, interrupted only by knowledgeable questions from Hitler, Porsche swiftly and in his broad Austrian accent, entirely penetrable by the leader, explained his car and his ideas.

Sufficiently briefed, Hitler ended the meeting without commitment but with a remark that admitted some hope: 'You will hear from me.' Three days later von Oertzen was informed that the Auto Union project would receive government support. The executive had no illusions about the reason why: 'Hitler supported the construction of our racing cars. But he did that not for liking me, but rather for liking Porsche.'

With remarkably little change in its basic concept the P-Wagen, which was known as the A-Type Auto Union in 1934 and progressively the B-Type and C-Type in subsequent years, raced successfully from 1934 through 1937. The team's ace drivers Hans Stuck, Achille Varzi and Bernd Rosemeyer battled the Mercedes-Benz battalions with brio, scoring 32 victories in both Grands Prix and hillclimbs and setting speed records to boot.

More than a successful racing car, the 10 March 1933 encounter between Ferdinand Porsche and Adolf Hitler created a remarkable meeting of minds. It initiated a partnership that would have fateful consequences for both men.

Although Willy Walb was the first man to drive the new Auto Unions in 1934, Ferdinand Porsche also tried out his creation. The radical cars successfully challenged the best efforts of Alfa Romeo and Mercedes-Benz.

CHAPTER 9

THIRD REICH ASSIGNMENTS

Half a year after his unrolling of the drawings of the future Auto Union racing car in Adolf Hitler's office Ferdinand Porsche found himself in Berlin again for another meeting. In the autumn of 1933 the engineer was called by Mercedes-Benz man Jakob Werlin, who urged, nay insisted, that he come to Berlin for a 'very urgent' meeting the following afternoon at the Hotel Kaiserhof.

Opening pleasantries in Werlin's suite were agreeable. Born in Austria's Graz, the Nazi had begun his apprenticeship at the Puch factory there and was entirely familiar with Porsche's glittering Austrian career. When the Daimler-Benz man came to the point, Porsche knew at once that he was a confidant of Hitler, for he, Hans Stuck and Baron von Oertzen had kept their meeting with the Chancellor entirely secret:

> You see, Dr. Porsche, since Herr Hitler met you in connection with the Auto Union racing-car project, he has gained an even higher opinion of your professional capacity as a designer. Let me come straight to the point. Hitler is very interested in the possibility of small cars. He will be here any minute now and perhaps you can enlighten him on the subject. You told me that you have been working on problems associated with small cars for some time.

Before he could react to this revelation Porsche saw a door to the suite swing open and Hitler enter. After tea was served and the niceties observed, the dictator seized the floor. Adolf Hitler held forth at length and in detail about the kind of car he had in mind, something to suit the German family with three children, a proper car but not too fancy, economic to run and repair, a real 'Volkswagen' – a car that would suit his people.

Hitler, auto designer *manqué*, detailed his thoughts. This was to be no crude three-wheeler or cyclecar; a genuine car for the German working man was wanted. It should be four-wheel-drive, Hitler suggested, with a three-cylinder air-cooled diesel engine, preferably front-mounted. This gave rise to two conclusions: first, that he was still reading his car magazines and second, that he had not overlooked the vehicle's military potential.

This was familiar territory to Porsche but Hitler's answer to the designer's question about the desired selling price was not: 'At any price, Herr Dr. Porsche – at any price below 1,000 marks!' This was staggering to the engineer. The small cars he had designed for Zündapp and NSU would have cost much more than that just to produce and would have retailed for around 2,200 marks. Porsche never claimed to be a production expert; he had not yet worked with high-volume car projects. But this seemed a chimerical goal.

At the start of Germany's 1934 2,000km Trial, in which cars were flagged off in pairs at intervals, a smiling Ferdinand Porsche waved to officials. The car behind him was a mid-engined Mercedes-Benz Type 150 sports model.

After a glance at his wristwatch Hitler left the suite's sitting room. Jakob Werlin was prepared for the next step. He asked Porsche to consider the matter and to put his thoughts about such a car on paper. The first internal Porsche discussion of the 'Volkswagen' problem was held in the last week of September. By January 1934 an 'exposé' on the subject was ready for Hitler's review.

As to his proposed car's selling price, Porsche was on the spot. He believed 1,000 marks to be out of the question – yet this was what Hitler wanted. Clearly the car had to be priced lower than anything on the German market. After much cogitation Porsche opted for the figure of RM1,550 in his exposé.*

All went quiet on the 'Volkswagen' front until the opening of the Berlin Auto Show on Saturday, 3 March 1934. Adolf Hitler opened it with an elaborately staged address in a swastika-bedecked hall. Porsche was in the military-uniformed chancellor's audience for the strongly politicised speech that began in the dying fanfare of an Army band. Hitler issued a call for action:

> Germany has only one automobile for every one hundred inhabitants. France has one for each twenty-eight and the United States one for each six. That disparity must be changed. I would like to see a German car mass-produced so it can be bought by anyone who can afford a motorcycle. Simple, reliable, economical transportation is needed. We must have a real car for the German people – a Volkswagen!†

Hitler urged his auto industry 'more and more to design the cars that will compellingly attract new buyers by the millions'.

One man in the audience knew that his exposé had reached the eyes of the Führer. After the show this was confirmed by Reich Transport Minister Brandenburg, who added the obvious point that Porsche's proposed selling price was much higher than the figure specified by Hitler and thus would require further

---

* At such a price, the author surmises, if criticised Porsche could have said that he had misunderstood the assignment, that he thought the *factory cost* was to be RM1,000 and he had simply added to that the various commercial discounts, profits and distribution costs to arrive at the figure of RM1,550.

† Sources are not in agreement as to whether Hitler used the word 'Volkswagen' in this speech; Hopfinger says he did not.

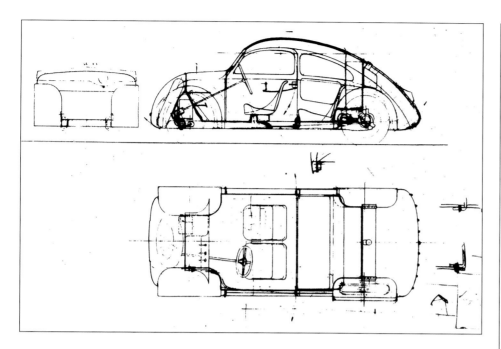

A sketch of 27 April 1934 set out a scheme for the future Beetle, showing how early its basic form was established. An important difference would be a final platform frame that relied on a central backbone for much of its stiffness.

study by the engineer. The meeting gave Porsche cause for some confidence; he checked with Brandenburg's office in early May but was told there was still no news.

The Werlin channel opened again in the last week of May 1934. Visiting Porsche in Stuttgart, Werlin said, 'You will shortly receive an official order to proceed with the development of the Volkswagen. This order will come not from the Ministry of Transport but from the Society of German Automobile Manufacturers,' the RDA. The Mercedes official explained that this decision had been reached at the level of the Chancellor in order to ensure a commitment by the car producers to the project. If they were paying for the development, Hitler reasoned, they would be more likely to exploit its fruits. They were expected to unite to build the car as a joint effort.

Official notification to Porsche from the RDA's Robert Allmers followed in early June, after which a contract was hammered out. Signed on 22 June, the document was relatively brief. So was the time for its realisation. It gave Porsche only six months to design the Volkswagen and four months to build its prototypes. When it entered production he would be entitled to a royalty of 1 mark per car.

In February 1936 the first of three prototypes was ready for testing. With this new project on their books, on 18 May 1937 Porsche and his lawyer son-in-law Anton Piëch founded the Company for Preparation of the German People's Car, which took over the fourth floor of the Ulrichsbau. In German this was the *Gesellschaft zur Vorbereitung des Deutschen Volkswagens m.b.H*, which they abbreviated as GEZUVOR. This had the agreeable and appropriate meaning of 'Go ahead!'

The prosperity of Porsche's enterprise became such that it could build offices of its own in the suburb of Zuffenhausen, north-west of Stuttgart's centre. The new headquarters offered complete facilities for designing, building and testing motor vehicles. Moving overnight to avoid the loss of a single working day, the Porsche men were installed in Zuffenhausen in June 1938.

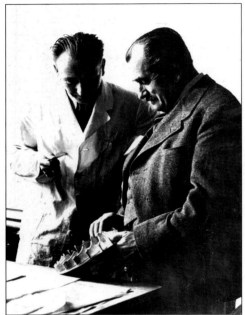

In 1938 Ferdinand Porsche conferred over an impeller with Franz Xaver Reimspiess, whom he commissioned at the eleventh hour to design an engine for the VW-to-be. His engine would perform yeoman service in the Second World War.

The year 1938 saw the completion of the final design of the KdF-Wagen, later known as the Volkswagen. Although its creation was a team effort, orchestrated by Karl Rabe, this historic car's creator was incontrovertibly Ferdinand Porsche. Inventors galore had proposed and built small air-cooled rear-engined cars, but it took a Porsche to make such a car a high-volume production reality.

Key elements of the Type 60's architecture were Karl Fröhlich's suspension using Josef Mickl's torsion bars, Erwin Komenda's body shape and Franz Xaver Reimspiess's engine. Frustrated by constant problems with two-stroke engines, perilously close to an agreed delivery date for the first Type 60 prototypes, Ferdinand Porsche gave in to Reimspiess's plea that he be allowed to design an engine, on the strict condition that it work well straight from the drawing board. Within 48 hours he had its basic design ready. So pleased was Porsche with the result that he gave the engineer a 100-mark bonus. The Reimspiess boxer four powered the first prototypes of 1936 and went on to propel Volkswagens in their millions.

Although the exigencies of its low price and mass production kept the Type 60's final design from having features that specifically pointed towards military use, such as Hitler's favoured four-wheel drive, its potential for wartime service was never overlooked. During its gestation Hitler introduced conscription, in 1935, and scrapped the name Reichswehr for his armed forces in favour of *Wehrmacht*, or 'Defence Force', of which the *Heer* or Army became a constituent. Its *Heereswaffenamt* (HWA) continued to define and source its equipment.

Motorisation was a priority for the new Wehrmacht, its leaders proclaimed, and the HWA had already made up its mind about the kinds of vehicle it required. It looked favourably on four-wheel drive and rugged solid axles. For general troop-carrying and field service its vehicles were open-topped with bucket or *Kübel* seats that let fully equipped troops jump easily in and out. Such a vehicle was a 'bucket-seat car' or *Kübelsitzwagen*, which was referred to simply as a *Kübelwagen*, whatever its make or size.

Lighter vehicles had proved their value to the German Army in service in the First World War: motorcycles with and without sidecars. 'As a result,' wrote historian Hans-Georg Mayer-Stein,

> when the Reichswehr was reorganised as of 1927, [the motorcycle's] range of use in reconnaissance, communication and supply units and in tactical service was expanded considerably. Cycles with armed soldiers were soon organised into their own cycle companies as very-fast-moving attack units. When the Wehrmacht was organised in 1935, the motorcycle riflemen became a service arm of their own.

In April 1934, a month before Porsche was authorised to proceed with the Volkswagen project, the Reich's military proffered advice to the putative designers. The car-to-be, recommended its Major Zuckertort, should have space for 'three men, one machine gun and ammunition'. This was the first contact between Porsche and the Reichswehr under the new regime. A comprehensive review with the military of the state of his Volkswagen effort was conducted by Ferdinand Porsche in Berlin on 15 March 1935.

Porsche's Volkswagen was viewed askance by Hitler's armed services. Its lack of generous ground clearance was considered a fundamental drawback. Nor was the rear engine regarded as an advantage; a *Kübelwagen* version of the rear-engined Mercedes-Benz 130 had been hugely disappointing in the many off-road sporting events organised by the Nazis' National Socialist Motoring Corps (NSKK) headed by Hitler crony Adolf Hühnlein. The Wehrmacht strongly supported this *Geländesport*, whose many participants were their reservoir of future manpower.

One might expect the enthusiastic members of the NSKK to have been recruited to drive the 30 Beetles in the V30 prototype series that began testing during 1937. However, Hühnlein felt, as he expressed privately, that the effort to build a car for the masses would lead to a weakening of Germany's automotive bloodline. As well, in 1934 Hühnlein and General Blomberg told Germany's auto makers that only a 'field-service-capable' vehicle could be considered a proper Volkswagen. Recognising that this would not be a priority in the car's design after all, in spite of his recommendation, Hühnlein kept his NSKK at arm's length from the evolving Type 60.

In fact Porsche himself rejected this criterion in a technical report about the capabilities of the vehicles used in the NSKK's winter trial of 1935. He stressed that in order to cope with off-road driving in all weather conditions, special vehicles were needed – special cars that 'naturally can only have a somewhat limited civilian use'. In fact Germany's car makers were obliged to produce extra-rugged machinery that could handle the NSKK's often-exaggerated cross-country trials.

In the event the 30 Beetle prototypes were test-driven by members of an SS troop, under the supervision of Ferry Porsche. The SS men fell under the spell of this new-fangled automobile. Encouraged by the reactions of these beguiled SS motorists, the Porsche/GEZUVOR test workshop under Rudolf Ringel modified a test Beetle to improve its off-road mobility. Special tyres were fitted and ramp angles at both front and rear were increased by shifting the spare wheel and muffler upwards. Tests from December 1937 showed that this made the little car quite agile across country.

The modified Bug was seen in action by an HWA officer, who blessed an effort to extract 'as much as possible for military application from the existing vehicle'. But in January 1938 he was unable to make much more of a commitment than that to Porsche, simply pointing out that a military version would need lighter-weight bodywork if it were to be able to carry four troops plus all their equipment – a desirable objective.

In typical stance Porsche looked on as one of the V30 prototypes of 1937 went out for its trials. One of the results of its evaluation was the decision to make the final design of 1938 larger, a full four-seated car.

By the time of 1937's Berlin Show a Type 60 chassis was on display. Porsche explained its attributes to Adolf Hitler watched by Robert Ley (left) and his deputy, the tall Bodo Lafferentz, behind them.

Considerably more enthusiasm was generated at the higher levels of the SS that same January when the leader of the V30 test cadre, Captain Albert Liese, proselytised on behalf of the Beetle's military utility to Lieutenant General Josef 'Sepp' Dietrich. An intimate of Hitler's, the influential Dietrich could see the merits of the VW's low profile, light weight and, particularly, potential for low cost which would facilitate a rapid conversion of Germany's military to more modern vehicles when the balloon went up.

'Even if Dietrich led the negotiations with the Volkswagen works primarily to ventilate his resentments over the conservatism of the HWA, he supported the Porsche concept particularly on the grounds of technical modernisation and thereby contributed substantially to the development of a long-term co-operation among the SS, the Volkswagen Works and Porsche.' So wrote historians Hans Mommsen and Manfred Grieger about the intervention of Dietrich and his SS in the fortunes of a military version of the budding Beetle.

A dollop of fertiliser for the flowering of this relationship was spread by Ferry Porsche in his stewardship of the prototype testing. At first, he said, he looked on the SS drivers sent to him as snoops and spies:

> Every little detail, each day, had to be recorded, and at first this was intolerable and struck me as quite absurd. It was often difficult for me to contain my anger. But after a while I discovered that few of those SS men were stupid thugs. Some could be approached and spoken to. Some would listen and a few even approved of our own suggestions about testing and performance evaluation.

Ferry's forbearance at a crucial juncture was to be richly repaid. In addition to the rest of his work on the VW project, Ferry became closely involved with the development and testing of military versions of the Type 60.

The knotty problem of funding a factory to build the Type 60 was resolved by having it manufactured and sold by the national labour union, the German Workers Front (DAF). Regular contributions to Robert Ley's DAF were made by both workers and employers in relation to the sizes of their pay envelopes and turnover respectively. The funds were banked by the Bank of German Labour (BdA). Money also flowed in from workers' subscriptions to the lay-away schemes of the immensely successful *Kraft durch Freude* (KdF) or 'Strength through Joy' movement, which organised holidays and cultural events for workers and their families.

Robert Ley received the idea of DAF involvement with 'approbation'. This was not surprising, for Ley delighted in exploiting his beloved DAF, a workers' organisation, in the world of capital enterprise. By the end of 1936 the DAF and its leader had decided to take on the responsibility for building and selling Hitler's dream car. The price could be held down, it was thought, because the usual taxes would not have to be paid by a quasi-state enterprise and the car's sales through regional offices of the DAF would incur minimal distribution costs.

On 20 February 1937 Adolf Hitler opened the Berlin Show with a speech in which he made clear his determination to achieve the production of a car for the people. That evening the dictator invited 400 car-industry workers to dine at

THIRD REICH ASSIGNMENTS   119

Mufti-garbed Porsche joined his Führer for the ceremonies attendant upon the May 1938 cornerstone-laying at Fallersleben. He was close to restoring his former Austro Daimler role as the head of a car-producing company.

The conversation of Lafferentz (left) and Ley with a party out of shot seemed to leave Porsche cold. The two top officials of the German Workers Front played an essential role in the creation of a factory to build the Volkswagen.

The laying of a cornerstone for the Volkswagen factory on 26 May 1938 marked an important step in the car's realisation. It was done with respect for both tradition and the full array of Nazi panoply.

The mood was jovial at the May 1938 laying of the VW factory's cornerstone as Hitler tried his Beetle's rear seating. Robert Ley and Ferdinand Porsche were the most prominent of those behind him.

Berlin's Kaiserhof Hotel. There Hitler announced his assignment of the VW project to Robert Ley and the DAF. Ley in turn named his deputy, Bodo Lafferentz, as his representative on the board of the Volkswagen company. Also appointed to the board was Ferdinand Porsche. In this wholly unexpected manner Porsche achieved his personal goal of becoming a car manufacturer.

In a typically lurid swastika-bedecked Nazi ceremony Adolf Hitler laid the factory's foundation stone on 26 May 1938. A vast works was to be constructed at Fallersleben, near Hanover, on the east–west Mittelland Canal. To show his appreciation for what Porsche was well on the way to achieving, the *Führer* authorised a bonus payment to him from his 'Disposition' fund under the administration of the chief of staff of his chancellery, Hans Lammers. Porsche's 1938 bonus was the not-inconsiderable sum of RM600,000, the equivalent of $240,000.

On 16 August 1939, astonishingly only 15 months after the laying of its foundation stone, the mammoth new KdF-Werk at Fallersleben came to life. Ferdinand Porsche personally turned the huge valve, the size of a massive steering wheel, that initiated power and heat generation by one of its coal-fed Borsig turbines. It seemed that completion of the plant and its workers' city could continue unabated in spite of the outbreak of war. Italian labourers borrowed from Mussolini were building them; they would not be called up by Germany.

But higher priorities intruded. Construction was being accelerated on Germany's West Wall, on Hitler's transformation of Berlin into a capital city for the new Greater Germany and on facilities dedicated to war materiel. The allocation of raw materials came under the control of the General Construction Inspectorate (GBI) headed by Albert Speer, the Reich's fast-rising technocrat.

Under the gaze of board colleague Bodo Lafferentz, on 16 August 1939 Ferdinand Porsche watched the relevant gauge as he opened the valve that fed steam to one of the plant's power-generating turbines.

A KdF-Wagen cabriolet was a gift to Luftwaffe chief Hermann Goering, shown its controls at his Karinhall estate by Ferry Porsche while Ley and Porsche took a back seat. Lafferentz leaned in benevolently at the left.

Early and late Beetle prototypes were in the courtyard of the new Porsche headquarters and workshops in Stuttgart-Zuffenhausen. In this 1939 image a sign designated it as a branch of the Volkswagen Works.

Wooden formers were shifted from place to place on the construction site at Fallersleben for the pouring of reinforced concrete to form the scalloped roofs of the VW works. The workforce was predominantly Italian.

Viewed from its powerhouse at completion in 1940, the KdF-Werke was an awesome sight. In this, the first of its three projected stages, the plant would have some 2.2 million square feet of productive area.

Although close to Hitler, Speer did not possess the easy intimacy with the dictator that Ferdinand Porsche enjoyed.

Only 35 years old in 1940, Speer sought to propagate a youth culture into which he felt the elder Porsche fitted poorly. But the great engineer was untouchable. 'At most, a half dozen men in all of Germany dared to speak their minds openly before Hitler, and my father was one of them,' wrote Ferry Porsche. 'The situation, in fact, was in some respects as though my father were also Hitler's father.' Albert Speer could not stop the work of Hitler's favourite vehicle designer – but he could slow it down.

A bitter battle of wills followed. Speer gave little encouragement in a meeting at KdF-Stadt at the end of March 1940. Instead of the requested 3,000 tonnes of steel for the works and 1,200 tonnes for the city, a colleague told the Porsche men, only 300 and 400 tonnes respectively would be provided. Not even the new city's pressing need for a water supply could be granted a priority. Another hitch was that a number of specialised machines ordered from America had not arrived – and in fact would never arrive.

The plant's priority would rise if it were a designated war-production site. Opinions differed sharply over the merits of this. Some in the government's Economics Ministry thought such a huge factory could hardly be overlooked in time of war. This was the opinion of Colonel Thomas who, said Reinhard Osteroth, at a January 1938 meeting stated his view 'that in any case the factory's size and significance make it important to the war effort and must be exploited'. Others, viewing it from the military-ordnance standpoint, saw Fallersleben as a plant dedicated and equipped for civilian production that was and would remain unsuitable for wartime use.

Recognising the risk that the latter view posed to the near-term future of their factory, Porsche and his colleagues suggested as early as November 1938 that parts of it – preferably parts yet to be built – should be used to produce aircraft engines, propellers, vehicle engines of 200 to 300hp and electrical equipment for the Army and Luftwaffe. This view was supported by Erhard Milch, deputy of Luftwaffe chief Hermann Goering for production. Discussions to this end went so well late in 1938 that on 4 January 1939 Goering named Porsche a *Wehrwirtschaftsführer*, a Leader of the Defence Economy.

In charge of technical procurement for the Luftwaffe, First World War super-ace with sixty-two confirmed kills Ernst Udet supported Porsche's initiatives but could not force approval of the engineer's proposals for aero-engine designs.

An X configuration was chosen for the four eight-cylinder banks of Porsche's 1935 Type 70 proposal to the Luftwaffe. Its 17.9 supercharged litres were forecast to deliver more than 1,000bhp at the high crankshaft speed of 4,100rpm.

In their ambitions Fallersleben's managers were supported by the Luftwaffe's technical procurement office, which was headed by First World War ace and Porsche ally Ernst Udet. Udet was already co-operating with Porsche by approving his use of the advanced and ultra-secret 12-cylinder Mercedes DB 603 aero engine in a World Land Speed Record car that Porsche was designing for Daimler-Benz.

A few years earlier the Porsche office had already knocked on Udet's door with aero-engine design ideas. Having already pioneered important advances in aero-engine design, Ferdinand Porsche deserved to be taken seriously in this genre by the Reich. In 1935–36 the Kronenstrasse produced two impressive studies for water-cooled aero engines that betrayed a close relationship to the Auto Union Grand Prix car's successful V-16.

One concept, the Type 70, disposed of 17.9 litres in 32 cylinders deployed X-fashion from a central crankshaft running in nine plain main bearings. Top and bottom pairs of eight-cylinder banks were at a 45-degree vee, above each of which was a single shaft-driven camshaft operating overhead valves through pushrods and rockers exactly as in the Auto Union. With roller-bearing connecting-rod big ends, the supercharged and fuel-injected Type 70 X-32 was forecast to produce 1,040bhp at 4,100rpm – fast for an aero engine – geared down to drive the propeller at 1,700rpm.

No less ambitious in its way was Porsche's Type 72, another 1935 design. Although having half the number of cylinders of the Type 70 it had more swept volume at 19.7 litres and was expected to produce 900bhp at 3,700rpm. Its roller-bearing bottom end required the use of a Hirth demountable crankshaft. Mounted inverted, the V-16 was to have the unusual vee angle of 52 degrees. Rocker arms and pushrods again drove inclined overhead valves from a single shaft-driven camshaft. Its reduction gear would give propeller speeds of 1,500 to 1,700rpm.

Porsche recommended building twin- and four-cylinder test engines, its Types 73 and 71, to validate these designs for the German Aviation Experimental Establishment (DVL) that commissioned these studies. His hopes that the DVL would carry them through to fruition were reflected by that entity's explicit permission to work with Porsche as granted by his consulting contract with Daimler-Benz. However, the DBAG, BMW and Junkers had the inside track in

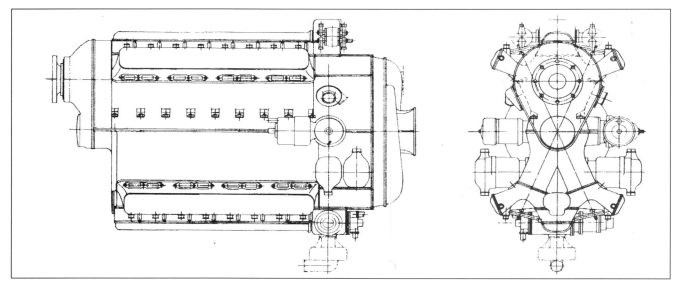

the supply of aero engines of their own designs to the Third Reich.

In spite of positive noises from the Luftwaffe in 1938 and 1939, Porsche's engine-design initiatives were not progressed, in part because the Reich Ministry of Labour was not prepared to allocate the workforce needed for their manufacture. However, the KdF-Werke could easily be making the engines of others. In fact this was the view of Junkers chief Heinrich Koppenberg, whose Junkers Flugzeug- und Motorenwerke AG was based in Dessau, at the southern apex of a triangle whose northern tips are Berlin and Fallersleben.

At first Koppenberg made an outright grab for the Volkswagen factory-in-embryo. Hitler asked him for a monthly production of 300 twin-engined Ju 88 fighter-bombers; Koppenberg would seize production capacity where he could find it. But with a little help from their friends, including Georg Thomas and Ernst Udet, the Porsches and Anton Piëch managed to ward off this thrust. Instead they inveigled to be identified as a valid supplier to the aviation industry, a much more advantageous business position that preserved the economic integrity of the factory.

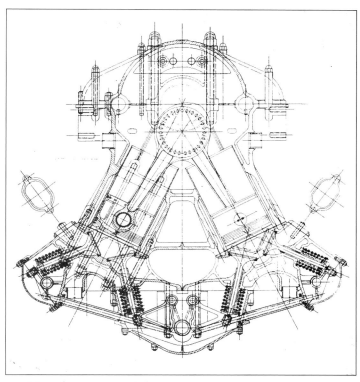

Designed to run inverted, Porsche's Type 72 V-16 of 1935 resembled the Auto Union engine in operating all its valves by pushrods and rocker arms from a single camshaft. Some 900hp was expected from its 19.7 litres.

On 18 September 1939 Hermann Goering decreed that 'for the carrying through of the Ju 88 programme, including the bomb production associated with it, the VW works is to be placed at the disposal of the Luftwaffe'. By March 1940, however, Porsche and Lafferentz were complaining to Albert Speer that almost half a year had passed since they had made their factory 'comprehensively available' to the Luftwaffe which, frustratingly, had failed to place any firm orders with them.

These were practical problems for the plant's managers. Although Robert Ley's DAF and its BdA bank were cheerfully carrying the cost of building and maintaining the works, the factory's need to start generating compensating cash flow by booking production orders was intensifying by the week. The KdF-Werke partners wanted to take every step possible to ensure that their plant would be completed and equipped so that car production could begin as soon as peace was declared.

In the summer of 1940 the general sense in Germany was that the war would soon be over. Surely the British and French would see the logic of reaching accommodations with Hitler, who had already signed a surprising non-aggression pact with the Soviet Union. Austria's Steyr, in fact, was already preparing new cars for its post-war market with the help of Ferdinand Porsche and his designers. Although the new Steyr 70's water-cooled V-8 engine was the work of the company's Oscar Hacker, Porsche's team designed its platform frame and all-independent suspension.

Also folded into the projects assigned to Porsche on 12 July 1940 was the design of two truck chassis for Steyr, the Model 170 with rear-wheel drive and the

Austria's Steyr, by 1940 part of Hermann Goering's industrial empire, commissioned Porsche to design both cars and trucks. The engineer took a practised look at a scale model of the planned Steyr Model 70.

As their Type 147 the Porsche team laid out this robust four-wheel-driven chassis for Steyr in 1941. Its front wheels independently suspended by torsion bars, it was destined for production in the thousands for the military.

270 with four-wheel drive for military use. Both laid out by early 1941, the first was Porsche's Type 146 and the second its Type 147. As well, Porsche tackled the conversion of Steyr's V-8 to air cooling, using a pair of blowers above the central vee. Pushrod-operated inclined overhead valves gave the 3.5-litre engine peak power of 85bhp at 3,000rpm. A 2.0-litre four-cylinder version of the engine was also created.

A rectangular-section tubular frame was at the heart of the Steyr 270. While rear suspension was conventional, with a leaf-sprung live axle, springing at the front was independent, using the enclosed drive half-shafts as the lower wishbone arms. Long torsion bars extended back from the upper suspension arms in an innovative layout. Behind its transmission a transfer box with centre differential distributed the drive to all four wheels.

The result, a versatile machine with 1.5-ton capacity uprated to 2.0 tons in 1944, was ranked by the HWA as its Standard Chassis II for heavy vehicles. Bodied to take on many tasks, it served as a maintenance truck, searchlight carrier, telephone truck, personnel car, ambulance, reconnaissance car and support for flak units. More than 21,000 left the Steyr factories to serve with the Wehrmacht.

As for the Steyr 70 passenger car, it was stillborn because peace had not broken out after all. Instead, production for war was accelerated during 1940. Some of

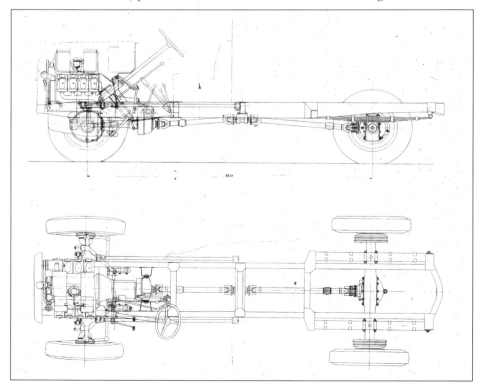

that commerce came to Porsche's Fallersleben factory. By February output was under way of wooden 300-litre drop tanks for aircraft, ironic in view of the plant's first-class sheet-metal-working facilities. In March Fallersleben started making 550lb bombs.

September 1940 saw generous recognition of the amazing career of Ferdinand Porsche on the occasion of his 65th birthday. Readers of leading journals were reminded that three days after his 63rd, on 6 September 1938, Porsche was a guest at the Nazis' annual Nuremberg celebrations, together with Willy Messerschmitt, Fritz Todt and Ernst Heinkel. All received from the hand of Adolf Hitler the German National Prize, recognition of the highest order in the Third Reich. The year 1940 brought another honour, the title of professor at Stuttgart's Technical University.

In mid-1940 Junkers finally placed some bigger orders with Fallersleben. The factory became a key site for the repair of damaged Junkers Ju 88 aircraft; throughout the war this remained its largest single task. A special workshop was opened in a hangar at the nearby Braunschweig Airport, where the plant's auto-production experts set up a dismantling/assembly line to speed Ju 88 refurbishment.

Phased in as well was the manufacture of new wings, tail assemblies and stabilisers for the Ju 88 and most of the components for the new Junkers Ju 188,

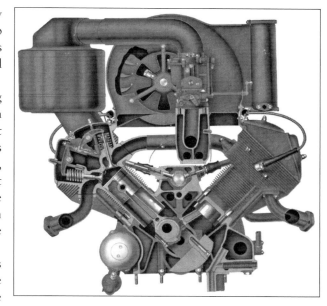

Based on Steyr's liquid-cooled V-8 the Zuffenhausen team created an air-cooled version with two sirocco fans and inclined overhead valves operated by pushrods and rocker arms from a central camshaft.

The Porsche-engineered Steyr 270 was designated a Standard Chassis II by the Army and by 1944 was rated at a 2-ton capacity. The Austrian-built vehicle was a purposeful machine of great versatility.

At the Nazi party's Nuremberg celebrations on 6 September 1938 Germany's highest civil honour, its National Prize, was awarded. Recipients being briefed by Josef Goebbels were (from the left) Heinkel, Messerschmitt, Porsche and Todt.

whose fuselage was made by Opel. This was a substantial and ongoing contract to mid-1944, when the plant began producing the major parts of the Ju 388 as well. Other KdF-Werke products were torpedo hulls and portable furnaces – of these one and a half million – to warm Germany's troops in the Russian winter.

A burst of business came in the autumn of 1940 from the production of 'swimmer' kits for tanks that allowed them to float and power themselves across rivers and estuaries. Of obvious value for an invasion of Britain, these also loomed large in the programme for 1941. From May this would be overseen by Anton Piëch, who took over from Otto Dyckhoff as the works manager. The ambitious Piëch accelerated efforts to bring more business to the plant, with the result that engine parts began to be machined for the Junkers Dessau works. Teller mines were produced in high volume in Hall 1, which had originally been set up as the tool and die shop.

Such was Porsche's reputation by now that speculation was rife abroad about his next contribution to Germany's war effort. His relationship with Junkers led to Britain's *Sunday Post* of 4 January 1942 revealing 'the silent bomber which Dr.

Repair and production of components for the versatile Junkers Ju 88 became the KdF-Werke's staple business during the war. This was its Ju 88A-15 version with an enlarged bay for carriage of a heavier bomb load.

Every bit as awesome as a set for the movie Metropolis, *this was the vast heavy-press hall at Fallersleben. It was put to work on the fashioning of stampings for the production of a wide range of weaponry.*

Ferdinand Porsche claims to have invented at the Junkers branch factory near Leipzig'. Porsche's public profile was proving to be as much a war-winning weapon for Josef Goebbels and his pantheon of wonder weapons as it was for the expansionist ambitions of Adolf Hitler.

Almost as ambitious as its Volkswagen effort was another Reich-supported Porsche endeavour, the *Volkspflug* or 'people's plough', created to motorise the German farmer. This was a small yet versatile tractor designed to do the tugging and tilling that oxen normally performed on small farms. For Robert Ley, the fanatical Nazi whose national labour union paid for the Fallersleben plant, this

Plans were advanced for the construction of a huge factory at Waldbröl in Robert Ley's Gau or District to produce the People's Plough, designed by Porsche under Types 110 to 113. War's outbreak rendered them redundant.

was a pet project. Plans were in train for a huge factory at Waldbröl near Cologne to make as many as 300,000 tractors a year.

Conceived late in 1937, Porsche's first Volkspflug design was the Type 110, followed by successive type numbers to 113 as development continued into 1940. Its proving ground was Porsche-owned acreage at Zell am See in Austria. Although in 1944 the plans for this chimerical effort expired, Porsche's engineers continued work on a version of the two-cylinder tractor powered by gas generated from wood, their Type 113G. Conversion of civilian vehicles to wood-gas generators to save precious petrol was in full swing late in 1944.*

In December 1941 Porsche received the news from Fallersleben that all construction work on his vast factory had stopped. The same applied to the adjoining company town, the KdF-Stadt, on the south side of the Mittelland Canal. That month also witnessed the Japanese attack on Pearl Harbor, after which Hitler unaccountably declared war on the United States. December 1941 saw him arrogate the function of Commander-in-Chief of the Army. His new role found the Führer taking a more direct personal interest in the technical development of armaments. Here Porsche would have an important part to play.

---

* This was the subject of a major decree from Albert Speer on 6 January 1945.

CHAPTER 10

KÜBELWAGEN TAKES THE ROAD

'For my father it was solely a matter of keeping the firm intact.' That was his son's verdict on the wartime activities of Ferdinand Porsche. The engineer had not forgotten the way his little company clung to solvency by its fingertips in the Depression years. Thanks to windfall earnings from his contract with Auto Union and the demands of his engagement to design the Volkswagen, Porsche was prosperous in the later 1930s. But neither Porsche nor his business manager Baron von Malberg was complacent. Security meant more and longer contracts, and if the Army was offering them, well, that would be nothing new for Porsche.

With Porsche's consultancy still contracted until February 1940 to Daimler-Benz, its chairman, Wilhelm Kissel, took care to ensure that the engineer was well acquainted with the Army Weapons Office (HWA). This was the all-important body charged with the task of providing Germany's demanding – indeed spoilt – generals with the optimum in equipment. Wrote historian Richard Overy,

> The officers who selected and developed weapons were highly educated, critical and demanding. The result was that the armed services expected the best military equipment they could get. Not only did they expect it to be of the highest quality in terms of the materials used and its technical efficiency, but also they insisted that the weapons should be constantly changed and modified to keep abreast of current research and the demands of the front-line soldiers. The effect was, of course, to produce weapons of remarkably high standard. But this faced the economy with difficulties. It meant that the weapons took a lot of development work and research and were very expensive.

At the HWA Porsche met an old friend, General Karl Emil Becker, four years younger, whom he had known and respected during the First World War. In March 1938 Becker became the army's chief of ordnance. This was promising, wrote Richard von Frankenberg: 'The first encounters with the HWA took place in an atmosphere of mutual consideration.'

The Porsche-HWA relationship clouded over, however, for quite soon it emerged that the HWA had been using Porsche patents for some time without having so acknowledged or having made appropriate licence payments. 'Above all,' von Frankenberg continued, 'it had to do with the torsion bar, of which, responding to Porsche's reproaches, an engineer in the HWA claimed to be the inventor – whereby it is scarcely credible that he, in good conscience, unaware of the existing Porsche invention had reinvented the torsion bar. Who in vehicle design was unaware of the Porsche torsion bar!'*

---

* Porsche did not 'invent' torsion-bar springing which, as its name implies, uses a long steel bar or bundle as a road spring, anchored at one end to the chassis and twisted by a lever at its other end by the movement of the wheel's suspension. In 1878 in Norway a patent for a torsion-bar suspension for wagons was obtained by Anton Lövstad. In 1923 and 1924 both Marlborough-Thomas and Thomas racing cars were given torsion-bar front suspension by J.G. Parry Thomas, who had patented such a concept in 1919.

Porsche's accomplishment was the practical realisation of torsion-bar springing through the mastery of stress calculation for durability and the provision and patenting of distinctive suspension linkages. Josef Mickl brought his analytical skills to the design and testing of reliable torsion bars. Porsche used torsion-bar springs for the Auto Union racing car and in a chassis he designed for Mathis in France in 1933. He undertook a similar project for Röhr in Germany in the same year. Thus he was well up on torsion-bar technology when he built the NSU small-car prototypes using this springing means and, of course, the first VW prototypes.

That torsion bars were not a Porsche monopoly was illustrated by their use by MG for its all-independently-sprung R-Type racing car of 1934 and later in the 1930s by BMW for rear suspensions. So far as the author is aware none of these applications required the payment of licence fees to Porsche.

The key protagonist in this conflict with Porsche was the engineer to whom von Frankenberg referred, Ernst Kniepkamp. In 1936 as a civilian Kniepkamp was placed in charge of Panzer design for the HWA. He introduced a number of innovations including his controversial torsion-bar springing, which immediately placed him at odds with such established tank constructors as Krupp. Ferdinand Porsche's HWA involvement had the effect of diminishing Kniepkamp's influence, an affront to his expertise for which he never forgave Porsche. Henceforward he would have little good to say about Porsche's initiatives.

After the torsion-bar contretemps, said von Frankenberg, 'Porsche was very critical of the HWA and made no secret of exposing its technical weaknesses. While the HWA seemed to give advantages to firms with which it had good personal relations, the general view of Porsche was that he was an unwanted critic, an outsider. Finally matters reached the point where the HWA only gave contracts to Porsche because it was ordered to by Hitler.' 'Pronounced controversies developed between Dr. Porsche and the Ordnance Department over these years,' added engineer-historian Walter Spielberger, 'and were never entirely resolved.'

Having made other arrangements, unconvinced of the KdF-Wagen's suitability for military use, the HWA ignored the potential that SS testers had discerned. Before the war began in earnest only 51 Volkswagen passenger cars of the final design had been built, and not at Fallersleben. All were hand-tooled and fabricated by Porsche and Daimler-Benz at a cost per copy said to be 'about as much as a Rolls-Royce'. They were used as development, press, show and display vehicles and toys for the Nazi elite. One, a rare cabriolet, was the factory's gift to Adolf Hitler. Ley and Goering were also presented with cabriolets, the model best suited to glorious public display.

Meanwhile the main plant had begun making a few Beetles. In all, 630 civil-bodied VWs were made to Porsche's KdF-Wagen design at Fallersleben between the slow beginning of series production in 1941 and its termination on 17 August 1944, two weeks after the fourth and final bombing of the plant. These rare and much-admired Beetles served as wartime transport for Nazi party functionaries, allocated to the regional *Gau* leaders and others in the hierarchy. Many were used for development purposes by key members of the Porsche office.

The man who had catalysed the relations between Hitler and Porsche, Jakob Werlin, was not forgotten. Nor were such key Nazis as Robert Ley and Albert Speer. A car from 1942 production was allocated to Emperor Hirohito as a gift from German foreign minister Joachim von Ribbentrop. Some of the early Beetles also served to cement business relations. When Ferdinand Porsche gave one to Willy Messerschmitt in November 1943 it was to remind the great aircraft designer of Fallersleben's production capabilities.

Helped by a nod from Sepp Dietrich, the HWA gave Porsche an order to build a military version of his KdF-Wagen on 26 January 1938. On 14 March Hitler was hailed in Vienna after Austria's annexation by Germany. For many in Austria this provided the economic integration with its powerful neighbour that had long been sought. Later that year, on 5 October, Ferdinand Porsche's home town was subsumed into the Reich by Hitler's annexation of German-speaking areas of Bohemia and Moravia in Czechoslovakia.

Before receiving the official go-ahead, the SS troops and the GEZUVOR workshop at Zuffenhausen had built a crude Kübelwagen that consisted of little more than a VW platform frame with angled-sheet fenders, three bucket seats on the floor and a mount for a massive machine gun. This resembled more a breadboard feasibility study than a serious design.

The first Porsche proposal, pictured in a Karl Rabe layout drawing of 15 May 1938 showing the Type 62, presented a distinctly 'civilian' aspect with rounded wings and engine cover and luxurious pleated-leather bench seats. A clue to the reason for this may be that the Type 62 project dated from 1936, according to the Porsche type-number list. This suggests that the number was assigned quite early to what the project list calls an 'off-road vehicle'.

A military version of Porsche's Wanderer chassis well illustrated the origin of the term Kübelwagen as used for most German military people-carriers with their Kübel or bucket seating.

In 1936 the Porsche people had been thinking of making a version of their new small car that would be suitable for the off-road trials promoted by the NSKK. These popular and well-publicised events would have helped spread the word about the capabilities of the VW-to-be. The car they built as the Type 62 looked ideally suited to that application, its body details modified to suit the final chassis design.

An alternative body was built in 1938–39 for the same chassis. This was a low, aggressive-looking and exiguous vehicle whose side-mounted spare wheels fooled some commentators into calling it a 'six-wheeled' prototype. Nicknamed the *Stuka* after the famous dive bomber for its pugnacious looks, it was as militaristic as Porsche's Type 62 was civilian in appearance. The *Stuka* took part in comparison tests with the Type 62 and other vehicles.

The body of the official Type 62 was commissioned on 17 May from traditional Porsche panelling partner Reutter of Stuttgart. Its spare wheel was

In 1938 enthusiastic SS troops connived with mechanics at Porsche's Zuffenhausen workshop to make this 'breadboard' sample of a possible military version of the Volkswagen. A sectioned V30 prototype was in the background.

Dating from 1936, Porsche's Type 62 project was first envisioned as a sporting car to take part in trials organised by Adolf Hühnlein's NSKK. Presented to the Army in 1938 it was found capable but not sufficiently 'military-looking'.

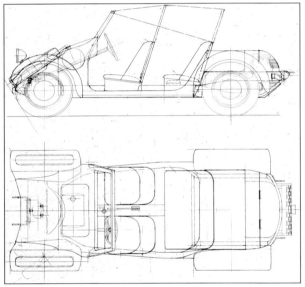

inset into the front deck and its sides were completely open, a few straps deployed to keep the occupants from spilling out. This prototype was ready for presentation to the HWA on 3 November 1938, showing its kinship with the Beetle in the shape of its windscreen and its rounded lines. Ferdinand Porsche presented it to General Becker and others of the HWA, including Lieutenant Colonel Sebastian Fichtner, head of the vehicle-test section.

The Army put the Type 62 to the test at its Münsingen Troop Training Grounds that same November, pitting it against one of its Class I military vehicles, the smallest 4x4 model in its inventory. The Porsche people brought along a Beetle prototype for comparison purposes. The open-topped Type 62 fared well enough, although the Army assessors thought it looked too 'civilian' and asked for more 'military elements' in its design.

While further tests were being conducted on the first Type 62 the Porsche engineers produced a more 'militarised' version. This Type 62 K1 kept the rounded wings and recessed spare tyre but had a more angular main body made of flat sheet steel ribbed for stiffness. One version resembled the first in having open sides with canvas doors; another had proper doors with side screens. The car with open sides was commandeered by the DAF's Robert Ley in October 1939 for a tour of Poland.

Tests of this new type in comparison to two of the standard Army vehicles at St Johann in the Tyrol in March 1939 showed it to be promising but still lacking the ample ground clearance needed for military duty. To increase the clearance 18-inch wheels were fitted but these were not the answer, especially because they raised the car's overall gear ratio when what was actually needed was lower gearing.

A different ring gear and pinion were tried in the Type 62 to give a lower axle ratio, but with a pinion that was too small this was a major and risky departure from the standard VW design. Ferry Porsche explained the problem:

> You had to be able to go at about the walking speed of a soldier carrying his full backpack, so that he could keep pace with the vehicle. This was about 4km/h (2.5mph). Thus there was one serious drawback to overcome. Low gear in the regular transmission produced about 8km/h (5mph). This was adequate for civilian use but too high a speed for cross-country military purposes.

As finally realised in 1939 and redesignated the Type 82, Porsche's military version of the VW looked tough enough. Ingenious engineering gave it parade-speed pace plus the added ground clearance that the Wehrmacht required.

Porsche's solution was typically ingenious. A pair of reduction gears was installed at each rear-wheel hub. The gearing raised the vehicle by 2 inches while reducing the overall gear ratio to give more pulling power at a lower road speed. The hub reduction gears had a ratio of 1.40:1, which combined with the standard final-drive ratio of 4.43:1 gave a ratio of 6.20:1. At the front the wheel-spindle carriers were modified to increase ground clearance there.

Now the Porsche engineers, well on their way towards a completely new vehicle design, awarded the VW-derived Kübel project a new number in 1939: Type 82. Another year was destined to pass after the presentation of the Type 62 before the first two samples of the Type 82 were formally accepted by the Army High Command in December 1939.

This new version reverted to 16-inch wheels for which tyres were more readily available. Now the engine was idling at 780rpm at the required walking pace of 4km/hr, so Ferry's goal was successfully achieved. At 3,300rpm in top gear the Type 82's maximum speed was 50mph.

This striking image of the Type 82 Kübelwagen with its complement of five soldiers showed the benefit of adopting bench rear seating. The use of doors was considered a radical departure for this class of vehicle.

Eight of the first pilot batch of twenty-five Kübelwagens produced in April 1940 paraded in the courtyard at Zuffenhausen, where they were produced. Manufacture at the Fallersleben factory commenced in May 1940.

Canvas hoods and side curtains were part of the Type 82's equipment, seen on the vehicles as they neared the end of the production line. A small area of the Fallersleben plant was at last producing Beetle-based vehicles.

---

* The body of the Type 60 Beetle made use of American Budd Corporation steel-pressing techniques under a licensing agreement under which Budd reserved the right to make the bodies for the first Volkswagen derivative. They had expected this to be a cabriolet but it was the Kübelwagen instead. Thus the bodies for this vehicle were made by the Berlin works of Ambi Budd. When Volkswagen began to build the Kübelwagen, Germany and America were not at war and Budd insisted upon producing the bodies as agreed.

No doubt remained about the vehicle's military bearing. It had square-rigged, corrugated, high-sided coachwork built by Ambi-Budd in Berlin to Porsche's designs.* Its spare wheel rested on top of the sloping front deck, simplifying the body. A serviceable hood and side screens were provided. A key decision was to fit the vehicle with doors. Doors kept the soldiers inside, meaning that tight-fitting bucket seats were no longer needed. Flat front seats and a wide bench rear seat could be used instead, offering more flexible carrying capacity. It also rendered the 'Kübel' nickname completely inappropriate – but it stuck.

The Porsche men deployed several secret weapons in the design of their Type 82 that contributed to its military success. One was lightness. The original Type 60 Beetle of 1938 had a design weight of 1,510lb and ready for the road it weighed 1,545lb dry. Compared with the small passenger cars offered by Adler (1,810lb), DKW (1,720lb) and Opel (1,700lb), its lightness helped the civilian VW gain a power-to-weight advantage over its rivals.

The same philosophy helped the Type 82 shine compared to the heavy Class I military vehicles. Rigorously controlling weight in every aspect of the military Beetle, the Porsche engineers brought this version in at the same design weight as its saloon counterpart, 1,510lb. At 1,600lb with all its skid plates and other battle gear, it weighed only 3.6 per cent more than the road-ready civilian car. This was an astounding accomplishment, even given the Type 82's open bodywork.

Another secret weapon in the car's design was its limited-slip differential developed by the ZF company. This used a central cage to drive a ring of sliding pawls which engaged wavy-cam surfaces that drove each of the rear-wheel axle shafts. When one drive wheel started to slip and spin, friction in the unit rapidly built up and began transmitting driving torque to the wheel that was not spinning – the one that had better traction.

The Porsche office exploited its early access to this ZF invention, first in a passenger car it was designing and then, in 1936, in the Auto Union racing cars. In these they stole a march on rival Mercedes-Benz until the Stuttgart firm twigged what they were doing. It was only natural that Porsche would use the ZF limited-slip differential to help compensate for their Type 82's lack of four-wheel drive.

As a form of protection for the project's future, four-wheel drive was also explored. This was done under Porsche's Type 87 designation, work on which continued into 1941. A drive shaft was taken forward from the front end of the secondary or output shaft of the gearbox to axle gears between the front wheels.

The front-axle gearing was given a drive ratio of 6.20:1 in order to match the gearing at the front to the double-reduction ratio at the rear wheels. In a clever design tweak, Porsche used hypoid gearing for its ring and pinion. This allowed the pinion to be offset upwards from the centre of the ring gear, helping to improve ground clearance at the front. Some Type 87 versions had self-locking front differentials and some did not.

Also added to the Type 87 was an extra-low gear for off-road use. A supplementary lever controlled both functions. Normally the front-wheel drive was not selected. Pushing the extra lever one notch forwards engaged it, allowing four-wheel drive in all four normal forward speeds. Pushing the lever forward to the second notch engaged the extra-low gear as well. Thus extra-low was only available when the Type 87 was operating in 4x4 mode.

Two Type 82s and two Type 87s were tested by the Army in February 1940 in company with a wide range of other vehicles including trucks. They were driven south from the Berlin Kummersdorf test ground to the winter test site in the mountainous Tyrolean Alps at St Johann, the town reputed to be the coldest in the region. 'The two Type 86 [sic] vehicles attracted much interest from all the factory and Army drivers present on account of their speed, manoeuvrability and good roadholding on the icy superhighways and country roads,' wrote VW historian Dr Bernd Wiersch. He quoted a report of 5 April 1940 on the Tyrolean trials:

> The Wünsdorf Test Centre is in general very enthusiastic about our vehicles. In the prevailing slippery ground conditions in the mountains, for example, our four-wheel-drive Types 86 and 87 cars without snow chains were vastly superior to the Army Uniform Personnel Car. For example, our Type 87 climbed the approximately 25-degree slope of the Hungerberg without trouble, while the wheels of the Uniform Car began to skid after a stretch of about 30 metres. Even our Type 82 with snow chains was better than the Uniform Car without chains, since the Uniform Car, with its inherent weight of 1,700 kg [3,750lb], reached its wheel limits too quickly despite its off-road gears.

The remarkable capability of the two-wheel-driven Type 82 had been confirmed back in January 1940 by Army winter tests in Eisenach involving both versions of the military Beetle. The traditional Class I Uniform Chassis was a tough bird to kill, however. The first wartime production plan for vehicles, intended to take effect on New Year's Day 1940, made provision for a derisory dribble of production from Fallersleben: a paltry 200 Kübels a month.

The year 1940 was one of desperation for Germany's entire motor industry. The Army seemed uncertain about its vehicle needs in both qualitative and quantitative terms, unsurprisingly in view of Hitler's fast-changing priorities for fronts on which to fight. Meanwhile the car makers' workforces were being gnawed away by conscription.

One of the most frustrated car makers, in view of the proven performance excellence of his light and low-cost military vehicle, was Ferry Porsche. Tests had shown the clear advantage of the Type 82 over the heavy Class I Uniform Chassis, but no decisions were being made on the basis of this unambiguous evidence, wrote Porsche the younger:

> Weeks went by in this way, and I finally became so annoyed that I openly proposed we take a VW Kübelwagen and run it hard for several more weeks, under all conditions, against the 'Jeep' designed by the Military Supply Office [HWA]. The offer was accepted, and far from weakening our case, the test decisively broke the deadlock in our favour. The Kübelwagen under many conditions came off better than the military version. But still the military held back giving us clearance to manufacture . . .

With the saloon and a display chassis, the Type 82 in its final form was featured at the 1940 Vienna Spring Exhibition. The HWA was thawing, albeit slowly. Pilot manufacture launched at Porsche's Stuttgart plant produced 25 units in April and thereafter 100 in May at Fallersleben. In June, when 200 were made, deliveries commenced to the Army. It paid RM2,945 for each of its Type 82s.

This early production 'involved mostly an assembly job,' wrote historian Art Railton, 'the line occupying a small portion of the huge factory. Joe Werner's shop was producing the engine-transmission, but the rest of the vehicle was shipped in from suppliers. The foundry was still not finished so castings came from a supplier in the nearby Harz Mountains.'

Joseph Werner was one of a handful of German-American engineers whom Porsche had recruited in Detroit in the 1930s to help mass-produce the KdF-Wagen. An ex-Ford man, he knew the art and science of volume production, especially for engines. So his instructions on the Kübelwagen job were especially frustrating: 'I was ordered to build no special tools and not to make the mistake of having any tooling left over for use in building KdF-Wagens when we finished the "jeep" order.'

There was as yet no demand for hordes of Kübels, the thousandth not built until 20 December 1940. The war was still expected to be over soon. The cover of the KdF magazine of May 1941 showed a Type 82 in front of Sicilian temple ruins with an inset view of a peacetime KdF-Wagen, the headline 'He too drives against England' and the caption: 'Mr. Churchill would certainly never have dreamed that so soon he would again meet the KdF-Wagen, which his radio

Although late to the party the Kübelwagen was soon to make its mark. Pictured outside the works was a swathe of the model which was designated the Wehrmacht's sole Light Uniform Passenger Vehicle by the end of 1941.

Part and parcel of the launch of the Type 82 was the need to train military personnel in its servicing. A class gathered for instruction in such matters as its rear-hub gear cases, whose lubrication was often neglected.

broadcasts long ago consigned to the graveyard. With unshakeable confidence in "their KdF-Wagen" hundreds of thousands of citizens continue saving, hoping to receive them soon after the war ends victoriously.'

Military interest in the little vehicle began to grow after the invasion of Russia in mid-1941. The original Uniform Chassis I for light Army vehicles had been replaced by a new design in 1940, made exclusively by Stettin's Stoewer. Under the challenging conditions of the Russian Front, the Stoewer vehicle exhibited 'severe defects in the frame, wheel suspension, clutch, drive shaft, steering, etc.'

In contrast KdF Kübels, just becoming available in Russia, performed much better. On 1 November 1941 an Army report stated, 'All special designs of the Wehrmacht in the realm of wheeled motor vehicles were weeded out (some of them are still being concluded) and, specifically, special designs for the Light Uniform Passenger Vehicle were replaced by the VW (two- and four-wheel drive).'

Before this decision could be reached the Army set one more exam for the military Beetle, one they were confident it would fail. They sent two samples to the North African desert, where the HWA and Uniform Chassis builders were certain the Bugs would simply sink into the sand and never be seen again.

'If they proved right,' Ferry Porsche reflected, 'this ordeal would do us serious damage and cause Hitler's men to lose confidence in our capabilities. Our detractors were to be deeply disappointed, however. The military version of the Volkswagen performed without giving the least trouble, despite the desert heat, the sand and the brutally rough strains imposed on them. On the contrary they seemed to thrive on this kind of treatment!'

The ascendancy of the Type 82 was affirmed on 19 March 1942 by an Adolf Hitler decree that gave Fallersleben a monopoly on the production of light military vehicles. The Type 82's merits had finally triumphed. In his table conversation among friends on 9 April Hitler elaborated on his thinking:

> After the war we must, for military reasons, limit the German motor industry to the production of a dozen models, and the primary objective of the industry should be the simplification of the engine. Higher power must be achieved by increasing the number of standard cylinders rather than by the introduction of a variety of new cylinders. But the most important task will be the design of one single engine which can be used just as well for a field kitchen as for an ambulance, a reconnaissance car, road-haulage or a heavy artillery tractor. The twenty-eight-horsepower engine of the Volkswagen should be able to meet all these military requirements. The ideal standard engine which I envisage must have two characteristics: (a) it must be air-cooled; (b) it must be easy and swift to dismantle and change. The latter characteristic is particularly important because, as this war has shown, it is more difficult to get spare parts than to get a complete engine unit.

Production shifted up a gear; in 1942 the delivery of the 5,000th Kübel could be celebrated. At that time its price to the Army was RM3,457. Production, which had been flat through most of 1941, began to rise late in the year and kept climbing well into 1944. Tropical battle zones opened up to the versatile Kübel after proving trials were conducted successfully in occupied Afghanistan and Greece.

Traditional celebrations were observed for the completion of the 1,000th and 5,000th Kübelwagens. The latter milestone vehicle was equipped with the wide-section 'aero' wheels and tyres best suited to desert use.

Equipped with large-section (200 x 12) 'aero' tyres on special Kronprinz wheels that let them take full advantage of their light weight, the Type 82s proved their merit by skimming over the desert sands of North Africa. And if one did get stuck in a ditch it was easy enough for a few soldiers to heave its light chassis out and send it buzzing onward. It would be an exaggeration to say that the Kübels in North Africa were trouble-free. Said one German workshop report from the front, 'At first this vehicle gave very good performance, but after 5,000 to 6,000 kilometres every possible type of trouble appeared.' One weak spot was the rear-hub reduction gears, a design feature which was unique and, as a result, tended to escape the routine maintenance so essential to its reliability.

Engineers from Porsche were quickly on the scene to diagnose such problems. They suggested field expedients and recommended changes to the production cars. Porsche's people considered it vital to build a co-operative relationship with the Army vehicle-service personnel; generally they succeeded. They realised that it was essential that the military version of the KdF-Wagen gained an excellent reputation among the soldiers and their families they were counting on to keep saving to buy the civilian edition. To this same end the technical manuals produced for the Type 82 carried the civilian KdF-Wagen service emblem on their covers. This, soldiers were to understand, was the wartime version of the same car they would be driving after their victory.

Field Marshal Erwin Rommel first made use of a handful of the fast-moving Kübels in the invasion of France. Later, saying that they could follow wherever a camel could go, the bold and brilliant Rommel exploited their capabilities to the full in his African campaigns. 'With its black, white and red command flag on its fender,' wrote historian David Irving, Rommel's 'own Volkswagen Kübel car was clearly visible. From it, he set the angle and tempo of the attack. If his car was shot up or ran over a mine, he simply commandeered another.'

'You saved my life,' Rommel told an astonished Ferdinand Porsche when they met during the war. 'I was using one of your Kübelwagens,' the Field Marshal

'Aero' wheels and tyres underpinned this Kübel in Tunisian service. During his invasion of France Field Marshal Rommel first appreciated the Type 82's value, which he proved decisively in his North African operations.

explained, 'which went through a minefield without setting it off. The big Horch that was following me, with all our luggage, went sky-high!'

'It was a great car,' Rommel's technical adviser John Eschenlohr told Art Railton. 'Everybody drove them, officers and men. You could trust it because you knew you would get back if you went in a Kübelwagen.' If they received cars without the big 'aero' tyres that coped easily with the sand dunes, they installed their own from aircraft supplies.

Kübels were ubiquitous in all German theatres. They were personnel carriers, munitions carriers, fuel carriers, ambulances, siren cars, cannon tractors, engineer

This Kübelwagen was booty for the British Army in North Africa. It is one of the Type 82s that was sent back to the United Kingdom for detailed evaluation to gain insights into the technological threat posed by the Volkswagen.

vehicles and communications cars. Able though they were, however, the Type 82s could do little without fuel. Lack of fuel, especially in the Wehrmacht's more extended and remote fronts, meant that many serviceable vehicles of all kinds, including Kübels, were abandoned in the field.

In North Africa many Kübels were left behind with only vapour in their fuel tanks when the rest of Rommel's army withdrew after the British counterattacked successfully late in 1942. A decade later 350 wartime Volkswagens were still being driven happily by the local population in Libya and other Northern African nations. The situation was no better on the Russian front, said Hans-Georg Mayer-Stein:

In Greece, from which British, Anzac and domestic forces were swept by the Wehrmacht in the spring of 1941, the German flag fluttered over the Acropolis on 27 April. A Kübelwagen posed there triumphantly.

> Since the German troops in the east were more and more involved in losing battles, at least since 1943, there was scarcely time for vehicle maintenance. In the front-line repair shops, only makeshift repairs were made, or damaged vehicles were cannibalised, in order to keep their time out of action short. As a rule, the lifespan of a vehicle lasted only three weeks. The German military vehicles of 1944–1945 thus made a miserable impression: bashed and bent body panels, missing parts, wrong wheels and tyres, etc. Many a Volkswagen was simply left by the roadside as the German troops retreated.

Many examples, captured intact, were commandeered for use by the Allied forces, who valued their mobility. In the Sahara their saying was that one Kübel was worth two Jeeps. Several were liberated for study back home in Coventry and

Observed by horse-drawn troops, the driver of a Type 82 peered around his door to find a way forwards through the muddy terrain that plagued the Russian front. Its lightness was a huge asset to the Kübelwagen in such conditions.

Most prestigious among the VW's wartime adaptations was the Type 287 Command Car, which unlike the standard Kübel had four-wheel drive. A Beetle look-alike, it was equipped for use in the field by senior commanders.

Detroit. The Allies were deeply curious about the design of these agile German vehicles, based as they were on Hitler's famous yet mysterious and indeed notorious people's car.

Under Ferry Porsche's supervision, Porsche's designers created a bewildering array of variants based on the Type 60 and 82 designs to meet the war's exigencies. Dubbed the Type 82E was a high-ground-clearance Kübel chassis carrying a KdF-Wagen body. Two were built in 1941 and from 1942 to 1944 the output was 564 of the Type 82E saloon plus two cabriolets. They were chiefly used by SS units to fly the flag of the Third Reich in occupied lands.

Among the most confusing of Fallersleben's wartime automobile products – and at the same time among the most interesting – were those with Beetle bodies on four-wheel-drive chassis. Dubbed the *Kommandeurwagen* or Command Car, this Beetle look-alike had the full 4x4 propulsion equipment under its floorboards. Its dry weight of 1,740lb made it one of the heaviest VW derivatives of the period. Only the much more elaborate Type 166 Schwimmwagen, described in the next chapter, was heavier at 2,005lb empty.

Designed as Porsche's Type 287, the Command Car seated the 'commander' behind the driver with free space to his right, while an adjutant occupied the front passenger seat with a map and writing desk that folded out into position. Rifle racks, a machine-pistol holster and first-aid kits were built in. Floors were wooden-slat-covered for service in the mud of Russia.

Output of this highly specialised Beetle was 134 units in 1942, 382 the following year and 151 in 1944. We can envision them being built by the VW factory to be sold on to the Army on an advantageous basis, a car that looked just like the famous peacetime KdF-Wagen yet was fully equipped to cope with wartime conditions. This was a heady combination that granted exceptional status to any officer fortunate enough to be assigned one of these rare military Beetles.

Ordinary rear-drive Kübels had their own terrain to conquer; twin tyres at the rear of the Type 82 were tried for improved traction. Porsche went even further with half-track-like treads driven from the rear axles in several configurations. Officially the Type 155, this was the *Schneeraupe* or 'snow-caterpillar' to cope with conditions on the Russian front. High cost and complexity stifled its production.

In 1943 another version of the wartime VW, the Type 157, had flanged steel wheels to allow it to be driven on railways. That same year the wartime Beetle's engine was enlarged from its original 985cc to 1,130cc by expanding its cylinder bore. Horsepower rose only from 23½ to 25 but torque or pulling power was usefully improved.

The larger engine was better suited to the more demanding service imposed by heavier KdF-Wagen variants such as the Type 166. As well it helped compensate for the drop in power that accompanied the use of generator gas as fuel. In Porsche's Type 230 project VWs were powered by generator gas from a wood-burning plant installed under their distended noses.

Among the many other studies carried out during the war were automatic-transmission experiments; the design of an exhaust-driven turbocharger;* a version of the engine using the disc-type valves developed by Felix Wankel, later famous as a rotary engine inventor; a six-wheeled off-road vehicle powered by two VW engines; a five-speed gearbox; the use of fuel injection and versions of the VW with power by acetylene and electricity.

Tests with alternative materials were important as well in the straitened circumstances of the German wartime economy. Trials were conducted with cast iron instead of aluminium for the VW engine crankcase and gearbox casing, but this was found to be counterproductive. The added weight at the rear overstressed other components and disastrously impaired the vehicle's handling.

The Porsche organisation was not conducting all this design and experimental work out of its affection for the Third Reich. Through 1941 an accounting showed

Among evolutions of the Type 60 devised under Ferry Porsche was wood-gas-generation equipment to replace petrol fuel. Less ungainly than many installations, these two variants carried generator units in their noses.

---

* The design placed the turbocharger adjacent to the engine with the drive turbine below the compressor wheel in a position where it could be cooled by air blowing across its face after it passed through the engine. Such cooling was essential with the relatively poor turbine-blade alloys that were available to the German engineers. Porsche patented this design by Walter Becht.

that its expenditure on the design and development of the KdF-Wagen amounted to RM8.4 million ($2.8 million), paid from various sources. On average, through the subsequent war years Porsche was billing Volkswagen 300,000 marks monthly for engineering work or some RM3.6 million each year, equal to $1.2 million. In 1944 the turnover of the Porsche office reached RM5,825,000, of which Volkswagen-related engineering accounted for more than half. Working hard for the Reich, Porsche was also rewarded well. Its staffing in 1944, by then scattered throughout Greater Germany in Gmünd, St Valentin and Zell am See in Austria, in Stuttgart and at Fallersleben, amounted to 588 people – 299 technical staff and a workshop force of 289.

Parts for engines and vehicles built at Fallersleben were sourced throughout Greater Germany. Castings and forgings came from Czechoslovakia. The Peugeot plant at Sochaux, which became a subsidiary of the Volkswagen Works in June 1943, supplied 20,000 connecting rods, 3,000 flywheels and 5,000 forgings for crankshafts – the last a rush order. Type 82 body panels made in the press works of Ambi-Budd in Berlin were shipped west to the plant for assembly.

Of the Type 82 'bucket-cars' the plant's total output was 50,435. Included in this total were such specialised versions as radio cars (3,326), intelligence cars (7,545) and repair vehicles (2,324). This made it Germany's most abundant light military vehicle, well ahead of the wartime version of the Mercedes-Benz 170V, of which 19,000 were built.

Substantial though they were by German standards, these Beetle numbers were an order of magnitude smaller than America's production of 650,000 Jeeps for use in the Second World War. Design and development of both vehicles had taken place at about the same time, but the American military was quick to see the value of its Jeep, which had been pioneered by its Quartermaster Corps and car maker American Bantam. It shifted both Willys and Ford into high-gear production of the tough and versatile Jeep.

Kübel production slumped in the last year of the war when conditions worsened. Supply of vital components was fitful. Raids on Berlin kept Ambi-Budd from supplying bodies regularly. Only in the last weeks of the war did the KdF staff succeed in bringing the Type 82 body dies to Fallersleben, where they stamped their own panels for the last 665 bodies made.

After attacks from the air, assembly of the engines and vehicles was moved to the spacious cellar of the plant, originally intended only for services and piping. The workforce was composed largely of Italians, Russians, Poles and Frenchmen, held there against their will. As early as 1943 only one auto worker in eight in the plant was German.

In spite of shortages on the Eastern Front enough fuel was available for one Kübel to make a long journey in March 1945. The Army's Commander-in-Chief decided to leave his Berlin headquarters and travel east in an inconspicuous Type 82 to a castle near the Oder River where his Ninth Army commanders were based. There he urged them to do all they could to resist the Russian Army's powerful attack on Germany and Berlin. Then Adolf Hitler returned to Berlin, chauffeured as usual by Erich Kempka. During the trip back, wrote John Toland, 'Hitler sat silently, deep in thought.'

CHAPTER 11

MASTERPIECE SCHWIMMWAGEN

At leisure on the Kraft durch Freude holiday liner *Robert Ley* in the summer of 1940, Ferdinand Porsche reflected on yet another commission for his designers. On 1 July 1940 the HWA asked him to design and build an amphibian version of his military four-wheel-drive Type 87. Here, Porsche well knew, was a clear sign of future maritime missions for the Wehrmacht. On the day he received the commission German troops disembarked at the Channel Islands. An invasion of the main British isles could not be long in coming.

The order for an amphibian Beetle followed an initial request prepared on 18 June that said such a vehicle was 'urgently' needed by the Army's engineers. The HWA's contract made RM200,000 available for the project. The machine was to be able to achieve 6mph in the water and 50mph on land. As well its transition from water to land was to be achieved without the occupants having to leave the vehicle. A concession on the HWA's part was that the vehicle would not have to have doors.

In a project led by Ferry Porsche, the Zuffenhausen engineers adapted the 4x4 Type 87 platform to this new application with remarkable speed. With Swabian practicality their first step was to weld a Type 87's doors shut, attach a provisional propeller and experiment with it in a fire-fighting reservoir on their Zuffenhausen property. These tests, said Karl Rabe's son Hans, 'produced fundamental knowledge about the behaviour in water, the weight distribution for trimming, the power unit, the draught and the freeboard, also during water entry and exit. They also showed the unfeasibility of a simple reworking of the Kübelwagen for operation in water.'

The end of September 1940 found Ferry Porsche at the wheel of the first Type 128 prototype as it surfed the Max Eyth Reservoir with a full complement aboard. Its Drauz-built body was doorless.

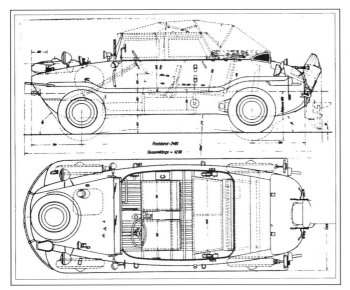 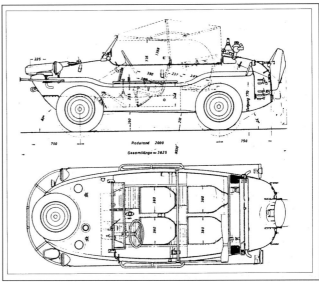

Side and plan views show the proportions of the Type 128 (above) and the final Type 166 Schwimmwagen. The latter sacrificed stowage space at its rear to preserve seating for four fully equipped servicemen.

In the final design of what became their Type 128 the Porsche engineers worked with coachbuilder Drauz in Heilbronn to fashion a suitable doorless body on the Type 87's floor pan. They fitted water-tight seals to the axles and speedometer-cable aperture. Water-tight material covered the cables of the mechanical brakes. An engine-driven propeller blossomed at the rear. Simply enough, the front wheels acted as rudders when turned, a solution foreseen by the HWA in its guidance to Porsche.

Completed on 21 September 1940, at the end of that month the first Type 128 plunged into Stuttgart's Max Eyth Reservoir* in a bend of the Neckar River east of Zuffenhausen with Ferry Porsche at the wheel and four colleagues checking for leaks. It worked well enough, but it needed a healthy push from those colleagues and a few more to get it up the muddy bank and out of the reservoir. October saw revisions and improvements in the three prototypes built.

The Army's HWA lost no time in putting its experimental 128s to the test after it received them in early November. For six weeks they were driven on Autobahns, across country, on minor roads and in water, using the military proving ground at Kummersdorf near Wünsdorf. Their tests included comparisons with a front-engined amphibian that had been under development for six years by Hanns Trippel, most recently at the former Bugatti factory at Molsheim, and, improbably, one of the Light Uniform Chassis vehicles swathed in a watertight linen covering.

'During this testing,' wrote VW historian Bernd Wiersch, 'the vehicles covered between 3,207 and 3,496 km, 1,400 km on the Autobahn, 1,270 km on roads, 180 km in rough country, 200 km in very rough country and 300 km on very rough mountain roads. Every vehicle spent 18 hours in the water.' These figures did not apply to the linen-sheathed Uniform Chassis amphibian, which 'was constantly defective and dropped out as unusable in the first water and off-road tests'.

While waterborne the fuel economy of the Type 128 was an astonishing three times better than that of the Trippel and on land it was twice as economical. They were similar in terms of road speed but the VW derivative was faster in the water at 6 versus 5mph. Even better, the air-cooled VW could maintain its peak water

---

* Born in Württemberg in 1836, Eduard Friedrich Maximilian Eyth was an engineer who joined England's Fowler, a leading producer of steam tractors, in 1861. He remained with Fowler until 1882, when he returned to Germany. There he led the movement towards increased mechanisation of farming.

velocity for an hour while the water-cooled Opel Kapitän engine of the Trippel boiled over all too quickly.*

After this successful debut the HWA approved the building of a pre-series of 30 Type 128s in 1941 for further tests and evaluation. But the speed with which this could be achieved would depend on the priority level granted to the project. Ferdinand Porsche seized the opportunity to ask for a higher priority when he presented the Type 128 to Hitler and General Field Marshal and Army Chief of Staff Wilhelm Keitel at the Berlin Chancellery on 26 February 1941.

'The Führer thought that this was do-able in view of the small quantities involved,' wrote Porsche afterwards in an aide memoire, 'and with respect to this he turned to General Field Marshal Keitel, who gave the assurance that he would make the "special level" [of priority] available to the Schwimmwagen.' Porsche reminded Keitel that in ordering materials for the Type 128 he was already operating at priority 'special level' SS; the Nazis managed priorities by adding new and more urgent levels at the top of the range as and when needed.

The priorities were secured and the 128s were built. The Army's Kummersdorf proving ground was the site for more trials in May and June. In August three of the pre-series cars traipsed 1,600 miles through the Alps. Some 15 per cent of the running was under 'poor and poorest' off-road conditions and 20 per cent over rough mountain tracks. As well 29 hours of floating trials used the Danube and the Bodensee. The testers said that this trip 'demonstrated to us as never before the Type 128's extraordinary off-road capability when driven sensibly. We drove on paths that had never before seen a motor vehicle, and the total cargo always amounted to almost half a ton.' The HWA's experts found plenty of faults, however. Much strengthening was needed, particularly in the propeller drive, whose durability they found unfit for purpose.

These detailed test findings were a tribute to the extensive and intensive testing of new vehicle designs conducted by the HWA, in spite of the urgent requirements of war. This was consistent with the German tradition of thorough engineering as well as with the concept that it was best to go into the field with proven equipment. This was not always the case, especially later in the war, but Germany set and maintained a higher standard in this respect than most of her enemies.

At the end of 1940 the aquatic-Beetle project received a sharp shove forward. On 3 December Berlin's Waffen-SS took an interest in the development of a new light armoured scout car, thanks to the quick thinking of Ferry Porsche. His timely suggestion during a meeting in Stuttgart with the regional chief of Heinrich Himmler's elite military staff brought his attention to the work being done on an amphibian version of the Type 87. The needs of the SS could merge well with those that had hitherto guided work on the Type 128.

What the Waffen-SS was seeking was a replacement for the motorcycles with sidecars used by its mobile guard units. Motorcycles were proving unequal to the demands of modern warfare in difficult terrain. The utility of BMW, NSU and Zündapp motorcycles on scouting and courier work in the easy conditions of Hitler's first campaigns in Western Europe was undisputed. But in rustic Poland and in Africa, where roads were poor or non-existent, they fared poorly. From Africa a German workshop company reported to headquarters as follows: 'In the light of recent experience, motorcycles are most unsuitable for the desert and are

* Hitherto Germany's leading exponent of amphibious vehicles, former racing driver Trippel hit tough competition for the first time in the Type 128. He bounced back, however, building and supplying some 800 of his Opel-powered SG6/41 amphibians for the Waffen SS through 1944. After the war Trippel would be famous for his Triumph-engined Amphicar, introduced in 1961. One made a successful Channel crossing.

the type of wheeled vehicle that appears most commonly in the workshops. There is extraordinarily high wear on pistons and cylinders. Clutches, carburettors and Bowden [cable] equipment full of sand were the order of the day.'

Motorcycle makers did their utmost to rise to this challenge. Their first step was to add a drive to the sidecar's wheel. Tests of whether a differential was needed were inconclusive. Zündapp came up with the concept of a special spur-gear differential that allocated the drive torque in proportion to the centre of gravity loading across the vehicle's track. This was found to work well; the ratio settled on was 2.2:1, motorcycle to sidecar.

Blower-cooled motorcycle engines were also being studied near the end of the war. As well the idea of designing a motorcycle powered by the KdF-Wagen's flat four was explored. This was not pursued, Ferdinand Porsche arguing successfully that even with enhanced power a motorcycle with sidecar would never be a well-integrated vehicle. To cap the argument against them, motorcycles cost more than their VW-based counterpart, at least in Type 82 form.

On 22 December 1940 Porsche received a contract for the development of a vehicle that would meet the needs of the Waffen-SS, sweetened with a fee of half a million marks for development and the loan of ten engineering draftsmen from SS ranks. While the Type 128 had been promising, it was not yet the complete article. Built as it was on the long 2.4-metre wheelbase of the Types 60 and 82/87, the Type 128 was not as gainly as it needed to be to fulfil its mission. This had been especially evident in its awkward entries and exits to and from water.

The solution was a shorter wheelbase of 2.0 metres (78.7 inches). 'We began work on the drawing board in April 1941,' wrote Ferry Porsche. 'By August of that year the first prototype was ready for testing.' All relevant sources identify this final Schwimmwagen design as Porsche's Type 166, which is described in the original list of Porsche type numbers as *VW-Krad-Wagen*, meaning 'VW motorcycle car' – a description more of its mission than of its design.

With its shortened wheelbase the Type 166 gained agility in entering and leaving the water as tests demonstrated. Its development was funded by the Waffen SS, which provided engineering draftsmen to Porsche for the project.

That autumn the first prototype had a high-level viewing. The younger Porsche was 32 at the time. 'I was asked to bring this car to Hitler's headquarters for a demonstration,' he said, 'and he appeared to be pleased.' The headquarters in question was the famous Wolf's Lair in Rastenburg, East Prussia. The request to bring the first Type 166 there was certainly made by Heinrich Himmler, whose SS had commissioned it. Looking even more obsequious than usual in the presence of his Führer, nervously twisting his gloves as he sought Hitler's approval, the odious Himmler was the man on the spot at the forest presentation of Porsche's spruce amphibian with its SS licence plates. Mufti-garbed Ferry showed the versatile auto to a platoon of uniformed military leaders including Keitel and Jodl of the Army and Wolff of the SS.

Many features of this first Type 166 differed from the final configuration. A design outline was confirmed in the late autumn of 1941 for the production, at the Porsche works in Zuffenhausen, of a pre-series of 125 vehicles. Such a large pre-series was a reflection of the urgent need for vehicles of this genre, especially in Russia. That November Ferdinand Porsche met twice with Hitler and his retinue at the eastern Wehrmacht headquarters at Rastenburg, the *Wolfsschanze* (literally 'Wolf's entrenchment', known in English as 'Wolf's Lair').

Work on prototypes of amphibious Beetles was in full swing at Porsche's well-equipped Stuttgart-Zuffenhausen workshops. They were in operation before the company's new headquarters building was completed.

Ferry Porsche, who led work on wartime development of VW variants, presented the T166 prototype to Hitler and – behind windscreen – Himmler at the Wolf's Lair in the autumn of 1941. Keitel and Bormann were at the right.

To one of these meetings Porsche brought a portfolio of photos that showed the undernourished condition of the 850-odd Russian prisoners who were part of his workforce at Fallersleben. Prepared at the request of works doctor Hans Körbel, it contained photos taken by Fritz Heidrich. Just as he had at Austro Daimler in the First World War, Porsche urged better arrangements for nourishment of the

---

* Here was a clear example of Porsche's ability to bring unpalatable facts to the attention of Adolf Hitler. Although Porsche would not stint in so doing, Hitler's ability to take advice would diminish as the war ground on.

'He appeared to be pleased,' said Ferry of Adolf Hitler's reaction to the SS-backed Type 166 at its autumn 1941 Wolfsschanze presentation. It brilliantly married four-wheel drive to full amphibious capability.

Instead of the prototype's abbreviated windscreen the final Schwimmwagen had a full-width screen. Its silencer was well clear of the water from which it emerged with its propeller still in seagoing position.

Russians.* His intervention with Hitler led to the issuance of instructions to the Reich Food Ministry for the 'pampering-up' of those prisoners who were thought to be capable of working well.

After a winter's work on the Type 166 the Max Eyth Reservoir was again the basin for the formal baptism of the new model in March 1942. Experience in the field with the pre-series proved beneficial in improving the design of the final production version, which was accepted by the HWA on 29 May 1942, the end of the month in which it began to be made at Fallersleben. Like the Type 82s, its one-piece welded body-hull was produced in Berlin by Ambi-Budd and fitted at Fallersleben. The advantages of Ambi-Budd's press techniques were shown in the 166's one-piece side-panel pressings. Weighing 122lb when hammer-formed and taking 29 hours to make, when die-stamped each was produced in 65 minutes and weighed 60lb.

A major challenge facing the designers had been the provision of space for four fully equipped soldiers in spite of the shortened wheelbase. This was solved with style in the final Schwimmwagen design. To increase range it had more fuel capacity, storing a total of 11 gallons in two tanks of equal size. Inherently buoyant, the Type 166 combined manually selected four-wheel drive with a swing-down propeller at the rear driven directly by a dog clutch and roller chain from the engine's crankshaft. Its short wheelbase, extra-low gear and four-wheel-drive made it an even more effective off-roader than the Type 82.

The Type 166's final design for production differed in many details from its Zuffenhausen pre-series. It no longer had a pointer on the steering column that showed the driver the front-wheel angles, at first thought useful when waterborne. Also deleted was a buzzer that warned the driver if the fan-drive vee belt broke.

Many attachments differed, redesigned at the KdF-Werke for easier production. A tubular towing bracket at the rear was omitted. While the fat 'Aero' wheels and tyres suited the model, scarcity meant that many were built without them.

Before the Schwimmwagen could be released for service with the Pioneers and other forces it had to be type-approved by the respective authorities. As a vehicle this presented few problems, but the Type 166 was not just a vehicle. It was also a boat, said the HWA's approval commission, and boats have to display bow lights, red on the port side and green to starboard. 'The Porsche engineers thought the HWA people were crazy,' said Richard von Frankenberg, 'and had no thought of installing such lights. This bothered the HWA not at all – the vehicle was simply not accepted.'

Doubtless with some trepidation SS cadres had their first taste of motoring Schwimmwagen-style in the waters of the Mittelland Canal adjacent to the KdF-Werke at Fallersleben.

After series manufacture finally began, by 6 June the first hundred had been completed. The German military paid RM4,200 for each of its water Beetles. The price paid bore little relation to the real cost of the Type 166, which was so complex that it heavily burdened the largely Italian workforce that assembled it. For this reason its production was halted on 26 August 1944, in the wake of the final bombing of the factory, after 14,276 were produced. This fell well short of

With prototypes in the foreground, the Porsche workshop at Zuffenhausen produced the first 125 units of the Type 166 in the 1941/42 winter, using steel hulls fabricated in Berlin by Ambi-Budd.

A completed Type 166 Schwimmwagen was a thoroughly evolved vehicle, shown with its propeller and soft top deployed. Air feeding and cooling its engine was inducted well above the waterline.

Carrying only two troops, a Schwimmwagen's freeboard was more than adequate as it cruised down a French canal. Strong bow waves suggest that it was running at its maximum speed in water of 5½ knots.

the 20,900 orders that the factory had booked for this remarkable vehicle.

The Schwimmwagen was a prized asset in all the Wehrmacht and Waffen-SS theatres. When the Waffen-SS paraded in Paris in 1942 it was with a long line of Type 166s. Pristine or battered, armed or as transport, they were ubiquitous at the fronts in spite of their relatively small numbers. Especially in the east their versatility was prized. Porsche supported the model's development throughout the war, using the fifth Type 166 from the pre-series as its guinea pig.

After the war one British officer took a special interest in Porsche's Type 166. This was Major General Percy Hobart, who had been responsible during the war for all the 'funnies' in the Tank Corps: flamethrowers, mine-destroying flails and the like. He set up a visit to Wolfsburg especially to see and assess the Type 166 Schwimmwagen. Although the works could not then produce its four-wheel-drive system for lack of certain forgings, it still had the complex and costly dies for the steel pressings that made up its sophisticated hull, which like those for the Kübelwagen had been sent there by Ambi-Budd. The officer in charge at the factory, Major Ivan Hirst, took care to set these aside for possible future use.

'The General arrived in a Jeep,' Hirst recalled, 'and we had a contest with the Schwimmwagen. It easily went where the Jeep couldn't. As a result of these trials Hobart tried to get the British government interested in the Schwimmwagen, but he was told that no money was available for that sort of thing. So I had to let the dies go.' A vehicle that could justly be termed one of the masterworks of the Porsche office could no longer be produced.

In the spring of 1943 a Schwimmwagen navigated the waters of Brittany in a training exercise. Like the Kübelwagens surviving examples would become prized trophies for the Allies when they invaded the following year.

The occupying Allies evaluated the Type 166 at Fallersleben. Tragically the dies for its hull were trashed after the British government declined to accept high-level recommendations that it take advantage of this ingenious machine.

Still wearing its stencilled delivery address, a Type 166 was delivered by the Army to the General Motors Proving Grounds at Milford, Michigan, for evaluation by the Motor City's engineering elite.

The Type 166 was also of great interest to the Americans. A fully operational sample was shipped to Aberdeen, Maryland, and given an initial assessment in a Memorandum Report dated 30 October 1944. This included a translation of the technical specifications of the Type 166 and gave the first impressions of US Army Ordnance experts on this Porsche creation.

They liked it. 'The torsion bar suspension on this vehicle enabled it to perform extremely well over smooth and rough terrain,' the Aberdeen experts found, and 'the front and rear locking differentials were very effective for mud operation.' Its engineering attracted admiration: 'The simplicity in design of this vehicle lends itself well to mass production.' Its seaworthiness won plaudits too. 'The all-around performance of this vehicle in water was exceptionally good,' read the report. 'Front wheel steering in water instead of using a conventional rudder was found to be very effective.' Summing up the Aberdeen appraisal, Lieutenant Colonel G.B. Jarrett found the water Beetle worth emulating: 'Because of the excellent performance of this vehicle during limited tests and because of the simplicity of the design, it is recommended that the vehicle be further investigated with a view toward having our automotive industry adopt some of its salient features.'

With just such a technology transfer in mind, another captured Type 166 was forwarded by Aberdeen to the General Motors Proving Ground at Milford, Michigan, where the auto industry could test it on a rich variety of surfaces and gradients, including GM's 'Mud and Billy Goat Hill Test', and pilot it into and across Sloan Lake. All the members of the Overall Vehicle Sub-Committee of the Society of Automotive Engineers' Captured Enemy Equipment Committee drove and rode in the little vehicle blazoned with its stencilled Aberdeen address.

They liked it too. Their well-illustrated report, issued in August 1945, could hardly have been more complimentary to the Type 166 and the engineers who conceived it. Their General Observations were as follows:

On the surfaced road the smoothness of the ride and the way the vehicle hugged the road and floated along were noteworthy.

The vehicle covered the mud and hilly route with much greater ease and smoothness than the American Jeep, which followed it on each trip around the circuit.

The vehicle was impressive for the manner in which it was manoeuvred by its front wheels in the water, its steadiness in the water and the ease with which it entered and left the lake.

The general overall performance was highly satisfactory for the purpose of reconnaissance for which it was designed.

The Porsche engineers, still sheltering in a sawmill in Austria, would have been gratified by this assessment of their work by the enemy industry whose mass-production skills they had sought to emulate. But clearly the Type 166 was a vehicle of a type that the Americans would not have thought of building. As the engineers remarked, 'American vehicles because of their surplus power can always do more than they were designed for.' They added:

> However, in the judgement of the committee this vehicle has sufficient power for its intended purpose, namely that of a reconnaissance vehicle in which it may be called on to cross ponds, small lakes and rivers, soft ground, etc. More power would mean the necessity of larger component parts, which would mean that the vehicle would lose many of its outstanding characteristics. If this vehicle had the power plant of an American Jeep it would probably mean that its weight would go up to that of the standard Jeep. The American Amphibious Jeep weighs approximately 3,400 pounds compared to slightly more than 1,700 pounds [1,725] for the German amphibious Volkswagen.

When empty the two vehicles had similar ratios of power to weight, the SAE experts reported, although the Jeep weighed almost twice as much as its amphibious Beetle equivalent. 'There are important lessons in the reduction and saving of weight in this vehicle, especially in the engine, and the hull construction,' they added.

After their meticulous analysis of the structure and concept of the Schwimmwagen the American auto engineers gave it a rave review: 'The general architecture of the job offers an ideal combination of structural strength, maximum passenger space and low silhouette. Such items as the elimination of side-entrance openings, the relocation of exhaust system and the torsion bar method of suspension, while lending themselves to the nautical personality of the job, have definitely contributed to its light weight as well as to its performance and stability as a land vehicle.'

A dichotomy of the Type 166 design was noted in the report: 'The general feeling of the engineers present was that they would not design an amphibious vehicle for an air cooled engine.' With all that water around, they thought it should be exploited. The engineers had not been party to the German tests in which the Trippel amphibian's water-cooled engine overheated. They opined that 'the definite trend of the German designers toward air cooled engines

Although traditionally critical of 'foreign' developments, Detroit's senior engineers found much to praise in the Schwimmwagen's design and performance. They posed with their sample and Army colleagues.

resulted from the problem they faced in winter fighting on the eastern Russian front'.

A complaint common to all the early KdF-Wagen designs was expressed here as well: 'Brakes are inferior. When the vehicle comes out of the water there are no brakes. They are not kept dry. It was explained that the brakes had not been good since the vehicle arrived from Germany.'

Many of the American engineers – among them the technical directors of their respective firms – were having a first opportunity to experience Beetle-style handling. 'There was a feeling that the vehicle steers too fast – over steers – on land,' they reported. 'It was pointed out that even though the steering is fast in [comparison] with American practice it probably is a national German desire to have fast steering. It will hold anywhere on a curve.'

A detailed breakdown of the welded-steel hull of the Type 166 was performed. 'Efficient shaping of the body panels,' said the report, 'has resulted in a job having unusual roominess and a pleasing appearance. It has also contributed to light weight, structural stability and seaworthiness.' Specifically, the SAE engineers concluded:

> In the judgement of the committee the body or hull of this vehicle creates the impression of a well engineered product. It is composed of a minimum number of stampings of substantial size, calling for an elaborate and expensive tooling program. The entire vehicle would be costly to build from our standard for small vehicles. However, cost in a military vehicle is secondary to the man hours required in manufacture.

In this finding they put their fingers on some of the considerations – complexity and the manufacturing-manpower requirement – that had led to the VW works' decision to cease production of the Type 166 in August 1944 after the waves of bombings.

Their experiences in and out of GM's Sloan Lake impressed the assessors with the shrewdness of the Schwimmwagen's hull contours. They praised the way the underside of its 'bow' was configured to permit the front wheels to steer the craft so effectively. 'Road and water tests have proven [that] a great deal of thought has been given to the actual contour of the hull,' they found, adding, 'It may well be advisable to make a more complete study of this hull by making a female plaster cast of same, which could be used as a basis for further experimentation and development.'

Their careful evolution from the 128 to the 166 in the waters of the Max Eyth Reservoir had rewarded the Porsche team with enviable engineering success, not only in the eyes of the soldiers that used their creations but also in the eyes of their peers.

Another enthusiast of the basic Beetle was Adolf Hitler himself, who expounded on its merits to colleagues at midday on 22 June 1942. Saying that practical experience in the desert confounded the thesis of some generals that vehicles could only operate there on roads, he said that 'it has always been my wont to insist that theoretical theses of this sort must be tested practically, and it was on these grounds that I ordered the construction of the Volkswagen'. He continued:

> And it was this same Volkswagen, which is now giving so magnificent an account of itself in the desert, that convinced me of the futility of this particular thesis. The Volkswagen – and I think our war experiences justify us in saying so – is the car of the future. One had only to see the way in which these Volkswagens roaring up the Obersalzberg overtook and skipped like mountain goats around my great Mercedes, to be tremendously impressed. After the war, when all the modifications dictated by war experience have been incorporated in it, the Volkswagen will become the car *par excellence* for the whole of Europe, particularly in view of the fact that it is air-cooled and so unaffected by any winter conditions. I should not be surprised to see the annual output reach anything from a million to a million and a half.

Here Hitler, so often a fount of forecasts that failed to bear fruit, hit the nail squarely on the head. The car whose creation he inspired would become a favourite not only of Europe but also of the world.

CHAPTER 12

LEOPARDS AT NIBELUNG

On 3 September 1939, two days after his invasion of Poland, Adolf Hitler issued his second order for the conduct of the war. One of its provisions was that 'the conversion of the entire economy to a war economy is ordered'. This meant intensified armaments production, of course, but also attention to agriculture, raw materials and all the other concomitants of conflict. Priorities had to be balanced between the crucial importance of the short term, for Hitler foresaw Europe on its knees in 1940, and the lesser possibility – as it then seemed – of a longer war.

'Hitler was fully clear about the economic consequences that followed' his short-term focus, wrote Adam Tooze. 'He needed to force the Wehrmacht's economic experts to abandon their caution and to press immediately for the expenditure of all available resources in preparation for the 1940 offensive, regardless of the long-term consequences for the viability of the German war effort.'

Hitler's ukase of 1939 was by no means the first call to mobilisation of Germany's economy. Since 4 February 1938 the newly created *Oberkommando der Wehrmacht* (OKW) or High Command of the Armed Forces had replaced the old War Ministry in overall command of the Reich's military. In charge of its powerful military-economics office was Major General Georg Thomas, who in 1933 had been the HWA's chief of staff. By the autumn of 1938 Thomas and his allies were well on their way to implementing what Tooze called a 'convulsive acceleration of armaments activity'.

Eager to be seen as responding to Hitler's demands, at the end of February 1940 General Hermann Goering, in his role as master of the Reich's Four-Year Plan, appointed Autobahn-builder Fritz Todt a special troubleshooter to locate and smash armaments-production bottlenecks. Since December 1934 Todt had been the Nazi party's designated leader of engineering. Named plenipotentiary for all construction by Goering in 1938, Todt introduced a system of priorities and rationing of materials.

'Hitler trusted Todt implicitly,' wrote David Irving, 'and Todt had also won the respect of the party.' On 17 March 1940 Fritz Todt became the Reich's first Minister for Armaments and Munitions, with specific powers over the Army's activities. He immediately ordered a stop to government-backed experimental projects that would not yield results by October, an order which was widely disregarded.

Mustering of armaments went too slowly for Hitler, who in January and February 1940 'cursed General Becker's lethargic and inadequate Ordnance Office', said Irving. Well entrenched, Karl Emil Becker did not readily relinquish power. His countermove to Todt's appointment was a bid to set up an overall ordnance office for the entire Wehrmacht, including the navy and air force. Although accepted by Hitler on 8 April, this initiative was countermanded the same day after representations by a Krupp executive, who hinted at unspecified scandals in Becker's family. Becker shot himself that evening.

Sharing as they did a positive attitude to problem-solving, Fritz Todt and Ferdinand Porsche were good friends. This helped work proceed on the construction of the KdF-Werke at Fallersleben in 1939 and 1940 in spite of first Speer's and then Todt's constraints on many projects. That the two men were kindred spirits with experience in common was suggested by remarks in a letter that Todt wrote to Albert Speer about a dispute in which the latter was involved:

> In the course of such great events . . . every activity meets with opposition, everyone who acts has his rivals and unfortunately his opponents also. But not because people want to be opponents, rather because the tasks and circumstances force different people to take different points of view.

Here was a philosophical explication of the kinds of situations that Porsche had experienced many times during his long career, especially during his Daimler-Benz years. It was a mindset that helped both men cope with the jealousy and antipathy that their successes engendered.

Another new task for Ferdinand Porsche was his role, from March 1940, as chief engineer of the new Nibelung Works at St Valentin in Austria, north-east of Steyr towards Linz. Laid out for automotive-style flow-line production, this new manufacturing site was the only one that Germany built expressly for tank production during the war. Wearing the code name *Spielwarenfabrik* or 'toy factory', it was owned by the Montan GmbH, a creature of the HWA. Personnel and other supporting services were provided by nearby Steyr-Daimler-Puch AG, which in turn was under the control of Hermann Goering's burgeoning empire.

Turning 62 in 1937, Ferdinand Porsche was at an age when many began contemplating retirement. Instead he was about to embark on the most demanding phase of his remarkable career.

Production and assembly halls at the Nibelung Works were deliberately spread among forests that grew to provide concealment. The plant was managed by nearby Steyr-Daimler Puch AG, rump of the company that Porsche once ran.

Franz Xaver Reimspiess was chief designer at St Valentin, which was on a single level in nine open halls in the midst of a forest and well hidden from the air. On its grounds was a test area for experimental and completed Panzers that included a 45 per cent gradient. Built in four stages, the factory began operations in the autumn of 1940 with tank repairs. By the end of 1944 some 8,500 would be working at St Valentin, the majority Austrian but of many other nationalities and, at the end, more than 1,000 concentration-camp inmates.

The mission of the Nibelung Works was to build Panzers from Porsche's drawing board. Now he needed a concomitant assignment from the HWA. Although Wilhelm Kissel of Daimler-Benz, with Ferdinand Porsche on his books as a consultant, was always looking for ways to work through Porsche to gain assignments from the Reich, as related in chapter 10 the HWA's relationship with Porsche suffered from recriminations over rights and patents. Viewed as more than fully occupied with the Volkswagen project, Porsche was only slowly drawn into the most intense rearmament effort.

On 6 December 1939 an order for a Panzer design was finally booked with Zuffenhausen. Ferdinand Porsche was instructed to design and develop a tank of between 25 and 30 tons that could carry a gun of 75mm or a high-velocity weapon as large as 105mm. By 9 December Karl Rabe had made a first sketch of its design. After spending the first days of 1940 in Berlin, Porsche returned on 6 January for a meeting with Rabe to discuss the project. The Panzer would be Porsche's Type 100 – a significant number for a significant project.

On 9 December 1939, three days after his Type 100 was commissioned by the Wehrmacht, Karl Rabe sketched a scheme for Ferdinand Porsche's first Panzer since his Daimler-Benz GT I of a decade earlier.

Parachuted into the Third Reich's war effort at this high level, Ferdinand Porsche was destined to spend his next five years amidst the mud of military proving grounds and the tattoo of rivet guns in vast tank assembly halls instead of the racing circuits he loved so much. Porsche plunged his engineers into the design of armoured military vehicles with the same enthusiasm and commitment that had already made his firm famous in other fields.

Ever the autodidact, Porsche would bone up on a subject with which he had had little contact for more than a decade. His last adventure, with the Daimler-Benz *Grosstraktor I* or GT I, had been inconclusive for the engineer, who left the company before its testing began. Since then his focus had been on the KdF-Wagen project, at the opposite ends of several scales with its small size and high production volumes. However, Porsche and his team were nothing if not quick learners.

Difficulties with the HWA notwithstanding, that body was keen to emphasise that it made available all possible resources to Porsche, whose people had long been welcome observers at its proving grounds and test sites. They were given every chance to view examples of both domestic and foreign tracked vehicles, including a German electric-drive system, and were supplied with countless drawings of elements of armoured vehicles both macro and micro. Open discussions took place both in Berlin and Zuffenhausen over technologies and requirements.

Officially known as the VK3001(P),* the all-new Porsche design was paired by the HWA with a rival, the VK3001(H) made by Kassel's Henschel und Sohn, whose Erwin Aders had been designing Panzers since 1936. Thus Germany's neophyte tank designer was pitched against one of her more experienced. The project's objective was to replace the Krupp-built Panzerkampfwagen IV, known as the Pz.Kpfw.IV. First in service in 1938, the Maybach-engined machine evolved into Germany's workhorse medium tank. In its developed H-version it weighed 25 tons and was capable of 24mph.

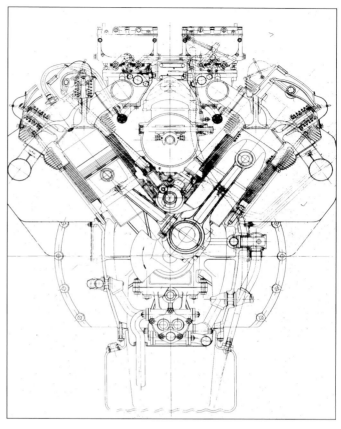

As its Type 101 Porsche designed a petrol-fuelled 11.4-litre air-cooled V-10 engine to power its Panzers. Pushrod-operated rocker arms opening its vee-inclined overhead valves were sharply disparate in length.

'In the field of Panzer development,' said colleague Walter Rohland, 'Ferdinand Porsche knew very well how to find favour with new ideas.' This applied especially to Porsche's relationship with Hitler, who had a passion for the novel and fascinating. This had helped win the Reich's financial support of the rear-engined racing car for Auto Union. Now Porsche paid attention to the Führer's passion for air-cooled engines, known to him since 1933. 'The water-cooled engine will have to disappear completely,' Hitler reminded Porsche and Josef Werlin in January 1942. 'I gave an order for them,' he added. 'It will be forbidden in future to build engines except with air cooling.'

In addition to being air-cooled, the Type 101 military engine designed for the Type 100 was a V-10, an all-but-unknown type. Its direct antecedent, designed but not built, was the 1½-litre V-10 of the Type 115 Porsche sports car of 1938–39. It embodied principles worked out and patented by Munich engineer Hans Schrön in 1935 for a five-cylinder in-line engine. In a 1939 study Schrön explored the balancing of multi-cylinder engines and arrived at 72 degrees as the best vee angle for a V-10. This was adopted by Porsche and his team for their Types 115 and 101.

A single camshaft in the block of the Type 101 opened inclined overhead valves through pushrods and rocker arms, the latter being very long for the exhaust valves and very short for the inlets. Individual aluminium cylinder heads were screwed to finned steel cylinders, which in turn were studded to the aluminium crankcase. Each cylinder assembly had a single long external stud from head to crankcase on the outside of the vee. A gear-driven axial-flow blower provided cooling. Plain bottom-end bearings carried fork-and-blade connecting rods.

---

* VK meant *Vollkettenfahrzeug* or fully tracked vehicle. The first two digits indicated the weight class in tons and the latter two a design series. A bracketed letter identified the maker.

Displacing 11.4 litres, the Solex-carburetted ten weighed 1,210lb and developed 210bhp at 2,500rpm.

With its design led by Paul Netzker and production in Vienna by Simmering-Graz Pauker AG, the Type 101 was to be used in pairs in the Leopards. Arriving in Zuffenhausen during July 1941, by the end of the month two V-10s for the first tank had been run. Although hopes for the design were high, the Type 101 was destined to be a leak- and wear-prone engine that significantly undermined Porsche's tank-building efforts.

The screwing of a steel cylinder into an aluminum head was not unusual in and of itself, in spite of the difficulty of maintaining a good seal when the aluminium expanded more than steel when hot. In this construction the screwed joint was a major structural element of the engine, holding each head to its cylinder and thence to the crankcase. In this respect the design was like that of most radial air-cooled engines, which dispense with the usual system of numerous long studs holding both heads and cylinders to the crankcase.

In a system that was radical for German tanks but mother's milk for Porsche, the Type 100 powered its tracks by electric motors. Driving through planetary reduction gears, they received current from generators at its twin engines. The electrical equipment was supplied to his specifications by Siemens-Schuckert. Here was an apposite rebirth of the Mixte power-transmission system that had served Porsche so well in his projects since 1901.

In a tank this drive gave exceptional flexibility and control because the speed of each track could be varied electrically for pinpoint steering. Another advantage was that the electric motors could deliver maximum torque literally from zero revolutions, helping a tank break free of almost any hindrance. Such was their torque, in fact, that in the Type 100 their drives included slipping clutches to prevent breakage in the event of a seized track.

One previous tank had similar advantages, the Saint-Chamond of 1916–18. Produced in the city of that name by the FAMH company, it was designed by Colonel Émile Rimailho, creator of the 75mm cannon it carried, also made by FAMH. Ingeniously as well as prudently, Rimailho adapted an existing electric drive developed for railcar use by Soc. Ets. Henry Crochat, where Paul Colardeau was employed. The result was the Crochat-Colardeau drive powering Holt caterpillar tracks from a Panhard-Levassor engine. The 377 Saint-Chamonds produced served well as heavy assault weapons in spite of their mediocre trench-crossing ability.

Riding on American Holt caterpillar tracks, the French Saint-Chamond tank of the First World War was a respectable performer with its Crochat-Colardeau electric drive from a Panhard-Levassor engine.

On a visit to the front in September 1916 Britain's tank major domo, Colonel Albert 'Bertie' Stern, was impressed by the speed and ease of ratio changing offered by the Crochat-Colardeau system. He ordered one for testing in a British tank while, wrote John Glanfield, 'with conflicting advice from other quarters Stern called for a range of transmissions to be evaluated in a field competition'. Other electric drives were ordered from the UK's Westinghouse and Daimler.

First to be fitted to a test tank was Daimler's system, which 'linked the engine directly to a generator which powered two electric motors,' said Glanfield, 'each driving its own track. Stern was so confident of success that he ordered 600 sets on 6 January 1917 before the system had been tested. Tests soon afterwards were disappointing – the tank could only pull out of a shell hole in a succession of violent jolts while racing its engine at 1,800 revolutions and suddenly shifting the brushes to deliver 1,000 amps. The order was cancelled.'

Porsche, at right in its conning tower, oversaw trials of his Type 100 tank. Built in the Nibelung Works where Franz Xaver Reimspiess was chief designer, only one of its two prototypes was made fully operational.

Neither did Bertie Stern's other two electric-drive orders meet his expectations. The French disappointed, not having their equipment ready, while the British Westinghouse Company's effort yielded 'a bulky transmission weighing five tons'. There is no indication that it was ever installed and tried. Success went to Walter Gordon Wilson, whose mastery of epicyclic gearing produced viable tank propulsion and steering.

Now it was Ferdinand Porsche's turn to adapt his Mixte drive to the Type 100. Its two engines and their generators were at the extreme rear. The Leopard's two electric motors were placed in line in the centre of the hull to drive the tracks' front sprockets through a planetary transmission, where steering was effected electrically. Supporting the tank on its tracks were unique twin-wheeled bogies, each of which ingeniously incorporated an internal torsion bar that took up no space inside the tank – a feature of which Porsche was justifiably proud.* Its design was called *Kniehebel* meaning 'knee-lever' or 'knee-action'.†

Equipped with solid-rubber wheel facings, three twin-wheeled bogies were on each side of the Type 100 aka VK3001(P). Officially known at Porsche as the

---

* The Henschel design also used torsion bars, but these crossed the complete chassis, taking up room and requiring much more meticulous machining of the hull. Porsche's jealous critics – of which there were many – said that the low-mounted torsion-bar housings of the bogies made them poorly suited to mud crossings.

† This was also translated as 'Porsche toggle' suspension. Tank expert Walter J. Spielberger described it as follows:
  'Each bogie unit consisted basically of a carrying bracket and a primary and secondary arm, each of which carried one road wheel. The bracket, which was fixed to the tank hull, was integral with a spindle on which the primary arm oscillated. The primary arm carried a fixed shaft, the outer end of which formed the spindle for the leading wheel and the inner end of which was a hinged pin for the secondary arm. This secondary arm was made of a hollow steel casting in which the torsion bar was carried. This torsion bar was splined on both ends. It was anchored to the trailing end of the secondary pin. The forward end of the torsion bar was connected with a torsion cam unit which consisted of a relatively long tubular member of which the load carrying cam was an integral part. This cam unit was free to oscillate in plain bearings, located in the forward end of the secondary arm. The cam reacted against the arm which was splined to a shaft. A tension helical spring was used for holding the bearing surfaces to a reaction arm and the cam. A substantial rubber bumper was mounted on the trailing end of the primary arm.'

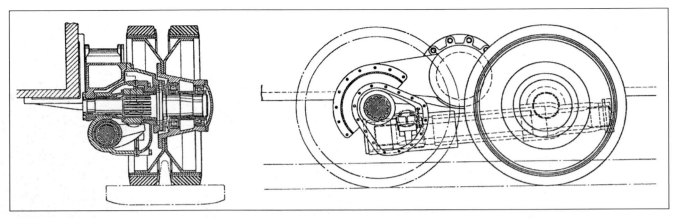

A robust torsion bar and its housing suspended the pairs of wheels of the bogie that Porsche's team designed for the Type 100 'Leopard' tank. This self-contained assembly was the Porsche Kniehebel *or 'knee-action' bogie.*

Sonderfahrzeug I or 'Special Vehicle I' for security, the tank had the unofficial nickname of 'Leopard'. Although two prototypes of the VK3001(P) were among the first tanks made at the Nibelung Works, only one was carried to operational completion. While its trials continued during the summer of 1940 its creator took the cruise break mentioned earlier aboard the motor ship *Robert Ley*. Obliged for once to relax, Ferdinand Porsche sat for a pencil portrait by artist Knudson.

Testing grounds at St Valentin were among those used to evaluate the Type 100 prototype in the summer of 1940. A Hitler Order suspended development of such long-term projects when the war seemed won.

In July 1940 Adolf Hitler reinforced Fritz Todt's order to halt all military research that was unlikely to bear fruit within a year. 'The war in the West is won,' he told his generals, only weeks after their invasion of the Low Countries and France was launched on 10 May. That same summer Todt recruited steel-industry expert Walter Rohland to chair a Panzer committee that was responsible for streamlining tank production. Taking the job in August, Rohland first convened his group on 29 November. Porsche of course was present.

Considerably to Ferdinand Porsche's surprise in the spring of 1941 he was asked to disclose any and all of his projects to a visiting delegation of tank experts from Russia, arriving as a corollary of Adolf Hitler's non-aggression pact with Stalin. Shown over all German tank factories and schools, the Soviets refused to believe that the Panzer IV, then at 23 tons, was Germany's heaviest tank and accused them of not revealing all their facilities as Hitler had promised. Their own

T-34, produced since mid-1940 but still secret, was a 30-tonner and they could not believe that the Germans lacked anything comparably heavy. Indeed, tanks in the Russian KV series already weighed over 50 tons.

Ferdinand Porsche supervised a demonstration of armoured equipment to Hitler on 8 April 1941 against a background of military successes in Africa and Greece and early preparations for *Seelöwe* or 'Sea Lion', the invasion of Britain. In Africa the latter's Matildas, although undergunned, had proved frustratingly resistant to incoming missiles. Adolf Hitler demanded a Panzer that was capable of decisive action against such well-armoured rivals by fitting the powerful 88mm cannon first introduced as an anti-aircraft weapon in 1933. As early as the Spanish Civil War the justly celebrated '88' proved its effectiveness in the anti-tank role as well.

With Hitler (left) and the tall figure of Albert Speer (right), Porsche oversaw Panzer demonstrations in the spring of 1941 that led to a decision to produce new heavier tanks with greater firepower.

This required a much more substantial tank, the long breech of the 88 needing a larger turret, which in turn forced an increase in the width of the hull. Special versions of the 88, with a barrel length just over 16ft, were developed for Panzer installation. Orders for suitable tanks were issued on 26 May 1941 to both Henschel and Porsche during a major meeting of all interested parties at Hitler's Berghof in the Obersalzberg. The aim was to create these on an accelerated programme with the first production to be in May/June 1942.

'In this year,' the Führer told his colleagues at the Berghof, 'we have had clear superiority over the English armoured weaponry. This superiority must never be lost.' He set out three key criteria to an audience that included Porsche:

- Greater ability to penetrate enemy Panzers;
- ours to be armoured more strongly than hitherto; and
- the speed to be no less than 40km/h [24.9mph].

Frontal armour of 100mm was necessary, it was decided, while 60mm would be sufficient along the sides of the new Porsche and Henschel tanks. The performance of the 88mm gun and its ammunition was to be enhanced with the goal of penetrating 100mm of armour at a range of 1.5km [0.93 mile].

Such tanks were intended to play a key role in continuing to implement Germany's successful *Blitzkrieg* style of warfare, making sharp and deep incursions into the enemy's territory that played havoc with their defences. In fact Adolf Hitler had asked for a Panzer equipped with the full-length version of the 88, with a barrel 20½ft long giving higher projectile velocity, that would more than menace any rival machine. He had been persuaded on this occasion that the shorter gun would be more practicable.

In light of the Führer's expressed desires, the HWA tipped a rare wink to Porsche. A Panzer equipped with the ultimate long-barrelled 88 might still be required. The ever-adaptable Karl Rabe opened a new folder for the Type 130 project, the erection of such a vehicle on the basis of the new weightier tank he was designing. It would use the same chassis albeit with the engine room shifted forward to clear space at the rear for the gunnery crew and their ammunition. Although mounted in a fixed cupola, the Type 130's gun would be able to traverse vertically and horizontally. For the time being both this electric-drive version and its sister, the Type 131 with hydraulic transmission, remained paper studies.

Ferdinand Porsche moved to the fore on 21 June 1941 when Fritz Todt named him chairman of the Panzer Commission of the Armaments Ministry.* This was a new entity, established as a buffer between the Wehrmacht and the suppliers of armaments, with the mission of moderating and interpreting the Army's expressed needs. When Porsche was away Oskar Hacker was designated to deputise for him. As chief designer at Steyr-Daimler-Puch AG, Hacker had developed numerous military vehicles during the 1930s.

Todt's elevation of his friend Porsche to this role was not without irony, because the Austrian engineer's relationship to the HWA, never easy, was under renewed stress. The point at issue was the turret-mount diameter of the new big tanks. Specifying a diameter of 2,000mm, the better to accommodate the longest 88 gun, Porsche went ahead and placed an order for such a turret directly with Krupp. The latter's engineers responded with creative and impressive suggestions that reduced the required diameter to 1,820mm, allowing hulls to be narrower.

Constructive though this sounds, it was in cheeky contravention of the status quo. Any and all orders to suppliers were only to be placed by the HWA, not by outside consultants like Porsche. The Zuffenhausen engineers had flouted this convention and not with a minor component. Turret size was a key determinant of a tank's design. That the outcome was a good one was of no interest to the angry HWA officials.

Adjudication of the issue fell to the Reich's chief engineer, Fritz Todt. His decision in Porsche's favour was a shock to the HWA, which – as was its wont – sent a comprehensive document to all parties defending its position. 'From this moment on,' wrote tank expert Mikhail Svirin, 'began the methodical "secret war" that the HWA subsequently carried out against Porsche. Now this uncomfortably stubborn Doctor became a deadly enemy for these men. They tried to suppress the obstinate Porsche, shamelessly fashioning numerous obstacles in his path.'

Postponing *Seelöwe* in favour of *Barbarossa*, on 22 June Adolf Hitler unleashed 120 divisions along a 2,000-mile front in the East, attacking Russia with the aim not only of swiftly humbling the hated Communists but also of seizing their oil, mineral and agricultural assets for the benefit of the Reich. With the war now engaged on a grand scale, on 7 November 1941 Hermann Goering convened a major conference on the reorganisation of production for war. 'Goering directed Todt to take new steps to increase productivity,' wrote Richard Overy, 'while the Four Year Plan was designated as the instrument for the newly proposed central planning office.'

This attempt by Goering to wield suzerainty over war production misfired. 'What turned the tide against him was the increasing unwillingness of the armed forces, industrialists and officials to accept Goering's direction,' added Overy. Todt and Luftwaffe procurement officers 'began to co-operate more fully in the last months of 1941 in an effort to get round the departmental fragmentation developed by Goering'.

---

* Daimler-Benz encouraged Todt to bring Porsche into his organisation in an advisory role. It was Todt's choice to appoint him to the Panzer Commission.

While the paladins jockeyed for command in Berlin, Germany's Russian offensive faltered with the onset of winter. At the end of July 1941 the Soviet T-34 tank first appeared on the battlefield. It proved a formidable adversary with its 500bhp V-12 diesel engine, 76.2mm cannon, high-speed suspension derived from the ideas of American engineer Walter Christie and sloping hull and turret shapes to deflect incoming charges. And soon there were lots of T-34s.

In November Ferdinand Porsche had a close-up look at the T-34. During that month Adolf Hitler's four-engined Fw 200 Condor and personal pilot Hans Baur took a group of officials to the Eastern Front for an inspection tour so they could see the state of affairs for themselves. Their invitation came from Panzer Leader Heinz Guderian, who felt that they had to visit the battlefield in person to grasp the challenges facing his tanks. Ten experts travelled, including Ferdinand Porsche, Walter Rohland, the HWA's Sebastian Fichtner and Ernst Kniepkamp – Porsche's *bête noire* in the department – and designers from the companies that produced Panzers.*

Leaving on 17 November, they spent the night at Smolensk before flying on to Orel and then travelling to Plawskoje. 'A trip to the workshops at the front gave a tragic picture of the effects of cold weather on our too-sensitive Panzers,' said Walter Rohland after their six-day tour of the front. 'The result was more than depressing. The Panzers could scarcely be deployed; if the engines and gearboxes were working the weapons failed, because they were frozen. To the contrary it was confirmed time and time again from the front that the T-34, with its copy of a MAN diesel engine, defied all the insults of winter.'

A week after their return from Russia the key Panzer players convened at the Reich Chancellery in Berlin for a major discussion on the subject of tank design. Porsche and Rohland joined a small pre-meeting of the principals with Hitler at 10:00 am on 29 November. Fritz Todt summed up his view of the position by saying,

'This war can no longer be won militarily.'
'How then should I end this war?' asked Hitler.
Todt: 'It can only be ended in a political manner.'
'But I can hardly see a way to end it politically,' Hitler replied. 'We'll have to discuss this another time.'

Formal discussions starting at noon did not shy from the suggestion that the Germans simply copy the impressive T-34 tank. Ferdinand Porsche favoured this highly practical yet heretical course of action. 'I am more convinced than ever that success lies in the direction of weapons and war materiel made as simply as possible,' he said. 'We should turn away from this mania of wanting to produce always more complex and sophisticated equipment.'

Evidence was marshalled against this course. The T-34 engine's cylinder block required a large amount of strategic aluminium, it was argued, while the steel used in the tank's armouring demanded alloying elements such as tungsten that were in short supply in Germany. Nor was its diesel engine an option at a time when Germany could not yet synthesise suitable fuel. Priority for the available diesel had to go to the Kriegsmarine for its submarines and pocket battleships. Petrol had to be used.

---

* Richard von Frankenberg related the charming story that the group originally planned to fly by one of Fritz Todt's normal Heinkel He 111 transports and crew but that Hitler intervened, saying to Porsche, 'I'll give you my airplane and Captain Baur – then you'll be safe. The generals can take this military plane. I have many generals but only one Porsche.' With Walter Rohland aboard, and all the group flying in the Condor, the story loses some of its validity.

One of Porsche's test benches was used to assess the V-12 engine of a T-34. The attending officers ridiculed the engine as giving only 400hp against the expected 600. 'The Russians are lying as usual,' they said. 'It's all propaganda.' But Ferdinand Porsche discovered that its fuel injector was governed in order to save fuel and extend engine life when full output was not required. After his adjustments the engine's output shot up to its full 600hp. Ferry recalled that 'the engineer-officers were left standing there with what Americans call "egg on their faces"'.

The upshot of the study trip and subsequent deliberations was confirmation, if such were needed, that the decision taken six months earlier to build larger tanks was correct. The skins of both existing Panzers and those planned were easily penetrated by the 75mm gun of the T-34. Larger tanks were needed with turrets that could accommodate the longer breech of the 88mm gun. Porsche's Type 100, which had yet to receive its Krupp-designed turret, remained at the two examples produced. Electricity cabled from their generators would power some of the trials of the prototypes of the next generation.

One of the reasons why Ferdinand Porsche chose electric drive for the Type 100 was his deep conviction that a mechanical transmission would wilt under the demands of such heavy Panzers. His experience with the Daimler-Benz GT I was relevant here. For his bigger *Sonderfahrzeug II*, known as the VK4501(P), he carried on with his electric drive while also commissioning a hydraulic transmission from Voith as an alternative.

Heidenheim's Voith was expert in the use of hydraulic torque converters to provide gear reductions, also in concert with mechanical systems. Their transmission for the new heavier Porsche tank had two engine-driven torque

A layout of Porsche's Type 101 tank depicted its two engines (O), two generators (G) and two electric drive motors (E) with their reduction gears and slipping clutches to avoid overloading the machinery. Its control strategy is also shown.

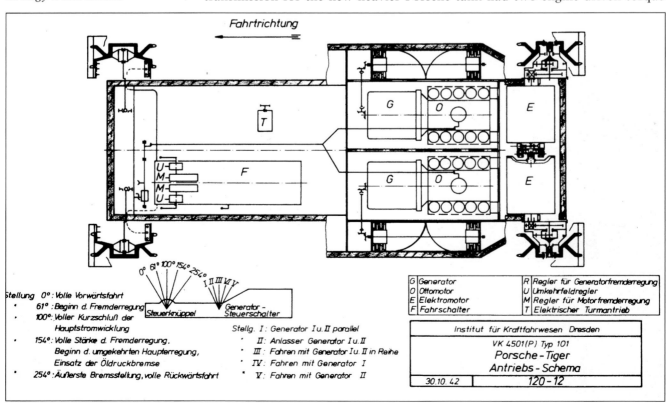

converters that were rapidly filled or emptied to shift the drive from one to the other. While one was propelling the tank the other was shifting to the next ratio needed. The two engines drove forward to a housing containing the torque converters, from which a single central shaft went rearward to the transmission with its forward and reverse ratios, emergency low gear and steering provisions.

The Voith-equipped tank's driver controlled the transmission with pneumatic systems that triggered the hydraulics. Its shifting was fully automatic. Steering was achieved by hydraulic control of a system of planetary gears. A Tiger (P) with the Voith transmission was Porsche's Type 102. Of the 50 units ordered from Voith only one was received at St Valentin and installed for testing. This covered some 1,200 miles on the proving grounds there and later around 150 miles at the Wehrmacht's Kummersdorf test site.

Again Siemens-Schuckert supplied the generators and motors for the electric-drive version of the VK4501(P). On Porsche's engineering team Otto Zadnik was his key electrical expert. The system's layout differed completely from that of the Type 100. At the extreme rear were its two 2,490lb drive motors, each rated at 230 kilowatts, placed across the hull with each driving a track through a planetary reduction gear and slipping clutch. Immediately forward of these were the two V-10 engines, each with its attached 275-kilowatt generator. The latter weighed 2,015lb apiece. In total the major electrical units required some 2,070lb of copper.

Powering both types of tank were two air-cooled petrol-fed V-10 engines of 15.0 litres, longer-stroke versions of the Type 100's power units. Fitted with Solex carburettors and magneto ignition, the enlarged ten continued the use of pushrod-operated overhead valves and developed up to 320bhp at 2,500rpm. During their development and production the single air-cooling blower was replaced by twin belt-driven axial-flow cooling fans. In addition shafts and bevel gears drove axial blowers on both sides of the hull to cool both the engine bay and the electrical components. Two 12-volt batteries used the generators as motors to start the

The V-10 engines were enlarged to 15.0 litres for the heavier Tiger (P) and equipped with twin axial-flow cooling blowers. They were tested complete with their generators which could be used to place them under load.

Seen in August 1942, the Tiger (P)'s two engines drew from twin-throated Solex carburettors that mated with air filters in the deck above them. Their combined output was 640bhp at 2,500rpm.

To the fore here is the gear train of the Type 101 Tiger (P)'s Voith transmission, which powered track-drive sprockets at the rear of the chassis. It received power through a shaft from the torque converters at the right and forward.

During some trials of the Tiger (P), whose wider tracks are evident, the Type 100 (right) supplied power through cables from its generators. Electrical development was overseen by engineer Otto Zadnik.

engines, one after the other. Inertia starters were added during early production.

Just as they had with 'Leopard', Porsche's engineers adopted a nickname, 'Tiger', for their bigger *Sonderfahrzeug II*. Picked up informally by both companies, this became the new tank's Führer-approved name. Ultimately it was exploited to identify work being done for the *Tigerprogramm* that had the very highest priority.

The Tiger (P) carried over the compact 'knee-action' suspension that had been promising in the Type 100's trials. A novel feature was the use of wheels with rims of steel instead of rubber, the latter being used in rings that 'sprung' the rims from the wheels. This effected a significant saving in rubber consumption. Crew capacity of the revised hull rose from four to five.

Built by Krupp to Porsche's design, the electrically traversed turret was set well forward where its weight counterbalanced the heavy masses of machinery at the rear of the hull. With his controls the driver could make use of either generator or both in series or parallel. Integral with the forward-speed control, braking was initially hydraulic with maximum retardation using the electric motors as well.

With 100mm frontal armour and 80mm on its flanks and rear the Tiger (P) scaled up to 57 tons complete. In both camps work on the Porsche prototype and its Henschel rival progressed at top speed and priority in the early months of 1942. The deadline was the presentation of both on Hitler's birthday, 20 April 1942. 'The time-scale imposed on the two competing design teams,' wrote Tiger historian Peter Gudgin, 'with less than a year from commencement of design to demonstration of a running prototype, was breathtakingly short.'

Shown with its barrel aimed rearwards, the Type 101 Tiger (P) underwent trials. Porsche strongly preferred drive through the rear sprockets, which left unstressed the upper return half of the caterpillar track.

Viewed from its front, the Tiger (P) manifested its 100mm armour, viewport for the driver and ball pivot for a machine gun. Krupp built its electrically traversed turret to Porsche's design.

Ferdinand Porsche's world changed on 8 February 1942. His friend and colleague Fritz Todt was killed in a plane crash. Albert Speer was named to take over all his powers as Minister for Armaments and War Production. Speer, who believed in the drive and imagination of youth, was disinclined to work well with Porsche. In

No armchair engineer, Ferdinand Porsche was on the spot to assess problems and find solutions in trials of the Tiger (P). A flat cap replaced his habitual homburg for work in the field with much younger military colleagues.

At St Valentin Porsche (right) conferred with colleagues next to a Tiger (P) with its 88mm gun pointing rearwards. Extensive venting in its rear deck was needed to cool its engines and electrical machinery.

an early meeting with colleagues he declared that no leading economic manager should be older than 40. 'Herr Minister!' Porsche piped up. 'I should advise you of my age. It is 66 years!'

While Porsche had found Todt a like-minded colleague who could comprehend and absorb technical information, he had a different view of Speer. 'Perhaps he's a good architect,' the engineer said in his family circle. 'I couldn't say, but that's not of interest here. In any case it would have been better if Herr Speer had remained an architect . . .' Not surprisingly such critical remarks reached Speer's ears.

'In one of his first discussions with Speer,' wrote Richard von Frankenberg, 'Porsche was to give his view on a planned electric transmission for Panzers. He reached this conclusion: "Herr Minister, in summing up I can thus say that this transmission is too complicated for our Panzer production." Whereupon Speer instantly rejoined, "And herewith I order this transmission to be installed."'

On Hitler's birthday in 1942 Albert Speer signed the ministerial decree that established a system of 'rings' for the producers of war materiel in particular departments. Some 30 leading industrialists chaired the rings and committees with ten heads of the main industry departments reporting directly to Speer. This was an increase from the three who reported to Todt. The five main rings oversaw iron and steel production, iron and steel processing, non-ferrous metals, engineering components and electrotechnical equipment. By the end of 1942 the Speer ministry oversaw 249 committees and subcommittees, rings and sub-rings.

'The fundamental idea,' said Joachim Fest, 'was to divide the whole field of armaments according to weapons systems, such as tanks, aircraft, artillery and tracked vehicles. Each section was to be headed by a *Hauptausschuss*, a central committee responsible for the final product. Soon there were thirteen of these central committees; attached to each of them were so-called "rings" responsible for delivering the raw materials, spare parts, fuel and special equipment.

'Speer appointed Karl-Otto Saur head of the central committees,' added Fest. 'Saur was a man of rigorous temperament who had already held a key position under Todt. Walther Schieber, who also came from Todt's staff, was given responsibility for the rings. According to his colleagues he had "the strength and energy of a bull".' Speer called his new system 'self-responsibility of industry', echoing a principle that had been implemented successfully by Fritz Todt. He continued the process that Todt had initiated of transferring more design responsibility away from the HWA and into the hands of the weapons producers.

'Our business was to build the tanks according to the requirements set by the army,' said Speer, 'whether these were decided by Hitler, by the General Staff or by the Army Ordnance Office. Questions of battle tactics were not our concern.' This was echoed in part by Ferdinand Porsche, said his secretary Ghislaine Kaes: 'Porsche took on every proposal, every order. He always tried to fulfil every task – he called that *durchziehen*, pull-through – and concerned himself neither with its effectiveness nor the possibility of series production. About these, he felt, the customer had to decide, saying that he was neither the General Staff nor the purchasing minister.

'This attitude,' Kaes continued, 'soon brought Porsche into conflict with the military on the one hand and on the other hand with Speer, who was unable to prevail with Hitler. Speer could not bear that again and again Porsche visited Hitler alone and was able to achieve what he had in mind. Hitler used to invite Porsche to dinner. On every occasion he had sausages and beer ready, the Professor's favourite combination. During such meetings Hitler himself used to draw vehicles on sketch paper to communicate better with the designer.'

Porsche found himself working at the nexus of the Third Reich's military effort, for Hitler gave priority to his army tanks above the needs of the Kriegsmarine and Luftwaffe, saying, 'Cost what it will, Panzers must be built.' With thoughts of copying the T-34 put aside, Hitler had approved not only the Tiger effort but also a new medium tank, the Panther. Heavier than the T-34 at some 45 tons, it was intended to get to grips with this feared adversary.

In his Panzer Commission role Ferdinand Porsche adjudicated in the important competition to design and build the Panther. At a conference with Hitler at Rastenburg on 23 January 1942 Porsche presented the competitive Panther designs of Daimler-Benz and MAN. The discussion gave the Führer yet another chance to flaunt his expertise.

As concerned armaments, said Albert Speer, Hitler had

> accumulated knowledge which was superior to that of his military staff. He was better informed than they about the characteristics of specific weapons and tanks, ammunition types and innovations. He could intelligently follow complicated reports, certainly better than his entourage. Actually he knew more than was good for one in his high position. It often seemed that Hitler took refuge from his military responsibilities in long discussions of armaments and war production. He engaged in them for relaxation, as he had in architecture before the war.

When Speer introduced the Führer to technical experts in industry he found that he accepted their views:

> He was amenable to their arguments and was even prepared to revise preconceived opinions of his own. I found that he readily accepted the judgements of qualified technical experts. Had his military entourage realised that in time, and had it given the experienced officers of the front a chance to speak to him as 'experts', his judgements would have been sounder in many cases. But they were afraid that such a procedure would only have revealed their own lack of knowledge.

Clear evidence of such a clash between Hitler's view and those of his experienced officers was the Führer's decision-making about Panzer armour. Heavy armour was the decisive factor in tanks, he decreed, for the new models being developed. Armouring must come before speed, he told Albert Speer. The two could not be combined. 'This decision by the Führer is to be regarded as final,' minuted Speer.

After the war Basil H. Liddell Hart discussed tank design with former General Hasso von Manteuffel, a leading Panzer commander. 'Tanks *must* be fast,' he told

Porsche was the centre of a group behind a Tiger (P) whose gun was aimed forwards. The tests depicted took place after the T101 had its first encounter with its Henschel-built rival the Tiger (H), as described in the next chapter.

the military historian. 'That, I would say, is the most important lesson of the war in regard to tank design. The Panther was on the right lines, as a prototype. We used to call the Tiger a "furniture van" – though it was a good machine in the initial breakthrough. Its slowness was a worse handicap in Russia than in France because the distances were greater.'* Here was flat contradiction of Hitler's opinion from a man with experience in the field.

In meetings on 5 and 6 March 1942 with Porsche in the chair the Panther proposals of Daimler-Benz and MAN were again compared. Daimler's was seen as superior. When the committee met again on 13 May it found faults in the Daimler proposals and prepared a fresh recommendation in favour of the MAN design.

Presented with this finding by Porsche, Adolf Hitler acknowledged that 'the quickest possibilities of manufacture, as suggested in the memorandum, are decisive, and in no case may two types be produced side by side'. Deciding to sleep on it, the Führer confirmed the next day that 'the proposals of Porsche are to be carried out. The MAN Panther is to be built.'

A similar decision awaited Ferdinand Porsche's own project, his Tiger (P), in competition with Henschel's more conventional offering. Powering the Tiger (H) was a single 21.4-litre petrol-fuelled Maybach V-12 of 650bhp, mounted in the rear and driving forward through a mechanical power train. Although promising fewer developmental headaches than the Porsche's Mixte drive, it lacked its compelling sophistication of control.

With his declaration that 'in no case may two types be produced side by side', Hitler had decreed that there would be only one winner in this competition to build the heavy Panzer that was expected to transform Germany's war fortunes. Prototypes from both candidates were expected at Rastenburg on 20 April 1942 for presentation to Hitler on the occasion of his 53rd birthday, a demand attributed to the hard-driving Karl-Otto Saur. Making a good first impression could be decisive.

---

* 'He considered that the Russian "Stalin" tank was the finest in the world,' Liddell Hart quoted von Manteuffel, referring to the JS or IS series that came on stream in 1944. 'It combined powerful armament, thick armour, low build, with a speed superior to the Tiger and not much less than the Panther. It had more general mobility than any German tank.'

CHAPTER 13

FATE OF THE FERDINANDS

'Early 1942 was the nadir of the Allied war effort,' wrote historian Michael Thad Allen. 'U-boats were preying on U.S. and British shipping with impunity; Rommel was chasing the Allies out of Africa; the Japanese were driving the Americans out of the Pacific; and tiny Britain remained very much hemmed in on its island.' Indeed, the Japanese had taken 70,000 British prisoners at Singapore. This was the hubristic background to Germany's deliberations about Panzers as the curtain rose on 1942.

In March 1942 the first of two prototypes of the Tiger (P) was ready at the Nibelung Works. Porsche's design was highly promising, Albert Speer told Adolf Hitler on 19 March. By October 1942, he said, Tiger production would amount to 60 Porsche and 25 Henschel types. Much depended, therefore, on the impressions given by the presentation to Adolf Hitler at the Wolf's Lair on 20 April.

Carried by a low-loading six-axled railcar, the Tiger (P) received its matt grey paint from loyal Porsche men during its journey east. German rail delivered both prototypes to a siding at Rastenburg, where they arrived on Sunday, 19 April. With no suitable ramps available, a 75-ton steam crane was used to transfer both to the ground. Because the Porsche entry's steering system was not yet finalised, it was unable to make the turn away from the siding. With some asperity Ferdinand Porsche rejected the Henschel representative's offer to use his Tiger (H) to tow the Tiger (P) into position to be driven away.* Instead the crane came to the rescue.

Meeting frequently both in the field and in lengthy conclaves, Albert Speer and Ferdinand Porsche both aimed to please their Führer. Often, however, their means towards that end differed sharply.

---

* In fact this was pure braggadocio because the Henschel entry had yet to move even an inch under its own power.

Porsche was a flat-capped passenger when Albert Speer took the controls of a Henschel Tiger (H) during a demanding evaluation. This was the turretless tank that Ferry Porsche ridiculed as a 'sports Panzer'.

After drives during which both proto-Tigers suffered breakdowns, the new and untested machines spent the night concealed in the forest. There was no rest for the Porsche engineers under Otto Zadnik, who used the time to complete their tank's control system to the extent possible. At 10:30 a.m. on the 20th both Panzers were front and centre when presentations of their features began. They were an odd-looking pair because only the Porsche tank carried its turret and cannon. Ferry Porsche joked that the much lighter Tiger (H) was nothing more than a 'sports Panzer'.

At 11 o'clock the Führer materialised to accept birthday salutations and introductions to the engineers present. He singled out one, Ferdinand Porsche, for a presentation of the *Kriegsverdienstkreuz* First Class with Swords, the War Merit Cross for distinguished civilian service. Score one for the Zuffenhausen team. Then Hitler engaged in a half-hour dialogue with Porsche at the Tiger (P), also with engineer Herrlein of hull- and turret-maker Krupp. He spent only a couple of minutes with the Henschel entry, asking about its cooling-air vents. Score two for Porsche.

'The actual demonstration took place at mid-day,' said Hans Rabe, 'in straight-line running on the country road that ran past the Wolf's Lair. Both vehicles quickly disappeared into the distance. Then it was made known that a second demonstration, to Hitler and Goering, would take place in the afternoon. Goering arrived around 3 o'clock and followed Hitler in mounting the Porsche Panzer. This time the Henschel vehicle received no attention at all.' Score three for the Tiger (P), inside which several electrical fires had been quickly extinguished during the trials.

'Goering, who had just fallen into disfavour, gushed tears of enthusiasm,' related Walter Rohland. 'Hitler ordered the production of 100 Porsche Tigers.' Score four for Porsche, although the force of the contest was dissipated by an evident intention to ignore the manufacturing guidelines set for the Panther so that two companies could build Tigers, such was the intensity of the Führer's craving to have many big tanks as soon as possible.

Among those present the least surprised by Adolf Hitler's order was Ferdinand Porsche. In the run-up to the birthday presentation he had met with the Führer on four occasions: 13 and 29 November 1941, 23 January 1942 and 19 March, when the fate of the Tiger (P) was the main topic of discussion. In certain circumstances, Hitler summed up, production of the Porsche Tiger could go ahead in advance of the critical comparison with the Henschel alternative since a 'pre-decision is possible' as long as the Porsche Tiger 'exceptionally satisfies' in the trials, wrote Volkswagen historians Hans Mommsen and Manfred Grieger.

Porsche acted quickly on receiving this green light. 'During a meeting on 23 March 1942,' stated Tiger historians Thomas Jentz and Hilary Doyle, 'Porsche requested that Krupp complete 50 Type 101 hulls for electric drive with air-cooled engines and 50 Type 102 hulls for hydraulic drive with air-cooled engines.' That other variants of the Type 101 chassis were discussed with Hitler was shown by another order to Krupp at that time: 25 engine-room covers for the Type 130, the Panzer-fighting version of the Tiger (P).

The road ahead would not be smooth. An important critic of Ferdinand Porsche's efforts moved into the front line on 1 June 1942, when former Opel production man Gerd Stieler von Heydekampf was named president of the Henschel & Sohn works at Kassel. He was to be a staunch defender of his company's products, thus automatically joining the HWA in opposition to Porsche.

A breakthrough development that summer allowed Germany to synthesise diesel fuels. At a conference on Panzer development on 23 June Porsche and Guderian heard Hitler call for 'the rapid development of an air-cooled diesel

As preparation for their production geared up at the Nibelung Works, two Tiger (P)s continued their trials in 1942 at both St Valentin and Kummersdorf. Porsche strode forward to address issues with one of them.

Two Type 101 Tiger (P)s rested between trials with their guns slung to the rear. Ultra-compressed production timing gave little opportunity for perfection of its engines, which continued to be troublesome in spite of modifications.

engine for tanks'. This challenge was taken up by the Porsche office, which had an air-cooled engine with which to experiment. Compensating for the lower specific power of a diesel cycle would be two more cylinders to create a V-12 and the use of a turbo-supercharger driven by exhaust gases. Forecast for this 18.0-litre engine was 500bhp at 2,500rpm.*

Late in June St Valentin's second Tiger (P) reached the proving grounds at Kummersdorf south of Berlin for tests of the competing designs during July. Mission rules there punished new vehicles over a demanding obstacle course to force failures of parts and designs. This did not take long, gloated Ernst Kniepkamp: 'The Porsche engines in the Tiger were totally shot after only 100 kilometres. In covering this distance one engine had used 85 litres of oil, the other 55 litres.' These findings and parallel tests of the Tiger (H) led to surprising guidelines from Speer, approved by Hitler, that this new generation of heavy tanks should be 'responsibly driven' both in tests and in action, piloted 'with feeling'. Their future service in France was thought less demanding than that in Russia.

Tank-production overseer Rohland was scathing about Porsche's effort in the Tiger competition. 'The Porsche Tiger,' he said, 'which in spite of the reservations of respected designers was quickly taken up with a program of 100, was a total catastrophe. The air-cooled engine was a little-tested experiment and for every technician it was clear that its drive through an electric motor could never produce the acceleration of a direct drive, not to mention the space demands of the generator, the poor efficiency and the additional weight burden.' Porsche's people countered these criticisms, citing the high torque, flexibility and indeed proven gradeability of their system.

Most importantly, Adolf Hitler had countermanded Speer with his initial order of 100 Porsche Tigers. 'During the discussion on the subject of the Porsche Tiger,' said Heinz Guderian about a conference on armoured vehicles, 'Hitler expressed his opinion that this tank, being electrically powered and air-cooled, would be particularly suitable for employment in the African theatre, but that its operational range of 30 miles was quite unsatisfactory and must be increased to 90 miles. This

---

* Before appreciable work had begun on this engine all Porsche's attention shifted to a new X-16 tank-engine concept, also combining air cooling with supercharging and the diesel cycle. This is described in Chapter 14.

was undoubtedly correct; but it should have been stated when the first designs were submitted.' Two tank detachments being readied for movement to support Erwin Rommel in North Africa, Panzer-Abteilungen 501 and 503, were designated to receive the air-cooled Porsche Tigers.

Nevertheless the critical findings of the Kummersdorf trials could not be ignored. 'Difficulties arose with the Porsche design,' said Hans Rabe, 'as a result of oil leakage from the engine and, as a result, substantially poorer engine cooling, as well as faults in the running gear.* Colonel Thomale from the Ordnance Ministry expressed himself as against the Porsche vehicle. Thereupon Porsche was granted three more months for experiments.'

This represented the first formal granting to the Porsche organisation of a period of time in which to conduct development work on its design. Such was the urgency of the *Tigerprogramm* that the tanks of both contestants had been introduced into production straight from the drawing board. Krupp had delivered its first four Tiger (P) hulls as early as December 1941, four months before a tank using them had been driven. Ferdinand Porsche had bent every effort to fault-finding and remedying but under the intense pressure to produce Panzers this had been only fitfully productive. Of the two Tiger concepts his was by far the more deserving of additional time for proving and modification.

By August 1942 the Nibelung Works had produced a total of six Tiger (P)s, well in arrears of its target of 66, building up to its monthly target of 15 tanks. Three of them were marshalled at St Valentin for inspection by Albert Speer on 27 August 1942. Driven in competition with them was the Wehrmacht's staple medium tank, the 25-ton Pz.Kpfw.IV. Its lightness gave it a hillclimbing advantage in the gravel pit on the factory grounds, where Speer personally took the controls of one of the Porsche Panzers. This time it was the Tiger (P)'s turn to be turretless.

After the visit five finished Porsche Tigers went to Northern Austria's Döllersheim training ground, where Panzer-Abteilung 503 had a chance to get used to the characteristics of this new kind of Panzer. In 1938 Hitler had ordered both that village and its adjoining hamlets to be cleared to provide a vast base for military manoeuvres that offered ample space as well as severe winter conditions.† Manifestly unready for combat, the Tiger (P)s were designated for humble training duty at Döllersheim.

Intending to curb the enthusiastic excesses of Germany's engineers, in September 1942 Albert Speer merged the armament-development bodies, including Porsche's, into his management structure. The expressed objective was to bring the military people together with designers to generate ideas for improved equipment. Porsche remained chairman of the Panzer Commission. His task was no sinecure; Hitler demanded an acceleration of tank production to 600 Panthers and 50 Tigers per month, his goal to be reached not later than the spring of 1944.

The future of the Tiger (P) production line at the Nibelung Works still hung in the balance. As of 5 October the factory's production schedule provided for 45 Tiger (P)s to be produced during the period from February to April 1943, using the improved Porsche V-10 engines and electric drive. Because Krupp's production was laid out to build equal numbers of hulls for electric and hydraulic drives, as requested by Porsche in March, a number of the latter hulls were returned to Essen for conversion to the electric-drive design.

---

* Rabe's report reflects the fact that a so-called air-cooled engine actually derives much of its heat dissipation from its oil circulation, which includes oil coolers. Early problems with the V-10s included foaming of the oil, which had been dealt with by added sump baffles and introduction of a centrifugal air/oil separator. In this instance the oil leaks resulted in loss of lubricant, hence loss of cooling.

† The affected area included the grave of Hitler's paternal grandmother Maria. Some thought its eradication could suppress information about Jewish ancestry on that side of his family. No supporting evidence for such ancestry has been found.

Ferdinand Porsche was often in St Valentin, working on problem-solving, while his engineers Rudolf Hruska and Emil Rupilius were at Simmering-Graz-Pauker in Vienna, attacking the V-10 engine's travails. On the latter point, a meeting between Porsche and Speer on 9 October reached the conclusion that the Maybach engine should be tried in the Tiger (P).

Suspension of activity at St Valentin became official on 14 October when Krupp was advised that until further notice no assembly was under way. With stocks of their completed hulls having reached 64 at the Nibelung Works, deliveries were suspended because Tiger (P) assembly had been paused for changes to be made to both engines and suspension.

Krupp also learned that further trials of the contesting Tigers, culminating in another Albert Speer demonstration, were imminent. These took place between 8 and 14 November at the Hainich Forest test area near Eisenach in Thuringia. Established during the 1930s for Panzer testing, they were known as the 'Berka' grounds after an adjoining village.

Porsches father and son were present for the trials, the findings of which would be adjudicated by a special Tiger Commission established by Albert Speer in October for the express purpose of deciding the fate of the Henschel and Porsche offerings. The 19-member Commission had two chairs. Speaking for the military was Colonel Thomale, whose personal opinion has been expressed above. The technical chair was Robert Eberan von Eberhorst, a Vienna-born engineer who on behalf of Auto Union had been responsible for the implementation of Porsche's racing-car designs. In this role he had not always been complimentary about the Professor's work.

After November 1942 trials in Thuringia's Hainich Forest the decision was taken to halt production of the Tiger (P). Instead its chassis became the basis of a self-propelled attack gun, the Maybach-powered Type 130 'Ferdinand'.

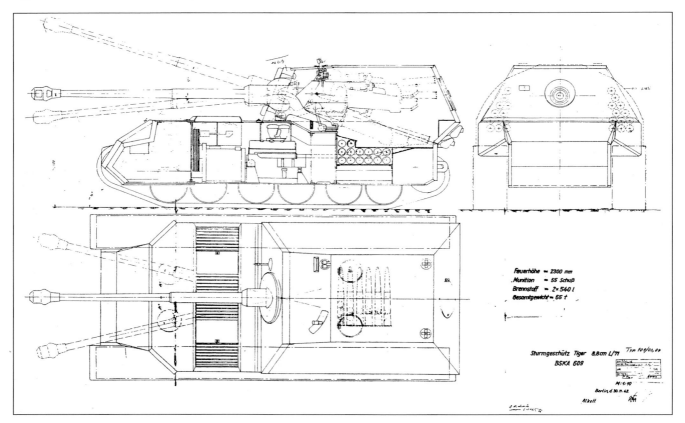

As so often, the Hainich Forest tests were not apples-for-apples. Henschel's tank again appeared minus its turret and with its Maybach specially tuned for the occasion.* Albert Speer delighted in driving this 'sports tank' over the forest's obstacles, this time completely ignoring the Porsche offering. Both entries, however, suffered mechanical failures.

Adjudicators from the HWA were not above a little chicanery to achieve their objectives, Porsche found. The trials in Thuringia also included captured Russian tanks. These put on a poor show compared to the latest Pz.Kpfw.IV. Checking the engines of the Russian machines, a sceptical Porsche concluded that they had been tampered with, not only to perform poorly but also to suffer failures from such sources as unsuitable lubricants.

Looming in the background of these assessments was the unfolding catastrophe of Stalingrad that witnessed the fatal encirclement of General Field Marshal Friedrich Paulus's Sixth Army by the end of November. Stalingrad's desperate city-centre battles were the inspiration for another Führer fantasy, said Heinz Guderian: 'For street fighting Hitler ordered the construction of three Ram-Tigers, to be constructed on Porsche's chassis. This tank, a product of Hitler's imagination, was designed with the purpose of ramming enemy tanks or breaking the walls of houses and other vertical obstacles.

'This "knightly" weapon,' Guderian continued, 'seems to have been based on the tactical fantasies of armchair strategists. In order that this street-fighting monster might be supplied with the necessary gasoline, the construction of armoured fuel-carrying auxiliary vehicles and of reserve containers was ordered.' Drawings and a wooden model showed a helmet-like carapace over the entire vehicle that would mow down obstacles. Although three Tiger (P) chassis were allocated to the project, to the relief of all, including Porsche, no Ram-Tigers were built.

'By and large the trials at Berka went well,' said Ferry Porsche on his return to Zuffenhausen late in the afternoon of 10 November. He left before the special Tiger Commission met to reach a binding conclusion. Arriving at noon on the next day, Ferdinand Porsche related its verdict. The Tiger (P) was halted in favour of the Henschel alternative.†

Further vehicles on the Tiger (P) basis were to be built as *Sturmgeschütze*, self-propelled attack guns. The core of this initiative traced its origins to an assessment by General Field Marshal Erich von Manstein of Germany's invasion of Poland in September 1939. Finding the infantry's contribution sub-par, he said that 'lack of armoured support was a fundamental reason why the infantry (except those in Panzer divisions which alone had tanks) was weak,' according to Kenneth Macksey. 'His proposal to reinforce infantry divisions with armoured assault guns was soon adopted. Artillery pieces with limited traverse were mounted on obsolete light tank and also modern tank hulls, thus enabling protected crews confidently to aim direct fire at the enemy.'

At Zuffenhausen Karl Rabe had only to reach into his Type 130 file, dating from early 1941, to complete detailed design work on a version of the Type 101 that would serve as a *Jagtpanzer*, hunter Panzer, or *Panzerjäger*, tank destroyer, as the *Sturmgeschütz* was also being dubbed.

'Overlooking the failure of the Porsche Tigers,' wrote historians Hans Mommsen and Manfred Grieger, 'Hitler remained loyal to his designer and in the

---

* Walter Rohland recalled the Porsche prototype arriving minus its turret. Recollections by others and photographs of the Henschel show it to be the one missing its turret.

† As related by Thomas Jentz and Hilary Doyle, the last complete Tiger (P) was the only one of its kind to see active service. Equipped as a command vehicle, it participated in the Berka trials and other tests. After replacement of its engines with Maybach units early in 1944 it was taken to the Eastern Front in April. Commanded by Captain Grillenberg, it was lost in action in July 1944.

In Ferdinand Porsche's implementation of his Mixte transmission systems in Second World War Panzers, Otto Zadnik was a crucial ally. It was Porsche's style to rely on key colleagues to help manage his many and overlapping projects.

---

* American doctrine in this respect would evolve significantly. After armoured manoeuvres in the Carolinas in November 1941, the view was firm that 'the antitank gun was the best antidote to the tank'. After the war, however, an extensive review led to the conclusion that 'tanks fight tanks'. This was a result of the trend, evident on both sides, towards tanks equipped with high-velocity weapons that were capable of destroying other tanks.

† As disclosed here and in Chapter 12 the conversion or production of some Tiger (P) chassis as tank destroyers had long been envisioned. Many interpretations intimate that the Type 130's creation was a rushed, last-minute makeshift effort. This was manifestly not the case.

run-up to 1943's spring offensive placed high hopes in Porsche's tank destroyers.' To a gathering of Army Group South's leaders, they said, he gave confidence in an address on 18 February 1943: 'At the beginning of May we will have 98 heavy *Sturmgeschütze* of the new Porsche design . . . Most of these new weapons are immune from damage. Their effect as weapons is unrivalled.' In fact the deadline he had given for their delivery was 12 May, in good time for the dry season in Russia.

Experience in Russia in the autumn of 1942 had already demonstrated the value and potential of tank destroyers.* The bespoke tank destroyer sacrificed some mobility of its weapon to mount the most powerful available cannon, capable of penetration and destruction at extreme range. In a defensive role the tank destroyer, lurking in a favourable position, could wreak havoc with an oncoming Panzer phalanx.

Accordingly the HWA placed orders for several types of *Panzerjäger* based on existing chassis. As long contemplated, this was the future as well for the available Tiger (P) chassis. Discussions with the Nibeling Works about the Type 130's manufacture took place as early as 27 July 1942.† 'Some of the Porsche Tigers then in production were to be changed into assault guns by having the revolving turret removed and being equipped with a long 88-mm cannon and 200 mm of frontal armour,' wrote Heinz Guderian about the meeting in which this was decided. 'The installation of a 210-mm mortar in this tank was also discussed.' While studies were made, this did not proceed.

Although the first two such units were built in Berlin by the Altmärkischen Kettenwerke GmbH, known as Alkett, the Nibelung Works implemented the conversion on 89 more hulls. A pair of 11.9-litre Maybach Type HL120 TRM V-12 engines, using dry-sump lubrication, replaced the still-troublesome Porsche V-10s. Each delivering 265bhp at 2,600rpm, this provided a total of 530bhp – by no means excessive for a vehicle that would ultimately scale some 70 tons equipped for battle. The electric drive gave three standard speeds both forward and backward and a maximum of 12½mph for a vehicle that was not expected to have to make sudden movements.

As drawn by Karl Rabe more than a year earlier, the Siemens-Schuckert motors and their generators were moved forward in the hull to make space at the rear for the artillery crew and 55 rounds of ammunition. These major changes were made possible by the flexibility afforded by Porsche's electric Mixte drive. Installed in a huge fixed superstructure at the rear was the Type 130's 88mm cannon, whose spherical mount allowed it to swing over a 31-degree range horizontally and 24 degrees vertically. The weapon compartment was manned by a gunner, two loaders and the machine's commander.

While steel plating at the sides and rear remained at 80mm, armour thickness at the front was doubled by bolting on external 100mm plates. From his familiarity with tests conducted by the Kriegsmarine Porsche knew that this was not equivalent to the protection of a 200mm slab, but a provisional solution had to be used. Armouring of 200mm protected the front of the gun compartment.

Fresh from the St Valentin factory, this Ferdinand still lacked a protective mantle for the base of its long-barrelled 88mm cannon. Although not the prettiest Panzer, it packed a decisive tank-killing punch.

Louvres in the front deck were inlets and exits for engine- and dynamo-cooling air while well-protected vents at the rear cooled the electric motors. Room was found for two fuel tanks of 119 gallons each for a total capacity of 238 gallons giving a range approaching 100 miles at moderate speed.

The crew used six circular access hatches, two of them above the driver and radio operator at the front, entirely separate from their colleagues at the rear with whom they communicated by intercom. A large hatch at the rear allowed installation of the gun and disposal of cartridge cases after firing. Protection had priority over visibility for both commander and driver in a machine expected to operate from selected fixed positions. For the same reason no supplementary armament was provided.

David M. Glantz and Jonathan M. House wrote that the Type 130 'reflected the German penchant for diverting resources from the mass production of mundane but essential weapons in favour of the illusory advantages of small numbers of super-weapons'. At least one person on the German side felt that such

Three fresh Ferdinands showed the extra frontal plating that gave them 200mm armour. Intended as long-range assault guns operating from the rear of an advance, they were not equipped with defensive machine guns.

advantages were anything but illusory. The notes of a Panzer conference of 26 May 1941 carried this paragraph, clearly a verbatim statement by the Führer:

> For the war in the period ahead technical superiority is of decisive significance. The Italians had to give way in Africa because they were unable to mount any adequate defence against the English tanks. Even small series of superior weapons can be decisive. A particular example of this is the use thus far of the 88 in the battle against fighting vehicles, which according to the available reports is of unheard-of value even in small numbers.

It has been said by some observers that in a conflict in which Germany could never hope to prevail by sheer numbers of men and machines, an emphasis on quality over quantity was not necessarily unwise. Moreover a case can be made that it would have been wasteful, not to say sinful, to allow so many heavy Panzer chassis to lie fallow – not to mention the electrical equipment that Siemens had continued to produce – when the Reich needed all the weapons she could muster.

To cope with the Ferdinand's mass, the Wehrmacht's experts allocated three Tiger (P) chassis to serve as recovery vehicles. When equipped for service they carried a small derrick to assist in extrications.

When trials of the first Type 130 began, one of the Porsche engineers found the standard designation *Sturmgeschütz mit 8.8cm PaK43/2 (Sd.Kfz.184)* too much of a mouthful. Addressing the massive machine, he called it 'Ferdinand'. Quite without Porsche's knowledge the name 'spread like wildfire'. In February 1943 the name was personally approved by Adolf Hitler for the Type 130.*

In the foreground of these developments was what Major General Walter Warlimont called 'the month of doom in modern German history, November 1942, when the enemy struck both in East and West'. The Allies were ashore in North Africa and the Führer's forces were facing stalemate at Stalingrad. In response Adolf Hitler launched a major tank programme on 2 and 3 December 1942. After review of a temporising report by Walter Rohland, Hitler shrugged off the many problems of factory space, raw materials and machine tools and commanded that his expanded Panzer production programme be carried out 'whatever the cost'.

On 17 and 18 January 1943 the Führer intensified his demands. Monthly armoured-vehicle production was now to be 1,500 to 2,100, which both Rohland and Porsche considered 'utterly fantastic'. In case a reminder was needed, on

* Some sources state that the vehicles were given the name *Elefant* or Elephant after their first engagement in July 1943. In fact an order of 27 February 1944 required these machines to be referred to as Elephants in the future – just another little annoyance directed against Porsche. By then, however, the name 'Ferdinand' had so embedded itself in the historical memory that the edict was widely disregarded. When questioned about the Type 130 after the war, Ferdinand Porsche said he was unaware of such a new designation, informal or otherwise.

6 February Hitler ordered that the 91 tank-fighting Ferdinands 'be made ready with all speed'.

Three of the Tiger (P) hulls were set aside for their preparation as recovery vehicles, a prudent precaution in view of the Ferdinand's tonnage. With similar power trains to the Type 130, these had a small rear superstructure and carried a derrick crane. After the initial revision of their engine rooms and drives by the Eisenwerk Oberdonau at Linz, the low-profile vehicles were completed at St Valentin. All three were ready by September 1943.

Hierarchical changes early in 1943 followed Germany's capitulation at Stalingrad on 6 February and the January conference at Casablanca in which the Allies agreed to impose unconditional surrender on the Axis. Walter Rohland stepped down as head of the Panzer Committee. Ernst Blaicher of Braunschweig tank-builder Mühlenbau und Industrie AG, known as MIAG, took his place. A prominent member of the SS who was expected to be more exigent than Rohland, Blaicher was the choice of Speer's deputy Karl Otto Saur.

Among new requirements in January 1943 was a tank-destroyer version of the Tiger, the *Jagdtiger*, to carry a 128mm cannon. This was produced at the Nibelung Works, which engineered it in co-operation with the Porsche office. On 20 October Ferdinand Porsche presented a wooden model of his planned *Jagdtiger*. Instead of Henschel's torsion-bar springing, whose bars went across the chassis, it was to be built with Porsche's 'knee-action' bogies with integral torsion bars.

So difficult was the struggle under wartime conditions to produce new prototypes that it took until April 1944 to prepare a sample of this *Jagdtiger*-to-be. It was the first of two equipped with Porsche's torsion-bogie suspension. According to the resident designers at St Valentin this would have resulted in a lighter vehicle overall and significantly less unsprung weight in its underpinnings.

'When they were finally ready to produce Tigers in the form of the *Jagdtiger*,' said Gerd Stieler von Heydekampf of the Nibelung engineers, 'they had adopted

Although these are Tiger (H)-based tank destroyers being built at St Valentin, two with the same superstructure were built on Tiger (P) chassis. They carried long-range 128mm cannon.

– in spite of Henschel's objections – the Porsche suspension. This failed to work satisfactorily and had to be replaced by the Henschel suspension. This changeover meant weeks – it may even have been months – delay.' This was a setback for the Porsche bogies, which would only see service in the Ferdinands.

Seventy-seven *Jagdtiger* were completed by March 1945. Few in numbers though they were, these were an unwelcome sight to the Allies in the Ardennes during the Battle of the Bulge. There they were fielded by Heavy Tank Destroyer Detachment 653.

The spring of 1943 found Ferdinand Porsche busy at the Nibelung Works. On 19 March he presented a prototype of the Type 130 Ferdinand to Hitler at Rügenwalde. He hosted Albert Speer's visit on 30 March to St Valentin, where the Reichsminister took the prototype's controls. A week later Speer was in Stuttgart for a half-hour tour of the Porsche headquarters at Zuffenhausen. Age difference notwithstanding, Speer was not above being beguiled by the creativity of Porsche and his team.

During evaluations of the Type 130 Ferdinand, Albert Speer enjoyed taking its controls, for its Mixte drive gave great flexibility. Among a gaggle of passengers Ferdinand Porsche perched next to him.

In February Heinz Guderian, acknowledged as the master of tank deployment in Blitzkrieg tactics, was recalled from relative obscurity to the post of Inspector General of the Armoured Forces. He participated in Adolf Hitler's planning of a counter-offensive against the Russians on the Eastern Front. Around the key city of Kursk the Soviets had thrust their way west in a salient that the Germans could not resist trying to lop off with a pincers assault. The effort was code-named *Zitadelle* or 'Citadel'.

Potentially to be launched in the spring of 1943, an offensive was mapped out. Its success depended significantly on the ability of the new-generation Panzers to shock and awe the Russians. Some Henschel Tigers were available, as were the Panthers and Ferdinands, all of which, however, were babes fresh from the womb and by no means field-proven.* Nevertheless these were the game-changing 'wonder weapons' about which the Führer had bragged to the officers of Army Group South. Ready or not, they would have to be deployed.

The crew of one of the last Ferdinands to be produced at the Nibelung Works showed their pleasure at handling the massive giant. At the left was one of the three recovery vehicles built at St Valentin.

'Apart from its single long-barrelled gun,' said Heinz Guderian of the Ferdinand, 'it possessed no other armament and so was valueless for fighting at close range. This was its great weakness, despite its thick armour-plating and its good gun. But since it had now been built I had to find some use for it, even though I could not, on tactical grounds, share Hitler's enthusiasm for this product of his beloved Porsche. A Panzer regiment of two battalions, each with 45 tanks, was set up with the 90 Ferdinand-Tigers.'

As they were completed the Ferdinands were delivered to Heavy Tank Destroyer Detachments 653 and 654, the last arriving in the east at the end of May. At the Ferdinand's March presentation to him Hitler declared complete satisfaction with the machine and ordered that they all be sent directly to the field from the production lines at St Valentin without further testing. 'This circumstance greatly troubled the father of the Ferdinand,' wrote Mikhail Svirin, 'who by his own admission was amazed that no complaints were received from the troops at the front about the manufacturing quality of these unsightly vehicles.'

On 24 and 25 May the 654th had a visit from Guderian at its base at Bruck an der Leitha in easternmost Austria. Allocation of the prized tank destroyers was to be even-handed. Half were to be fielded by Detachment 653 commanded by Major Steinwachs, part of Army Group Centre's attack on the north of the Kursk salient under Walter Model. The other half in Major Noak's Detachment 654 would support Erich von Manstein's Army Group South's thrust from that cardinal direction.

Originally mooted for May, Citadel was postponed to June and then again to its final date of 5 July. By then the Ferdinands were redeployed. On 15 June Detachment 654 was reassigned to the north, where it joined 653 to create the 656th Heavy Tank Destroyer Detachment within General Joseph Harpe's XXXXI Panzer Corps. In support of Model's attack Army Group Centre could deploy all 90 of these powerful machines against objectives in the north of the Kursk bulge.

Builders at the Nibelung Works showed their enthusiasm for the Ferdinand with their graffiti on the final chassis produced, number 150100. Built on 8 May 1943, it was about to be whisked to the Eastern Front.

---

* Of 1,866 tanks on the German side, only 347 were the latest types. The rest were the earlier Panzer II and IV.

This disrupted the plans of General Hermann Hoth in the south, wrote George Nipe: 'Hoth planned to distribute one company of the impressively armoured vehicles to each of the three Panzer divisions of XLVIII Panzer Corps. It seems probable that he intended to use them to furnish long-range overwatch fire to support the breakthrough operations of the Corps. With their powerful 88 the Ferdinands could have been a potent weapon against Russian strong points and gun positions.'

Should this have been the plan of 'Papa' Hoth, as he was known, it was a good one well suited to the Ferdinand's attributes. The phenomenal range of its long-barrelled 88 meant that it did not need to be in the heart of the action to have deadly effect. But in the eyes of Hitler and Manstein the Ferdinand was above all a *Sturmgeschütz*, a mobile attack cannon. They pictured these ungainly monsters in the thick of the action, blasting paths through Russian defences. Their interpretation was to prevail in the Ferdinands' deployment.

Meanwhile the Russians, familiar as they were with German initiatives, uncharacteristically laid low and built up their defences where they expected attacks. Both their own intelligence and that from British sources gave them a clear picture of the impending assault.* The local population pitched in to help Soviet sappers build successive rings of tank traps, minefields and defensive weapons. They had ample time to create wave after wave of menacing and deadly impediments.

Launched by the Germans on 5 July 1943, what became known as the Battle of Kursk has gone down in history as the greatest tank conflict ever fought. The open steppes surrounding Kursk, due south of Moscow, gave both sides manoeuvring freedom for some 5,400 tanks and assault armour, divided roughly equally between the protagonists. The Russians, of course, had done all they could to curtail that freedom.

In the midst of the July 1943 fighting this camouflaged Ferdinand was heading for battles around the key town of Ponyri. It and its sisters were assigned close-quarter attack roles for which they had not been conceived.

---

* The British used their Ultra sources, their codebreaking of radio messages, to gain a good appreciation of the German dispositions. Wrote Bletchley Park historian Michael Smith, 'between April and July 1953 the complete German plans for the forthcoming Battle of Kursk were sent on the link between the headquarters of the German Army Group South and Berlin. A sanitised version, disguising the source, was passed to Moscow.' Knowing well the suspicious nature of Josef Stalin, however, the British were sure he would not trust information that was transmitted directly. Thus they leaked the battle order to the Lausanne-based Lucy spy ring, which was known by the Russians to have good high-level contacts in Berlin. Coming from Lucy, Stalin believed the intelligence. As well, added Smith, 'Stalin was already receiving full transcripts. Despite the strict security in force at Bletchley Park, a member of one of the watches in Hut 3 was a KGB spy.' This was John Cairncross.

'My fears concerning the premature commitment of the Panthers were justified,' wrote Heinz Guderian. Many suffered technical failures from cooling problems and poor gasketing that led to fires. 'Few remained in action after a week,' wrote Richard Overy. In the south, he continued, 'Manstein's drive was hindered rather than helped by the tank in which such high hopes had been placed.' During the night of 12/13 July, for example, 93 Panthers were laid up in workshops for repairs – almost half their total strength.

As for the Ferdinands, at the launch of Citadel 83 were operational. Four were reported as repairable in a fortnight and two as requiring longer to be fixed. Nominally each detachment had 44 Ferdinands while one was allocated to its headquarters. Shortage of spare parts would keep many of the newest Panzers from returning to the fray. Panzer Corps XXXXI won a position at the centre of the front, its cadre of Ferdinands aimed straight at the important objective of Ponyri.

'Within twelve hours' of the start,' wrote Gregor Dallas, 'the fight for Kursk had been transformed into "a great glowing furnace"; by the following weekend it had become the largest tank battle in history: on the steppe south of Prokhorovka, studded with small plots, gardens and the cornfields about them; under skies of rolling storm clouds; in winds, lightning and a scudding rain, well over a thousand tanks fought in groups and in individual actions. The "slaughter at Prokhorovka", which left the ground strewn with burnt-out tanks, lorries and tens of thousands of dead, was completed in eighteen hours.'

Deployed as assault guns rather than tank destroyers, the Ferdinands were pitched into the thick of the fray. In the initial stages of the battle, said Terry Gander, they were 'divided into about fifteen sub-units to add spear-point punch and fire support to the initial stages of the attack. In effect the *Panzerjäger* were to be deployed as heavy *Sturmgewehr*.'

Wrote David Glantz and Jonathan House, 'As a recognised expert on defensive tactics, Walter Model was acutely aware of the vulnerability of his armoured vehicles – particularly the [Ferdinand] self-propelled guns – to short-range attack by determined infantrymen in prepared defensive positions. To reduce this vulnerability he had insisted that his armoured spearheads be accompanied by dismounted infantry.' Some of the Panzers even fitted a platform at the rear that carried five supporting soldiers. Crew in Noak's 654 Detachment became adept at opening their big gun's breech so they could fire machine guns down the 88's barrel at nearby targets.

Heinz Guderian provided a succinct appreciation of the situation in which the Ferdinands found themselves:

The Porsche Tigers, which were operating with Model's army, were incapable of close-range fighting since they lacked sufficient ammunition for their guns, and this defect was aggravated by the fact that they possessed

A graphic of Germany's Operation Citadel showed the existing front around the Kursk salient, the Wehrmacht's intended attack and the maximum penetration achieved. All Ferdinands were deployed in the north by the 9th Army.

Triumphant at Kursk after Hitler called off the attack, the Russians captured this Ferdinand in good condition. Such was their impact that the Soviets blamed Ferdinands for counterattacks across all their fronts.

no machine-gun. Once they had broken into the enemy's infantry zone they literally had to go quail shooting with cannon. They did not manage to neutralise, let alone destroy, the enemy rifles and machine-guns, so that the infantry was unable to follow up behind them. By the time they reached the Russian artillery they were on their own. Despite showing extreme bravery and suffering unheard-of casualties, the infantry of Weidling's division did not manage to exploit the tanks' success. Model's attack bogged down after some six miles. In the south our successes were somewhat greater but not enough to seal off the salient or to force the Russians to withdraw.

Guderian's assessment was that a tank killer that was best suited to stay in the rear to exploit the long range and flat trajectory of its 88mm cannon was thrust forward instead to perform the role of a 'spear point' for the attack. In this it had notable success, destroying not only tanks but also field fortifications to help Model's troops to some of their deepest penetrations.

Ferdinands were all but immune to attacks by enemy artillery. One who benefited from this was Captain Herbert Jaschke, experienced in the Ferdinand: 'It was the heaviest Panzer we had at that time, a vehicle that was superior to the Russian KV-2, which had a 12.2-centimetre gun. I even got hit from 1,200 metres distance but it only caused a dent in the steel. We carried up to one hundred rounds of ammunition. There was no room for us. We perched and knelt on the shells. But the ammo saved our lives.'

The opposition fared less well. Manned by experienced artillerymen, the Ferdinand proved that its vaunted range was no chimera. One report credited it with killing a T-34 with a single shot from a distance of 2.9 miles. Kills at lesser distances were routine. Surviving Citadel with 31 of its Ferdinands operational, Detachment 653 claimed 320 Soviet-tank kills. Another totting-up at the end of August showed some 500 enemy tanks destroyed against losses of 44 Type 130s. Not atypical for the Ferdinand, these totals gave it a kills-per-loss ratio of at least 10 to 1. Even allowing for some exaggeration this made it the Second World War's most effective tank destroyer – the job for which it had been created.

Such was the reputation of the Ferdinands that post-battle Russian reports vastly overestimated their numbers. The attackers had six Ferdinand battalions with 45 Ferdinands per battalion, said one review. Many Soviet reports stated that Type 130s were active in Army Group South at Kursk, although none was. In the latter battle, claimed one Russian summary, Manstein had no fewer than 253 new Ferdinand assault guns in July 1943.*

Although many sources refer to mechanical problems suffered by the largely untried Ferdinands, this had not been a criticism by the knowledgeable Guderian. Like the Panthers, the individual Ferdinands were all but untested, in battle or otherwise. It would have been astonishing if such complex and highly stressed machines had worked flawlessly in Citadel straight from St Valentin. Nor were their crews fully trained in their manipulation. David Glantz and Jonathan House remarked on this:

> There was every reason to assume that the new German weaponry, in particular the Tigers, Panthers and Ferdinands, would tip the scales in the Wehrmacht's favour. Soviet reactions to these new weapons clearly indicate that these German assumptions were correct. While in hindsight it seems easy to criticise the performance of these new weapons in combat, in early July few anticipated the technical problems these weapons would experience.†

During the attack phase, when XXXXI Corps had control of its battlefield, it was able to recover and rehabilitate broken Ferdinands. A minefield took its toll in the first days, but most damage was to tracks and bogies, which could be repaired. A post-operations Russian assessment was that ten of the 21 Type 130s abandoned around Ponyri were victims of anti-tank mines. Only three were thought disabled by infantry action.

Stood down by Hitler on 13 July after eight days of intense conflict, Operation Citadel was unable to roll up the Russians as planned. As of 29 July the two Ferdinand detachments reported outright Citadel losses of a total of 39 Panzers, 44 per cent of its starting numbers of 88. After the battle the 49 Ferdinands still alive and able to kick again were taken to depots at Zhitomir and Dnepropetrowsk for repairs and replacements from locally available stores. From there they regrouped near Nikopol in the southern Ukraine under the banner of Detachment 653, which had suffered the least losses. Some were deployed defensively at Zhitomir, an important rail junction, in the early stages of that winter's Russian attack on Kiev.

At the end of November the 48 Ferdinands judged ready for more action returned to St Valentin for refitting. There they received a ball-mounted machine

---

* Soviet reports also placed Ferdinands at Leningrad, Poltava, Kiev, Minsk, Bobruisk, Budapest, Prague, Königsberg, Berlin and even Stalingrad, Mikhail Svirin related. Any heavy German anti-tank weapon was now called a 'Ferdinand'.

† One who did have reservations was perhaps the most experienced of all, Heinz Guderian.

Forty-eight serviceable Ferdinands were recalled to St Valentin for refitting over the 1944/45 winter. They received a machine-gun aperture, commander cupola and rough-finished Zimmerit coating to shrug off magnetic limpet mines.

As overhauled and refitted for 1945 the Ferdinands were still awesome destroyers of enemy armour. They proved this in their deterrence of Allied advances up the Italian boot in the early months of 1945.

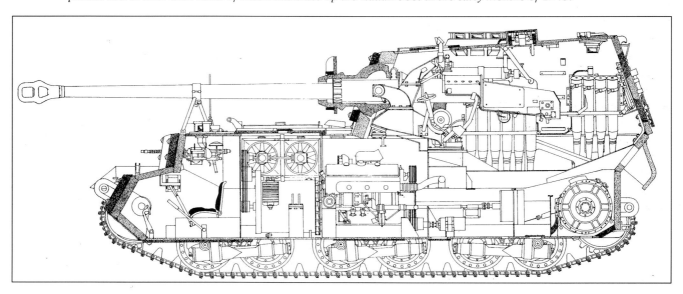

gun, a cupola for better commander vision, enhanced protection of exhaust pipes and anti-magnetic Zimmerit paste coating to discourage limpet mines. The upgraded Ferdinands re-equipped Heavy Tank Destroyer Detachment 653 and also formed a company attached to the 614th Tank Destroyer Detachment.

First assignment for the Ferdinands was for the 653rd in Italy, where the Allies were pressing northward. The Detachment mustered 14 Ferdinands. Although the terrain was less friendly to the big Panzers, for which bridge crossings were problematic, they were ranged to attack the landing Americans at Anzio and Nettuno south of Rome. 'There they showed their best side,' said Mikhail Svirin, 'successful in demolishing not only tank attacks but also infantry assaults. This Detachment's account even included a shot-down aeroplane.'

The next responsibility for the Ferdinands was the defence of Rome. 'When two Ferdinands took up the battle in a suburb of Rome,' wrote Svirin, 'according to various sources in ten hours they destroyed 40 to 50 tanks, assault guns and other vehicles of the Allies. They maintained their battle effectiveness in spite of constant mass attacks from Allied air forces and finally were blown up by their own crews when out of fuel and munitions.'

Withdrawing from Italy, the remaining Ferdinands joined the others in defence against the Russian Vistula–Oder Offensive of January 1945, which took the attackers to 45 miles from Berlin. In all a dozen of the big beasts mustered at Wünsdorf and Zossen, the Wehrmacht headquarters south of Berlin, as part of the hopeless effort to defend that capital in the final exigency. At the end of 1945, said Svirin, the Soviet Union had six more or less complete Ferdinands on its books, three of which were operational. One had been found at Wünsdorf with full fuel tanks and all its munitions – an indomitable trophy that the victors had learned to respect.

After sharp differences with Reich armaments minister Albert Speer, Ferdinand Porsche was moved out of his Armour Commission job at the end of 1943 to a more elevated and mainly ornamental post, that of Reich armaments councillor. His continued advocacy of exotic drives, engines and suspensions for tanks at a time when Germany needed mass production rather than the ultimate in sophistication earned Porsche a 'mad scientist' reputation in some circles, including Speer's.

Porsche could look back at a record of success during his stewardship of tank development. 'Germany increased its production of medium and heavyweight battle-tanks sixfold from 1941 to 1944,' wrote historian Dan van der Vat, 'delivering 2,875 in 1941, 5,673 in 1942, 11,897 in 1943 and 17,328 in 1944 – a result directly attributable to the special efforts of Speer, Rohland and such high-powered advisers as Porsche, urged on by an avid Hitler.'

Hitler biographer Joachim Fest wrote that 'this increase was particularly astonishing to Hitler, who had revealed a special regard for that weapon early on, possibly in remembrance of his triumphant French campaign, and had by a never-ending series of interventions totally disrupted all production schedules even at the development stage'.

On 28 September 1943 Porsche's personal tank factory changed hands. As part of the Reich's programme of transferring to private enterprises the facilities

The Eastern Front was a graveyard for more Ferdinands, this time much closer to Berlin as the Reich's forces struggled to stave off defeat. One of their last perimeters was around the Wehrmacht headquarters at Zossen.

it had commissioned in some haste early in the war, Steyr-Daimler-Puch acquired title to the Nibelung Works at St Valentin for a sum amounting to RM52,000,000. The Austrian plant's staple product remained Krupp's veteran medium-weight Panzer IV, which was Germany's most-produced tank during the war.

In his post-war assessment Ernst Blaicher was scathing about Steyr's management of the Nibelung Works. He said that its managing director, 'Georg Meindl of Steyr-Daimler-Puch, did not, in my opinion, considering the extent of his tasks, spend enough time on rectifying the inadequate output of his plant.* The labour conditions of the plant were unfavourable. There were not sufficient skilled workers on the spot; newly allocated labour consisted of workers from other labour-exchange districts which, if they had to lose any workers at all, only got rid of their bad ones.

'On principle,' Blaicher summed up, 'it can be said that in general an Austrian does not display the same energy as a leader as the so-called "Reichs-German".' That sin could not be laid at the door of the largely Austrian senior Porsche cadre, who now had even bigger projects on their hands.

---

* Remarkably, Albert Speer wrote in *Infiltration* that Meindl was the nominee of Hermann Goering to replace him, Speer, as the head of military production in due course. The Steyr Works was part of Goering's wartime accretion of industrial entities.

CHAPTER 14

VERSATILITY AND VARIETY

Had Ferdinand Porsche in his role as chairman of tank development been too ready to accommodate Hitler's interventions? If anyone were able to speak frankly to the Führer it was Porsche, whom his son Ferry described as having a quasi-fatherly relationship to the dictator. On one occasion, wrote Richard von Frankenberg, Hitler aide Martin Bormann took Porsche aside before a meeting and said, 'Now look, you tell Hitler how things really are. No one else can. Everyone who has made a presentation today has lied.'

In fact Porsche's removal from the decision-making process may well have contributed to the technological anarchy of the last two years of the war when, as Ferry related, 'no one contradicted Hitler any more. They took great pains, in fact, to pretend to agree with him so that the picture he got of what was going on around him no longer related to reality. He therefore reacted to a set of imaginary facts and cleverly contrived distortions and half-truths in a manner which he might not have done had he known what was really happening.'

Ferdinand Porsche's replacement as head of the Panzer Development Commission was Henschel's Gerd Stieler von Heydekampf, a Speer appointment. 'Although I was considered a production man,' said von Heydekampf, 'Speer entrusted me specifically with the task of maintaining a steady development of the types of tanks already in production, and to introduce mass production with the least possible disturbance.

'My association with Prof. Ferdinand Porsche was always that of a relation between competitors,' added von Heydekampf. 'As a member of the Opel management board [1936 to 1942] I had to endure the success that he had with his VW design. As the largest car maker outside America, Opel had also offered a "Volkswagen". But Porsche won. Thus we Opel people were bitter. Later in the war, in 1942, as managing director of Henschel and Son I took over the ongoing Tiger production. There Henschel had won, above all because the electric drive of the Porsche Tiger would have cost the war economy too much copper. This time the Porsche people were bitter.'

The change from Porsche to von Heydekampf was regarded as a success for Speer, who favoured young and 'dynamic' executives to press his programmes – at 39 von Heydekampf met his criterion – and who now gained control over tank development as well as production. It was of no financial import to Porsche; the job was strictly honorary. He and his team were still active in the design, development and manufacture of vehicles of all kinds, including tanks.

'Whenever we designed a new tank,' Ferry Porsche recalled, 'my father kept Guderian informed on the whole thing so that he came into our orbit not infrequently.' This did not mean that Guderian necessarily favoured Porsche's plans, Ferry admitted: 'We were under no illusions about this extremely clever man. Guderian was a blunt and forthright individual who had created entirely

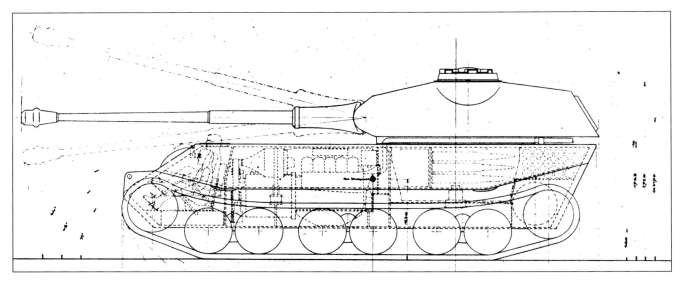

In February 1943 both Porsche and Henschel were asked by the HWA to suggest designs for an improved Tiger tank. Porsche submitted proposals for his Mixte-drive T180 with alternative front and, as here, rear turrets.

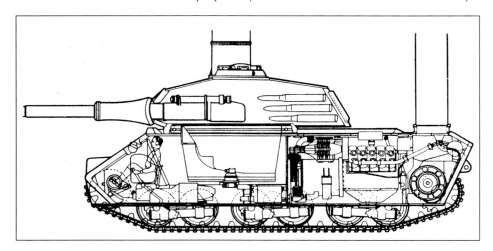

Engines and generators went to the rear in the front-turreted version of Porsche's Type 180 proposals for what the Army called its VK4202 programme. The distinctive and compact torsion-toggle bogies were again suggested.

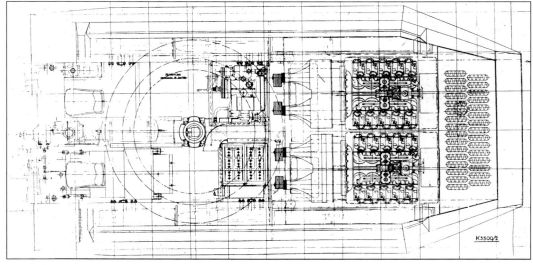

A plan view of the front-turreted Porsche T180 proposal of mid-1943 showed its doubled-up package of V-10 engines and generators as well as two crew members at the front, a driver and a radioman.

new tactics and strategy in the use of armour. He certainly was not afraid to stand up and be counted.'

The new team of Ernst Blaicher and Heinz Guderian oversaw the HWA's call for designs for an improved Tiger tank in February 1943. Its objective was to carry the formidable 88mm cannon and to house it in a turret that was better shaped to deflect incoming rounds in the manner of the Russian T-34. As with the original Tiger, Henschel and Porsche were asked to submit proposals for what was called the VK4502 programme.

Ferdinand Porsche put forward his Type 180 with Mixte drive and Type 181 with hydraulic transmission, officially the VK4502P and unofficially the Tiger II (P). Both concepts had rear-sprocket drive and 'knee-action' torsion-bogie suspension. An enlarged hull gave space for fuel capacity increased from 114 to 180 gallons, responding to criticisms of the range of the Tiger (P). Wading ability was also greatly improved. The electric-driven Type 180 offered two of the Porsche-designed V-10 petrol engines, for a total of 700bhp, while the Type 181 had a single Maybach 720bhp diesel driving through Voith's hydraulic transmission.

Layout alternatives from the Porsche office included placing the turret at the extreme rear and moving the engine forward, exploiting the unique design flexibility of the Mixte drive in a layout akin to that of the tank-fighting Ferdinand. Forward turrets were also proposed for a Panzer that would scale 70½ tons. On 9 October 1942 the Tiger II (P) received the green light, three prototypes to be ready by mid-December. The decision was reinforced in November, falling in favour of a rear-mounted turret, with production contingent on a successful prototype. With the press of other projects delaying the Type 180/181, in 1944 work on the Tiger II (P) was finally terminated.

With capacity available at the Nibelung Works, production there of 50 turrets of Porsche's design for the Type 180 had been given the green light. The turret's lines were particularly low and sleek. They were paid for, however, in a smaller internal ammunition capacity than Henschel's alternative because the Porsche design had more storage space in its hull. The turret also suffered from some armouring weaknesses. Nevertheless they could hardly go to waste so 50 of Henschel's new Tiger 2 – or Tiger B or II or King Tiger – Panzers wore Porsche turrets. In all 489 of the formidable new 68-ton tank were produced between January 1944 and March 1945.*

In 1943 Kniepkamp of the HWA launched studies into completely new tank concepts. Under the sobriquet *Sonderfahrzeug V*, by July and August 1943

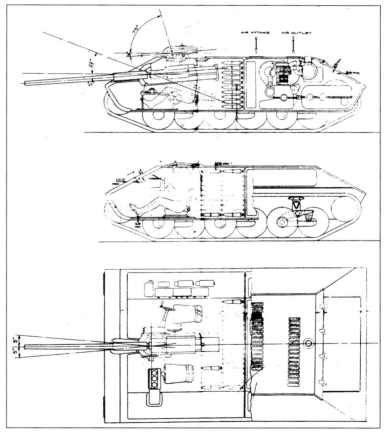

The advantage of an ultra-low profile was exploited by a Porsche Type 245 proposal for a new tank family, Sonderfahrzeug V, proposed by the HWA in the late summer of 1943. Crew members flanked a front-mounted cannon.

* Some sources say production was 487 or 484.

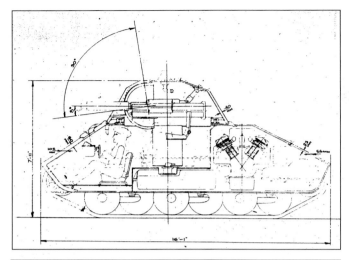

A model of Porsche's turreted Type 245 showed a two-man vehicle that would reduce demands on both manpower and factory capacity. There were concerns, however, about the penetrative ability of its fast-firing 55mm weaponry.

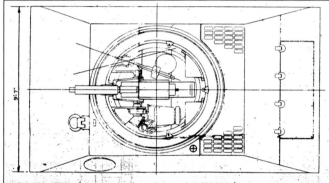

An alternative Porsche Type 245 proposal was compact, turreted and driven, like its lower sister, by a transversely mounted air-cooled engine. Bogies had new volute springs that were being developed for a concurrent project.

* This was an unofficial project in parallel with the HWA's official E25 programme, whose Panzer in the same category was being designed by Karlsruhe's Argus to carry a 75mm weapon. Argus proposed its own air-cooled engine or the ubiquitous Maybach V-12. Several hulls for this E25 variant had been produced by January 1945.

Ferdinand Porsche and his team were drawing up small and agile two-man attack vehicles that would economise on Germany's scarce mineral and petroleum resources. Their designs in this Type 245 series were low-profile with engines transversely mounted at the rear to conserve interior space.

In addition to a machine gun, common to all the Type 245 studies was a Rheinmetall Borsig MK 112 cannon of 55mm calibre. Originally designed as a fast-firing aeronautical weapon, this allowed a 350-round supply to be carried for a weapon that could fire 300 rounds per minute. Some Type 245s mounted it in a small turret while others placed it in the main hull, in the manner of a *Sturmgeschütz*, between a crew of two behind 60mm of armour plating. Posited for various versions of the Type 245 was Voith's hydraulic drive and the choice of Porsche's air-cooled V-10, giving 350bhp and 40mph, or the ubiquitous 11-litre Maybach six, whose 250bhp output was expected to drive the 15-ton vehicle to a top speed of 36mph. Evolved from the 'knee action' Porsche bogie, suspension used a volute spring instead of a torsion bar, the latter having been the source of some of the bogie's travails. None of these progressed from paper to metal.

In May 1944 Ferdinand Porsche launched design work on his *Jagdpanzer* E25 (Arta), developed from his smaller Type 245.* This too was a transverse-engined design that evolved from his 1943 proposals after direct consultation with possible users. When interrogated by the Allies Porsche said 'that in his position in the design of modern armoured fighting vehicles generally, 360-degree traverse of the main armament should be sacrificed with the object of incorporating a larger main armament and thicker armour, and possibly providing a better performance'.

These principles were manifested in the 27-ton E25 (Arta), which placed its 105mm cannon in the main hull and a 30mm anti-aircraft cannon – vital given Allied air dominance over Germany – in a small turret to be fired by the

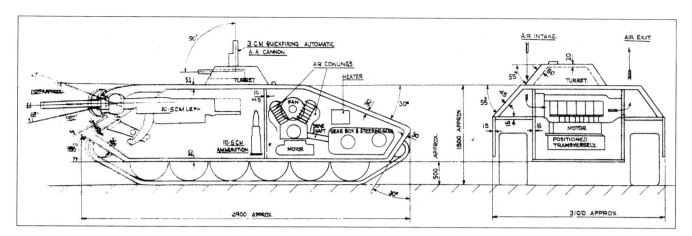

commander of a three-man crew. Proposed to power the E25 (Arta) was either the Simmering-built air-cooled V-10 or the planned V-12 version of the same, fuel-injected and turbo-supercharged to produce 500bhp. Voith's torque-converter transmission provided both drive and steering.

Although Porsche's E25 (Arta) remained a paper study, its features were at the cutting edge of tank engineering. An expert post-war assessment called it 'an example of the final stages of development in German tank construction. Heavily armoured and very mobile. Multi-purpose vehicle. Good AA protection and excellent shape. Very advanced mechanically.'

So troubled was the reputation of the Porsche-designed V-10 made in Vienna by Simmering-Graz Pauker that in 1942 that company scrapped parts being produced for 500 engines. Nevertheless, as mentioned, the ever-positive Ferdinand Porsche nursed hopes for a successful 12-cylinder version of that unit.

Moreover an even more ambitious Panzer engine was being created. In 1943 Simmering was commissioned by the HWA to create a new tank engine to be interchangeable, in both bulk and output, with the big Maybach unit. It was to be a diesel and air-cooled, as the Führer preferred. And the Vienna company was to work on its design with the Porsche organisation, taking advantage of its experience with air cooling and diesel combustion. The engine's configuration would be a rare one: 16 cylinders arranged in four banks of four to create an X-16.

Two such engines were its spiritual predecessors. In 1922 Britain's Napier produced six examples of its Cub, a 60.3-litre X-16. Considered to be the first

Ferdinand Porsche's mid-1944 E25 Panzer proposal aimed to satisfy needs expressed by commanders in the field. Using a Voith hydromechanical drive, it continued the concept of a transverse engine and carried a 105mm gun.

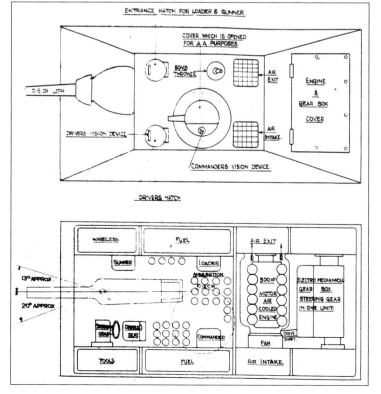

As drawn from memory by a Porsche engineer, plan views of the 1944 E25 (Arta) showed the V-12 version of the Type 101 engine, ammunition storage and places for a four-man crew flanking the weapon's breech.

Pictured at Vienna's Simmering-Graz Pauker works was the core of the Sla 16 designed by the Porsche team as its ultimate tank engine. Clustered at the X-16's output end were four four-cylinder Bosch diesel injection pumps.

A cross-section of the complete Simmering Sla 16 showed the shaft drive to its cooling blowers, its master connecting rod, diesel-fuel injection nozzles and prechambers and (upper right) one of its two turbochargers.

aero engine to develop more than 1,000hp, the Cub flew in several aircraft. Its future dimmed when the Air Ministry decided that an aeroplane would be safer with two 500bhp engines than with one of 1,000bhp. The other antecedent was an experimental Rolls-Royce X-16 of 1925. The 19.8-litre Eagle XVI produced 500bhp supercharged but was seen by aeroplane makers as hard to package. Only one was made.

The Porsche and Simmering engineers began their work by building their Type 192, a single-cylinder test engine for a planned X-16 of 36.6 litres. In Vienna on 6 November 1943 they celebrated its completion of a successful 48-hour test. The lone cylinder's output of 47bhp at 2,100rpm from 2.3 litres promised 750bhp from the complete engine – well on target for its mission. Porsche's Type numbers 212 and 220 both referred to the complete power unit, which was known as the Sla 16 by Simmering. Its four four-cylinder banks were spread at 135 degrees top and bottom and 45 degrees at the sides.

Designed by Porsche's Paul Netzker, the full-scale engine echoed some aspects of the unbuilt Type 70 X-32 aero engine of 1935. Unlike that unit it had plain bearings for its connecting rods and five main bearings for the forged crankshaft with its bolted-on counterweights. Each of the four reciprocating assemblies consisted of a master rod from whose four-bolt big end three link rods went to the other three pistons in its group. The crankcase was fabricated of sheet steel in upper and lower halves. Helical gearing at the crank's output end drove four camshafts, from which pushrods and rocker arms drove two gently vee-inclined overhead valves per cylinder.

Showing confidence in the fundamental concept it had introduced with its V-10, Porsche used the X-16's cylinders to retain their individual heads to the crankcase.

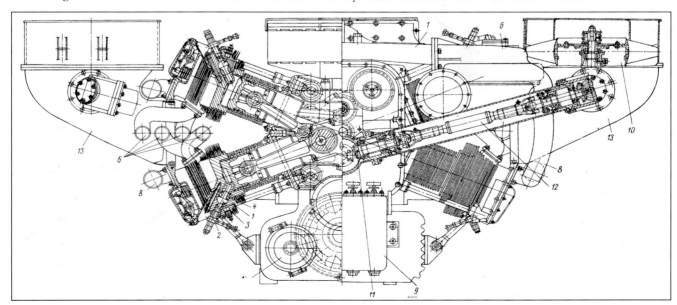

Screwed into its head, each of the forged chrome-steel cylinders with its fine machined fins was attached to the crankcase by 12 studs. Cast from an aluminium alloy called Hydronalium, the finned individual head had an inserted high-chrome-alloy pre-chamber for diesel combustion and gave a compression ratio of 14.5:1. While the prototypes had four Bosch oil-injection pumps feeding four nozzles each, the plan was to change to two eight-plunger pumps for production.

Source of the cooling draught was a pair of axial-flow blowers, one for each side. Driven by shafts and bevel gears, they turned at 4,100rpm for the crankshaft's 2,000rpm. Instead of the usual practice of blowing cool air over the cylinders, they sucked warm air out of the engine. Producing negative pressure in the engine bay, this avoided the problem of unpleasant gases leaking into the rest of the Panzer's interior. The price for this was the need for greater blower capacity than would have been the case with cool air under pressure. Two oil coolers flanked each blower.

Exhausts at the rear of the Sla 16 emerged from the twin turbochargers above the shafts driving its cooling blowers, which were flanked by oil coolers. The X-16 was planned for development to 700 shaft horsepower.

Installation drawings show the complete Sla 16 package, designed expressly to fit into Panzers. As the Type 212/220 this X-16 was one of the most remarkable engines ever to leave the Porsche drawing boards.

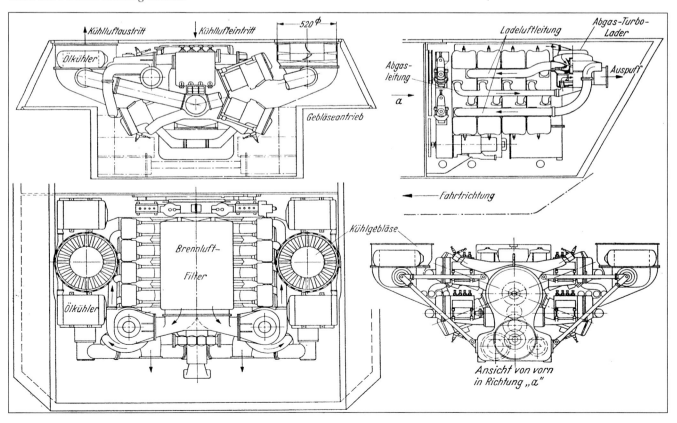

On the dynamometer at Simmering-Graz Pauker in Vienna the Sla 16 delivered net power of 685bhp at 2,200rpm. As pictured it was being tested in naturally aspirated form with slave cooling blowers.

While a drawing showed the Simmering X-16 in a Porsche-chassis Jagdtiger, its actual installation was made in a King Tiger at the Nibelung Works as the war ended. Most Sla 16 materials were taken by the Russians.

---

* At this stage the Porsche team were excluded from discussion of the manufacturing arrangements for the engines. The Simmering people were particularly concerned to be sure that they retained an important role in the engines' manufacture in spite of the necessary involvement of Steyr.

A particularly advanced feature was turbo-supercharging of the Sla 16 by its exhaust gases, a suggestion of Ferdinand Porsche. This was a technology that the Porsche engineers had been working with since the late 1930s. Each side cluster of eight cylinders had its own turbo unit, produced by Mannheim's Brown & Boveri, drawing from a central filter above the engine and reaching a maximum speed of 28,000rpm. This was a pioneering use of turbocharging for a high-speed diesel engine.

All this advanced machinery was ingeniously packaged into a compact assembly weighing 4,960lb complete. Testing of the first Sla 16 in 1944 yielded gross power of 770bhp at 2,200rpm, reduced to 685bhp at the engine's output shaft by the power diverted to drive the cooling fans. For this reason an increase in cylinder-bore size was planned for the series version to bring net power above 700bhp. Helical gears at the X-16's output stepped up its speed by 50 per cent to suit gear trains already being used with the Maybach engine, which revved to 3,000rpm. The gear train also positioned the output shaft lower to match existing transmission systems.

On 10 January 1945 the engineers at Simmering-Graz Pauker reported that two of these engines had completed a combined 300 hours of test-bench running.* In their view the design was ready for production. Planning began for a null-series of 20 engines plus spares to be built in co-operation with Steyr-Daimler-Puch. Foreseen was production of the first null-series engine in mid-June and volume manufacture of the X-16 from the end of July, after a first engine from the line had passed a 100-hour test. The Sla 16 would replace the Daimler-Benz DB 605D V-12 being phased out of Steyr's production lines.

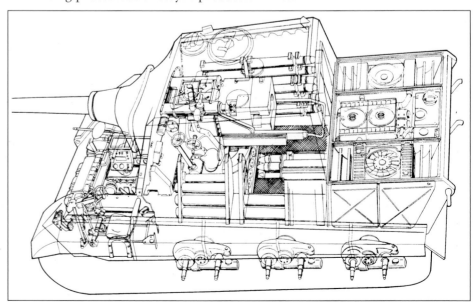

Much of the production of components and assembly of engines was to take place in the tunnel systems that the Organisation Todt was digging in Austria, exploiting slave labour from concentration camps established for this express purpose. Steyr would make components in the tunnels near Melk that were partially carved out and equipped during 1944. Code-named *Quarz*, these were already being used for ball-bearing production. Final assembly was scheduled for the *Zement* tunnels to the south-west, in hillsides near the Traunsee.

At the Nibelung Works a King Tiger tank stood ready for the first Sla 16 installation. After modifications to its engine compartment, undercarriage and hull, this was completed before the war's end but not subjected to definitive proving. Prolific engineer Wunibald Kamm, whose Stuttgart motor-research centre had designed the Sla 16's cooling blowers, proposed using 24 of its cylinders in a new diesel engine to power the biggest Panzers then on the Reich's drawing boards. This initiative, with its implied testimony to the validity of Porsche's basic design, was never realised.

The need for more power to propel progressively heavier Panzers led inevitably to consideration of the use of Germany's gas-turbine expertise to this end. A special advantage of the turbine in oil-poor Germany would be its omnivorous appetite, allowing virtually any liquid fuel to be used. This impacted Ferdinand Porsche's organisation in two ways. Porsche was asked to study the possibility of adapting the existing Junkers or BMW engines to tank propulsion. The aim was shaft horsepower at the 1,000 level.

Engineers Walter Becht and Josef Mickl were well qualified to make such an assessment. Their conclusion was that this was not feasible by virtue both of their high fuel consumption and of their poor torque characteristics, assuming that the

In its Type 305 programme Porsche undertook to fit a gas turbine engine in a tank. A single-shaft axial-flow turbine to Alfred Müller's designs was built to drive the tank through a hydraulic coupling that served a clutch function.

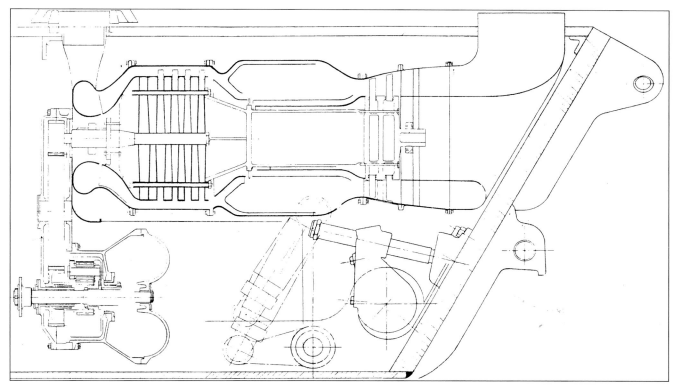

gas-turbine assemblies of the engines drove the vehicle directly. But in principle the gas turbine's potential was undeniable. Ferdinand Porsche told post-war investigators that for the same bulk and weight he could deliver 20 per cent greater output with a turbine in place of a conventional engine.

Porsche's team drew up a concept of their own. As their Type 305 they proposed a single-shaft turbine that diverted pressure air from its axial compressor to feed a separate combustion chamber whose hot gases spun the turbine that propelled the Panzer. This modality, they maintained, would multiply fivefold the available starting torque.

The Porsche concept was never put to the test because another player stepped onto the stage. This was Alfred Müller, who had form in this field as a designer of centrifugal superchargers for aviation. In 1943 Müller was at St Aegyd, in the forested hills west of Wiener Neustadt, at the Waffen-SS-supported Motor Vehicle Technical Research Institute.* From this impressive-sounding eyrie Müller suggested that a gas turbine be developed to power tanks.

Porsche engineered the Müller turbine's longitudinal mounting in the rear of a Henschel-built Jagdtiger. In spite of its risk to St Valentin's forests, it foreshadowed future turbine-powered tanks.

On 30 June 1944 the Reich informed Alfred Müller that it accepted his suggestion, whereupon work surged ahead on such engines. Also employed at St Aegyd was a colleague of Wunibald Kamm, blower designer Bruno Eckert. The Waffen-SS obliged him to undertake a parallel tank-turbine study. Alfred Müller's engines were to be built in three sizes: the GT-101 of 2,600bhp, the GT-102 of 2,400bhp and the GT-103 of 1,600bhp. The task of fitting them into vehicles was assigned to Ferdinand Porsche and the Nibelung Works, where the versatile Otto Zadnik was the relevant expert.

While Müller pressed ahead with his axial-flow gas turbine, Zadnik worked on ways of persuading the lengthy engine into a tank. On 25 September 1944 Ferdinand Porsche presided over the decision to concentrate on a gas-turbine installation for a Panther tank as a further effort under Project 305, using an adaptation of its mechanical drive train. He concluded that its output should be a minimum of 900bhp net of accessory power losses to achieve good performance in the 46-ton Panther. Any of Müller's turbines would more than suffice.

In mid-November Alfred Müller was able to show the basic design of his gas turbine to Otto Zadnik. After analysing the transmission requirements, Zadnik asked for improvement of the turbine's torque characteristics.† For experimental purposes he geared up to make the first test installation in the *Jagdtiger*, the 70-ton anti-tank version of the Henschel Tiger that was being built at St Valentin. By mid-February drawings for its turbomachinery were complete.

In spite of the exigencies of 1945 a workable prototype engine was produced in the relative calm of Austria. Remarkably, before the war's end they succeeded in

---

* *Kraftfahrttechnischen Versuchsinstitut*, part of Heinrich Himmler's intense wartime effort to establish an industrial organisation under the aegis of the SS. Typically the SS provided this laboratory with its own source of concentration-camp labour.
† The productive pairing of Müller and Zadnik did not end in 1945. Two years later they were in Britain at turbine pioneer C.A. Parsons & Company Ltd, where they and colleagues were inventing and patenting advances in the gas-turbine art.

fitting Müller's gas turbine into the *Jagdtiger*. Placed high and longitudinally at the rear, it drove forward and down through reduction gears to the transmission. Exhaust was vented upwards at the big tank's tail. 'Its exhaust temperature was 650 degrees,' Ernst Piëch told the author.* 'It burned up the trees around the tank!' No better testimony to the completion of this pioneering project could be given.

Although Ferdinand Porsche remained an outsider to the armament establishment and the old-line military, Adolf Hitler expressed his confidence in his engineer at the end of 1944 by naming him *Sachverständig für Rüstungswesen* or Armaments Expert for the Reich. While the post was more honorary than activist, it was appropriate recognition for a man who at the age of 69 was still endeavouring to serve the needs of his adopted country. 'Just as Porsche hated the public pomp of honours and celebrations,' wrote Richard von Frankenberg, 'so too was he delighted that his design activity was valued and encouraged more than before.'

This recognition for Porsche was some compensation for the naming of his adversary, Gerd Stieler von Heydekampf, as chairman of the Main Committee for Armoured Vehicles and Trucks on 1 February 1945. This brought all vehicle industries together under common supervision for the first time. In spite of steady attacks on its infrastructure the Third Reich's production of medium and heavy tanks had not faltered; indeed it rose by 46 per cent in 1944 to 17,328 Panzers.

Porsche's attention was frequently needed at St Valentin. It was also demanded at the KdF plant at Fallersleben, where his son-in-law Anton Piëch was in charge, and at Peugeot's factories at Montbéliard in France, which had been placed under the authority of the KdF-Werke during their German occupation. Porsche's presence was also demanded in Berlin and expected at least two days a week in Zuffenhausen, at the main design office which was then under Ferry Porsche's direction. In Stuttgart the family was drawn to the Ferdinand Porsche villa, with its steeply gabled tile roof, number 48 on the Feuerbacher Weg, high on a hill overlooking the city. To the south-east, near Klagenfurt in Carinthia, the 10-mile-long Wörthersee had been a favourite summer holiday spot for the Porsches since the early years of the century.

These were only some of the destinations on the crowded wartime itinerary of Ferdinand Porsche. He was under way more than at rest, whether on the railways or with chauffeur Josef Goldinger in his special streamlined KdF-Wagen, one of three prototypes for an abandoned race from Berlin to Rome. Ferry had a VW cabriolet with a Roots-supercharged engine. They kept their car radios tuned to the special frequency used by the air-defence authorities so they could shelter in tunnels or under bridges if attacking planes were sighted nearby.

The late winter of 1942 saw the first heavy bombing attacks on Germany by the Allies. On 5 March 1943 the RAF pounded the Krupp steel works in Essen with a force of 367 bombers. This was but a harbinger of more intense assaults on Germany that spring, abetted by high-level daylight bombing by the Americans.

'The mass attacks launched in the spring of 1943 came as a complete surprise,' said steel executive Walter Rohland. 'All the warnings given by the few men in industry who could form an objective opinion about the enemy's armaments potential had fallen on stony soil.' Among those men was Ferdinand Porsche, whose two research visits to America in the 1930s had left him in no doubt about the great depth and extent of America's industrial resources.†

---

* Given in degrees Celsius, this was equivalent to 1,200 degrees Fahrenheit.

† Hitler and Goering were aware of it too, but they could not conceive that the effete, pleasure-loving Americans would give up their beloved consumer goods to convert their industries to armaments. They were wrong.

Pictured in prototype form, the air-cooled flat four designed for the Volkswagen by Franz Xaver Reimspiess was a man of all work for the Third Reich's military in both two- and four-cylinder form.

The onset of the air raids impacted the way Ferdinand Porsche and his engineers went about their business. Hitherto they often used trains on which sleeping compartments were reserved for them by Mitropa on a permanent basis. This worked for a while, but increasingly they had to resort to the roads. Ultimately, as related in Chapter 17, a substantial swathe of the Porsche engineering cadre was relocated to a remote site in southern Austria for safekeeping.

In the summer of 1944 that witnessed the Russians advancing on East Prussia and on 20 July an assassination attempt on Adolf Hitler, Ferdinand Porsche's engineers managed to move their vehicle projects forward. They were working on their Type 293, a caterpillar-tracked personnel-carrying vehicle commissioned by the SS. With Stuttgart under bombing attack, components for its prototypes had to be made in Italy. Included in their remit was Porsche's adaptation of the VW engine to many other tasks. Compact, air-cooled and self-contained, the engine was a hugely versatile and useful prime mover. Fallersleben made many more engines than vehicles, both as field replacements in Kübels and for stationary power applications.

With magneto ignition and a speed-control governor, the flat four was a portable power package for generators, compressors and barrage-balloon winches. One of them drove the hydraulic pump that raised the V-2 ballistic missile into its firing position. Fitting Victor Derbuel's proprietary VD Roots supercharger increased the engine's output at altitude to power a portable cable lift for mountain transport. A flat-twin engine based on half a VW four produced 12hp for use as an auxiliary power unit for armoured vehicles and as a tank-engine starter.

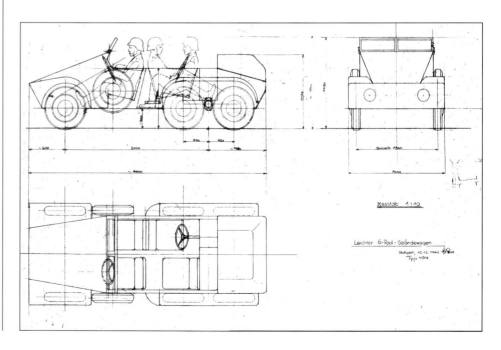

Porsche's Type 164 was a fascinating exercise of a twin-VW-engined six-wheeled scout car capable of instant operation in both directions. It would be tempting to try to build one in the twenty-first century.

During the war, said Ferry Porsche, 'it was vital that the Volkswagen be constantly reviewed, improved and kept up to date to meet changing conditions. So, since we weren't allowed to design a synchromesh transmission or a hydraulic brake system for private use, we designed them for "military use" instead – as "improvements" to the people's car.'

In the midst of the war the Porsche studio's familiarity with the VW engine was tested again. In their light attack boats the army's Pioneer troops, their field engineers, were having problems with water-cooled engines which picked up too much system-clogging debris. In 1943 they turned instead to the KdF-Werke to provide a suitable air-cooled engine.

More power was required because the four's standard output was inadequate, even at its larger 1,131cc. Porsche took up the challenge in its Project 170. The first approach was supercharging, using one of Victor Derbuel's Roots-type VD blowers. Driven by twin vee-belts at more than twice crankshaft speed, it brought power to 34bhp at 3,300rpm. Though this was enough, the package turned out to be too heavy. Another approach was needed.

Vee-inclined overhead valves were seized upon as the solution. The Type 171's valve gear used pushrods and rocker arms in a layout that required much longer rocker arms for the inlet valves than for the exhausts. Topping the individual cylinders was a single cylinder head with direct inlet and exhaust passages. Cooling came from a low-profile axial-flow design driven at twice engine speed by a

Testing as usual in the Max Eyth Reservoir, Porsche's engineers laboured to create a satisfactory engine for the boats of the Army's Pioneers, whose water-cooled engines ingested too much debris.

One solution for the Pioneers, Germany's engineer corps, was a Roots-type supercharger driven by twin vee-belts at more than double crankshaft speed. Although powerful, this was thought too heavy.

pair of vee belts. The Type 171 delivered 33bhp at 3,300rpm, a satisfactory result. Against it, however, was the myriad of special parts that it needed. This placed excessive demands on tooling and manufacturing in wartime, so the army requested a simpler version.

Porsche's final offering for an attack-boat engine was its Type 174. This was a back-to-basics design that relied on the Type 87 four as created for the KdF-Wagen with subtle refinements. While the valve gear was little changed, the heads were hot-rodded with individual inlet ducts from a larger central carburettor. Exhaust porting was more open as well. Cooling used the conventional sirocco fan.

At first the 174's performance was disappointing with only 25bhp produced. Given larger inlet valves, however, and a raised compression ratio, output of the 174A version rose to 30bhp at 3,300rpm. With the engine directly driving a long shaft and propeller, this met the needs of the Pioneers. Fully laden, their boats could break the waves at nigh on 11 knots.

A final power-unit variation for the KdF-Wagen, commissioned in the autumn of 1944, was a diesel engine. The Porsche team were asked to design a diesel that would be a straight replacement for the Type 60/82/166's flat four. Taking advice from Bosch, for the Type 309 they decided on two cylinders instead of four to suit the cylinder size that the existing oil-injection equipment could supply.

Diesel specialist Paul Netzker took charge of the design of a horizontal twin cooled by an axial-flow fan and scavenged by a Roots-type blower. Of 1,135cc, it was estimated to produce 22bhp at 2,000rpm, driving through a hydraulic coupling to moderate the fluctuating impacts of the two cylinders. With its cast-iron crankcase the twin posed a weight penalty of 17 per cent over the standard engine. Adding the required heavy-duty starter and battery hiked the gross penalty to 37 per cent. Its design only completed in May 1945, the two-stroke was never realised in the metal.

Tuned to 32hp, the KdF-Wagen engine took to the air on 25 January 1944. This was the first flight of the Horten H IIIe flying wing, essentially a powered glider. Very much in its element, the Type 247 air-cooled VW engine drove the craft's folding pusher propeller through five vee-belts. It was both more reliable and flusher-fitting in the aeroplane than the Walter Mikron engine previously used. One of the most beautiful of the tailless creations of Horten brothers Reimar and Walter, the H IIIe had a wingspan of 67.3ft. Capable of 87mph, the aircraft cruised at 74.

Porsche did not disqualify itself from the presentation of aeronautical ideas of its own. In 1934 it applied for a patent on an undercarriage for aircraft that took advantage of its torsion-bar know-how, making such bars an integral part of the mechanism that extended and retracted the landing gear. This was an attractive solution to a problem not then recognised by Germany's aero industry.

Easily as futuristic as any aircraft of 1944, the H IIIe flying wing of the Horten brothers was one of many prototypes they built and tested. Its Porsche-modified VW engine provided a smooth 32bhp.

Looming majestically in the Zuffenhausen forecourt, Porsche's Skoda-built Type 175 was created to duplicate on the Eastern Front the artillery-shifting achievements of Austro Daimler's Goliath in the First World War.

Other patents of 1938 suggested improvements in fuselage skinning and a method of using both charging and suction turbines to bench-test aero engines under simulated high-altitude conditions.

Firmly grounded was a job that the Zuffenhausen designers booked in January 1942. With the war in Russia dragging on into a second year, the HWA asked Ferdinand Porsche to design and build a modern version of the high-wheeled Austro-Daimler M 17 Goliath tug that had been so successful both in the First World War and afterwards. Fully briefed, Karl Rabe began to delineate its characteristics on Monday, 26 January. He entered it in his log as the Type 175, known as the *Radschlepper Ost* or 'Wheeled Eastern Tractor'.

Partner in the project was Skoda of Pilsen, reawakening an important First World War relationship. On 2 February Porsche and Rabe hosted a day-long discussion in Zuffenhausen with an 11-man delegation from the Czech company led by veteran design engineer and director Emil Řezníček. Skoda would produce the huge 9-ton vehicles to Porsche's designs.

Spaced on a 118.1-inch wheelbase, the same as that of the M 17, at just under 5ft in diameter the steel wheels of the Type 175 were even larger than those of its First World War counterpart. While small cleats were integral with the wheels, larger ones and ice studs were stowed on board to be attached when required. The use of steel was a direct response to the shortage of rubber, tyres of such size consuming far too much of this scarce resource.

Another shortage was of lead, for which the batteries in the submarines of Admiral Karl Dönitz had absolute priority.* Accordingly Porsche and Rabe provided a small crank-started VW-based parallel-twin engine to start the main four-cylinder unit. The latter was a long-stroke four of 6.0 litres with overhead

---

* The shortage of lead for vital submarine batteries was at one point so severe that the proposal was made that it be extracted from the keels of all the sailboats in Greater Germany. In a meeting Porsche rightly ridiculed this ludicrous idea after discovering that all the keels would make up only a week's supply of lead, saying, 'It would be better to end the war a week early and save the sailboats.'

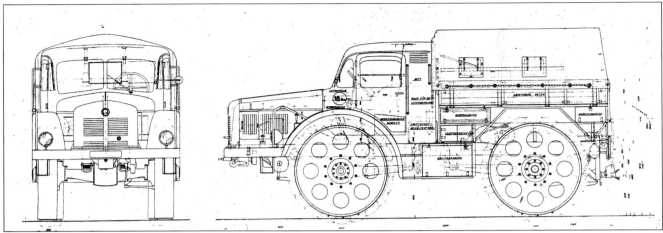

Approaching 5ft in diameter, the wheels of the Ostradschlepper *were cleated with high ground clearance to gain traction on the muddy Russian Front. Solid axles were carried by close-spaced leaf springs.*

After satisfactory trials 200 of the Type 175 were ordered. Production was halted at around four dozen units, however. It had turned out not to be the answer to the logistical challenge of the Eastern Front.

valves and air cooling by a blower driven from the nose of the crankshaft. Reached at a modest 2,000rpm, its peak output was 80bhp. A five-speed gearbox translated this into 11,000lb of pulling power, implemented when required by a chassis-mounted winch.

Wide leaf springs married with solid axles at front and rear that were guided by bracing triangles pivoted from the centre of the Type 175's ladder-type frame. Constant-velocity universal joints allowed steering of the driven front wheels, controlled by a gear that gave six turns of the wheel from lock to lock. The outer perimeter of the tractor's turning circle was 60ft. Conventional enough in appearance, the body and cab had seating for three plus space for a bunk.

On 1 October 1942 Ferdinand Porsche witnessed completion of the first prototypes of the high-wheeled *Radschlepper Ost*. At the end of the month they were being tested at Berka am Hainich, verifying their top speed of just over 9mph.

At 1.4 miles per gallon their petrol consumption from a centrally placed 250-gallon tank was judged high albeit not excessive for this class of vehicle. Nevertheless this was a constraint when fuel was in increasingly short supply.

On 20 November Porsche showed the *Radschlepper Ost* prototypes to Albert Speer. On 4 January 1943 Adolf Hitler witnessed a demonstration of Porsche's Type 175 near his East Prussian Wolf's Lair. Although the order to Skoda was for 200 units of the high-wheeled tug, its manufacture, begun in 1943, was halted somewhere between 40 and 50 units. Its fate was sealed by its excessive use of precious fuel and dissatisfaction with the functionality of its steel wheels. Porsche had given the HWA what it wanted, but it had turned out not to suit the ever-changing campaign in the east.

In the midst of these urgent activities Porsche took on another challenge, a commission from the Wind Power Research Company.* In July 1941, on a hill south-east of Stuttgart, a Porsche team erected their 736-watt Type 136 wind turbine on Hohenheim Agricultural Institute land. This was a brainwave of Fallersleben board member Bodo Lafferentz, who saw such turbines as ideal power sources for Germany's settlers on the steppes of Russia.

Encouraging work on this project for personal reasons was aerodynamicist Josef Mickl. The benign wind turbines suited this engineer because, as Ernst Piëch said, 'politically behind this was Mickl, who didn't like the war and everything. He was a little bit opposed to Hitler. He was so cross with him that during wartime he didn't have very much to do. He started this project more or less to gain his freedom. This was his form of opposition during wartime.'

Developed from a 130-watt prototype with blades of 11.2ft that had been tested on the works grounds, the wind turbine's four blades swept a diameter of 23.6ft. Mickl created a special hub designed to feather the blades automatically in the face of winds of excessive velocity. He also patented his design for the finning that kept its blades facing the wind.

Josef Mickl's Hohenheim turbine remained in operation for three years, only being taken down in 1944 when it was declared a menace to planes flying into nearby Stuttgart Airport. An improved and more powerful version, the 4,500-watt Type 137, remained unbuilt. Conditions in Nazi-occupied Russia, it seems, were not yet suited to the settlement of Germany's peasants.

Josef Mickl's artistry was manifested by his Type 136 wind turbine, built on Stuttgart's southern heights. With a generating capacity of 736 watts, it had aeronautical tail feathers to keep it facing into the breeze.

* *Forschungsgesellschaft für Windkraft.*

CHAPTER 15

THE MOUSE THAT ROARED

Adolf Hitler evolved the Big Idea during a meeting with Ferdinand Porsche at the Rastenburg Wolf's Lair on 29 November 1941. The Führer had in mind the building of a super-heavy tank, he told the engineer. 'Nothing could stop a land-battleship of this size and power,' he said, 'and it would prove a tremendous tactical surprise. Of course it would not travel alone but with an escort of smaller armoured vehicles, much the same as a convoy.' Hitler also saw his huge tank as being useful as a stopgap among pillboxes and bunkers.

With a generation gap between them Ferdinand Porsche was always ready to speak his mind to the frequent discomfiture of Albert Speer, who was Hitler's friend but not, like Porsche, his intimate.

Responding to his Führer, Porsche remarked on the need for extremely heavy armouring for protection and wide and long tracks to support such a tank's great weight. Hitler replied by giving his priorities for tank design: 'The first priority is the heaviest armament, second priority high speed and third priority heavy armour. However,' he added, 'heavy armour is unavoidably necessary.'

While many accounts credit Hitler's insatiable appetite for ever-bigger and more shocking weapons for the creation of the Panzer code-named '*Maus*' or Mouse, some logic did underpin his thinking. The Führer was doubtless aware of a desperate project near the end of the First World War to build a tank that was capable of crossing trenches as broad as 20ft.

'We designed a tank of colossal dimensions,' wrote August Horch, a member of the 22-man design team, 'with four engines, each of 600hp.* The vehicle could be dismantled, for this elephant could not be transported whole. It weighed over 120 tons and was a bizarre monster.' Its control system, from its captain to the crew and its weapons, was electrical, adapted from that used in submarines, while its drives were described as 'electric-magnetic'. At the Armistice two such monsters were being completed at the Riebe ball-bearing works in Berlin.

Adolf Hitler was also well aware of current trends in the Panzers of his enemies. Recent encounters with Russian armour had given no cause for complacency, the KV series carrying heavy guns and knocking on 50 tons when fitted with supplementary armour protection. Rumours about Soviet intentions hinted at machines as heavy as 120 tons, carrying 105mm cannon and two further heavy weapons as well as machine guns.

From the beginning of 1942 the Reich's Panzer makers reacted. Studies by Krupp envisioned tanks weighing from 130 to almost 200 tons. Ultimately Argus would be drawn in to support a project dubbed the E-100 that conceived a tank scaling some 150 tons.† Henschel too was looking at bigger and heavier versions of its Tiger (H), all in the name of carrying great firepower that could demolish

---

* A report in www.achtungpanzer.com credits the tank with two Daimler aero engines, each of 650bhp, and a weight of 148–150 tons.

† A single Maybach-powered E-100 was nearing final assembly by Adler near Paderborn at the end of the war.

defences. That competition would also come from Ferdinand Porsche was understood, for such a mission would not be denied the Führer's favourite engineer.

With plenty of other projects on its plate, including the Henschel and Porsche Tigers, the HWA dragged its feet in responding. 'When the project came to the HWA,' wrote Richard von Frankenberg, 'the technicians there were unanimous: building such an *Über-Panzer*, a Panzer-battleship, was not at all possible. Such a colossus could not be given sufficient mobility.

'And so they gave the job to Porsche,' von Frankenberg continued. 'He then gritted his teeth and was fully occupied for the next two years. Porsche did not question the operational validity of such a vehicle. That was for others to evaluate. He was given the design responsibility.' Indeed, it was just the kind of project that Porsche relished: one that everyone said was impossible, like shifting some 100 tons of howitzer up steep Italian mountain passes a quarter-century earlier.

The minutes of a meeting between Hitler and Speer over 21–22 March 1942 confirmed that 'Porsche is to be given the contract for independent design of a 100-ton Panzer.' Steps towards this goal, described initially as the VK.100.01, were erratic. Although such an instruction did reach Porsche at the time, he and his team were up to their ears in the preparation of their Tiger (P) for Hitler's birthday on 20 April. During that month Krupp was studying a turret that could suit both its own Mega-Panzer and Porsche's. Cannon with bores of 128 or 150mm were being considered, as was a coaxial gun of 75mm.

In a 13 May 1942 conference with Albert Speer the Führer gave his view 'that the heaviest Russian tanks will certainly appear by Spring. Therefore he demands that the heavy Panzers currently being designed be energetically carried out and holds the opinion that reducing the weight to 70 [metric] tons is incorrect. He has no qualms that instead of 100 tons one could even get up to a weight of 120 tons. Priority is to be given to the heaviest armour connected with a gun with the highest performance.'

Here was a specific example of Hitler's personal engagement with the heavy tank's design. 'In military conferences,' wrote Albert Speer, 'Hitler quite often presented as the fruit of his own reflections technical details that my experts, Professor Porsche or Stieler von Heydekampf, had instructed him on just a short while before. He also liked to claim that "last night" – although a conference might not have ended until nearly morning – he had "studied" a scientific or historical work of hundreds of pages. We all knew that he firmly believed in reading only the end of a book because everything important was to be found there.

'Of course we were all sticking together,' Speer continued. 'We had many talks about it and we tried, almost with intrigues, to attack Hitler's opinion from different sides and were winning over other officers who were coming with experience from the battles to tell Hitler. Well, he perhaps changed his opinion for a few days, but afterward he jumped back again. It was now the question that he wanted the heaviest tanks possible, which now were so slow that the tanks of the other side were far superior to them.'

After their preliminary studies, not until 4 June 1942 did the Zuffenhausen designers make their first definitive drawings of the big tank they identified as *Sonderfahrzeug IV* and logged as their Type 205 project. Leopold Schmid's outline

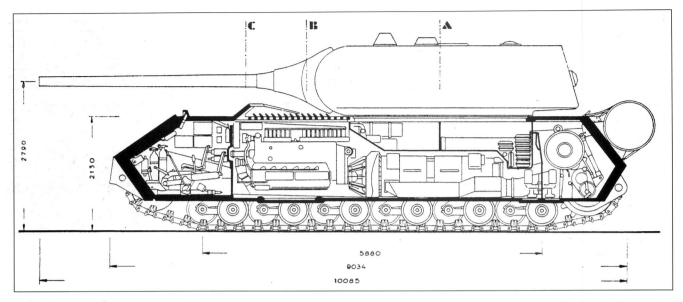

The size of its seats gave a sense of the dimensions of the Type 205 Maus, with its inverted V-12 aero engine driving directly into a tandem array of generators. Armour thickness is boldly portrayed.

for the design suggested 120 metric tons and a 150mm gun. That same day Hitler stated that he was 'in agreement that the super-heavy Panzer be a slow-moving vehicle (mobile fortress)'. Here he was echoed by Porsche, who said after the war 'that the vehicle should not be termed a tank in the strict sense of the word but a heavily armoured mobile pillbox'.

Four days later, in a meeting with Hitler and Speer, Ferdinand Porsche was commissioned to take steps towards completion of a super-heavy tank. The focus of the order was less the size and weight of the tank than its ability to carry very heavy weapons. Initially referred to as *Mammut* or 'Mammoth', the mega-tank was to be ready for trials to commence in May 1943.

Soon the early plans were reviewed by Wehrmacht higher-ups. According to Ferry Porsche, famed tank commander Heinz Guderian initially 'heaped scorn and ridicule on the whole idea' when he saw the proposals of 1942. The Type 205 was certainly not his idea of a machine to lead one of his fast-moving Blitzkrieg attacks. Later, however, in the hearing of Ghislaine Kaes Guderian said, 'With 100 Panzers like the Maus I'll hold the Eastern Front!' Its defensive potential was assessed as considerable.

From stem to stern the Maus, as it was definitively known, was rife with novel

In plan view the Maus echoed the view of an Allied investigator that 'it is filled almost completely with a mass of complicated machinery'. Any space not occupied by said machinery was packed with munitions.

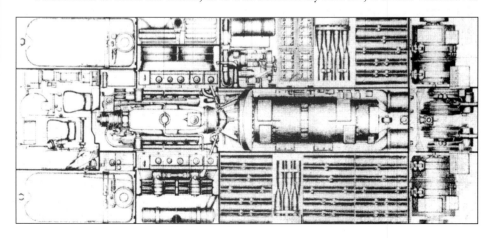

engineering solutions. 'The general impression given [by] the interior of the Maus,' said a post-war Allied report, 'in spite of its size, is that it is filled almost completely with a mass of complicated machinery.' Braking was by clutch-type discs. As usual for Porsche designs the drive was through the rear sprockets, instead of the front as on all other Panzers, because Ferdinand Porsche considered it best to prevent track shedding.

For such a massive machine the question of power was all-important. Although Maybach was promising a supercharged 1,000hp V-12, this was far from ready – and never would be. Porsche proffered the credentials of his Simmering X-16, but Speer rightly rejected this as unready. Porsche of course knew the right address: the Daimler-Benz AG and its aero-engine factory at Sindelfingen. Meeting with his Daimler counterparts, among whom were colleagues of old Otto Köhler and aero-engine designer Fritz Nallinger, Ferdinand Porsche settled on a suitable engine. This was the vehicle version of the newly developed DB 603 aero V-12, which displaced 44.5 litres. He knew this engine well as it had been the power source for his Type 80, a contender for the World Land-Speed Record that he had designed for Daimler-Benz in 1938–39.

Its vehicle counterpart was the MB 503 A, fuel-injected and developing 1,200bhp at 2,300rpm. This was gross output: 1,080bhp remained for traction after deduction of the power used to drive the radiator blowers. Designed to run inverted in aircraft, the engine was installed similarly in the Maus. An alternative engine would be its diesel counterpart, the MB 507 C with up to 1,000hp.* As used to start the *Radschlepper Ost*, two parallel-twin KdF-based engines were on board, one as a starter and the other as an auxiliary power unit for accessories. An acetylene-fuelled starter was also fitted for conditions of extreme cold.

Located on the tank's centreline, the engine was directly behind the crew compartment in the nose – an aural burden for its occupants. To its rear were two Siemens direct-current dynamos, placed in line. Each produced 400 kilowatts at 2,800rpm. Together the pair weighed 8,565lb. This reversed the layout of the Ferdinand, which positioned its single dynamo forward of the engine.

Scaling a total of 8,310lb, the two six-pole drive motors sat transversely at the rear. Each delivered 1,200bhp over a range of 1,500–2,400rpm through slipping clutches and planetary gearing that gave a choice of two drive ratios that were hydraulically selected.† They were also fully reversible and able to function as generators to provide braking which was augmented by the mechanical system. All these Siemens units were akin to those of the Tiger (P) and Ferdinand, asked here to run at higher speeds to transmit the increased torque.

'It was decided to build the Maus with electric drive,' Porsche and his engineers told Allied investigators, 'because of the large amount of power to be transmitted and the high tractive effort requirement of this vehicle. The decision was taken particularly in view of the short development time schedule originally established

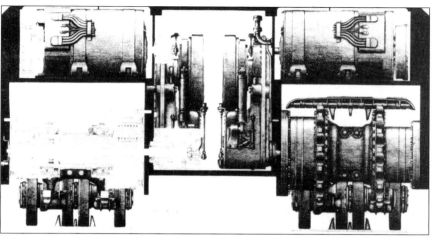

At the extreme rear of the Type 205 Maus were its drive motors, above the track sprockets and connected to them by two-speed transmissions giving a choice of overall ratios. These were problematic during development.

* Engines of these types also powered the 137-ton semi-mobile *Karl-Gerät* mortar built by Rheinmetall for the Reich's Second World War service.

† Difficulties with this control led to its replacement during trials with a mechanical shifting system.

Using volute springs made by Skoda to Porsche's designs, the Maus bogies were suspended from both sides of the sponsons containing them. Thickness of the surrounding armour was a major source of the tank's great weight.

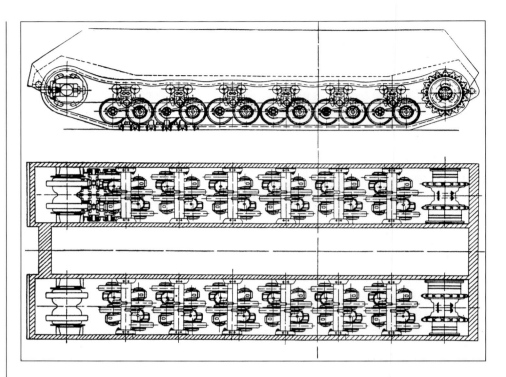

and the fact that extrapolation of constants appeared simpler for this type of equipment than hydraulic.'

Another advantage of the electric drive was that it could be used to ford even extremely deep waterways. With an air hose to a skeleton crew aboard a well-sealed Maus, the latter preparation thought to require three-quarters of an hour, electric cables to a sister tank provided the energy needed to complete a maritime passage. Here was Porsche technology used not only in the development of the Tiger (P) but also successfully in the C-Trains of the First World War.

Architecturally the Krupp-built hull was divided into three longitudinal sections, the outer ones acting as sponsons almost completely enclosing the bogies and their tracks, which were a generous 43.3 inches wide to cope with the Maus's mass. At combat weight, with main and disposable fuel tanks full, its specific ground pressure was 18.5 psi.* This was close to – and in action would quickly drop to – the HWA requirement for heavy tanks of a maximum of 17.0 psi.

The wide tracks were very close together by tank standards because the Maus had to be transported by the usual means – it had its own railroad flat car – and use normal roadways. Thus Porsche and Rabe kept its overall width down to 12ft. In fact it was narrower by a few inches than the Tiger in spite of the overhanging armoured skirts that carried its suspension bogies and protected its caterpillar tracks.

German tank designers and assessors took very seriously the ratio between the length of the track on the ground and the distance between the track centrelines – the L/C ratio. In their eyes a low L/C ratio, with tracks wide apart, was crucial to successful manoeuvrability, which was achieved by varying the respective speeds of the tracks to effect turning. Porsche's foe of old, HWA engineer Ernst Kniepkamp, 'was adamant that the L/C ratio should be kept down to 1.0,' he said in an interview. 'He strongly criticised Porsche's Maus which had an L/C ratio of

---

* This is the author's calculation. Porsche itself quoted a figure of 20.6 psi.

2.5. He could not think how any tank engineer could commit himself to such a manifestly unsuitable ratio.' Most shared Kniepkamp's view that the Maus would never be able to steer properly.

Planned for the Type 205 were Porsche's own 'knee-action' bogies with their longitudinal torsion bars. In spite of the difficulties experienced with these, it was considered to be a known and established design that should be used in view of the project's urgency. Supporting the tank on each track were six four-wheeled bogies, using the steel wheels with rubber-ring inserts created to conserve rubber. The bogies were a co-operative effort with Pilsen's Skoda, which designed and built the tracks in their entirety.

As in the Ferdinand, no internal movement was possible between the two-man driving compartment and the four crew members in the turret: the commander and three artillerymen. The Maus's driver's seat could be moved up and down to afford him ideal vision either enclosed or exposed, all his controls moving with him. Capable of rotating a full 360 degrees in only 16 seconds, the turret was mounted on a system of special rollers that were designed to be deformable, to absorb shock, and to be replaceable from the tank's interior.

Rear-mounted, the Krupp-designed turret was extremely large to accommodate rounds that were 60 inches long as well as a yard of recoil for the 128mm main gun. Total capacity was 68 rounds for the big bore and 200 rounds for the smaller coaxial 75mm cannon. The turret also carried a machine gun and smoke generator. In January 1943 Hitler personally confirmed this weaponry, while asking that guns of 150 and 170mm be designed for future installation.

Including its guns the turret alone weighed 50 tons, thanks to its size and frontal armour 215mm – 8½ inches – thick. Its roof was 65mm thick. The front of the hull was plated at 205mm while 150mm was not unusual for many other main surfaces. Placed forward athwart the crew compartment, internal fuel tankage was 365 gallons. Picking up an idea from the Russian T-34, a jettisonable cylindrical tank was attached to the rear of the hull. Weighing 518lb dry, it added 220 gallons. Fuel consumption was reckoned to be one-fifth of a mile per gallon in good going and half as far per gallon in bad.

The Maus was on the agenda when Porsche, Heinz Guderian and Dr Müller of Krupp met with Adolf Hitler early in December. 'Hitler said that he expected a specimen model to be ready by the summer of 1943,' related Guderian; 'he would then require the Krupp Company to produce five of these tanks per month.' Once commissioned, Porsche was not averse to encouraging his clients. From time to time he updated Hitler with news of super-heavy tanks being planned by the Allies.*

Porsche's Type 205 acquired added status when the HWA assigned it a project chaser whose remit was to follow up on the contractors: Daimler-Benz for its V-12 engine, Krupp for its hull and turret, Skoda for its tracks and suspension, Siemens-Schuckert for its Mixte drive and Alkett for final assembly.

Alkett was founded in Berlin in 1937 as a subsidiary of Rheinmetall-Borsig AG. Like its parent, Alkett was subsumed into Hermann Goering's industrial empire in 1941. 'Alkett's experimental shops under Hahne were famous for their versatility and unexcelled speed in putting out new types or variations of complete tanks or parts of tanks,' said Henschel's Gerd Stieler von Heydekampf. 'Alkett

* While the largest Soviet and American tanks of the Second World War scaled some 55 tons, the British built a 75-ton tank/tank destroyer, the A39 Tortoise, but failed to get it into service before the end of the conflict.

may be considered as *the* German development shop for armoured fighting vehicles.'

On 18 December Porsche made every effort to be civil to project-chaser Colonel Haenel at Zuffenhausen, where the HWA emissary 'issued orders that the Maus was to be completed and ready for trials on 5 May 1943', said a Porsche engineer. 'This was naturally considered as being more humorous than anything else and no notice was taken.' A follow-up visit by the HWA on the Maus's progress took place on 12 January 1943 at Zuffenhausen to set out in greater detail the assignments of the companies that would construct the colossal machine. All aspects of the project were discussed in a large conference dedicated to the Maus held in Berlin on 21 and 22 January.

'Porsche's Maus was to go into production and the figure raised to ten a month,' Heinz Guderian recorded. 'This gigantic offspring of the fantasy of Hitler and his advisers did not at that time even exist in the form of a wooden model. All the same it was decided that mass production was to begin at the end of 1943, that it was to be armed with a 128-mm gun and the eventual installation of a 150-mm gun was to be studied.'

Summoned to Berlin on 2 February 1943, Ferdinand Porsche learned from the HWA's Colonel Haenel that the Maus had to carry a flame-thrower with a fuel capacity of 260 gallons. The idea seemed bizarre – a close-combat weapon for a machine designed to kill at extreme ranges. 'Porsche said that it could not be done,' said one of his engineers, 'but was overruled and was told that the flame-thrower was considered essential.' Objections to this weighty addition to the super-tank were universal at a highly pressurised meeting among Porsche, the HWA and the manufacturing firms at Zuffenhausen on 10 February, but no relief was granted.

By April 1943 it was evident that the weight of the Maus was going to be higher than forecast. The flame-thrower installation alone added 5.4 tons, while build-ups of tolerances and armour thicknesses plus enhanced ammunition stowage contributed an equal amount. On 2 April Porsche announced to all concerned his plan to change the suspension bogies from his torsion-bar design to new bogies designed in collaboration with manufacturer Skoda. All were identical and interchangeable, an advantage in the field. Remarkably, the new running gear itself was credited with a weight reduction of some 4 tons per vehicle.

Porsche's rationale for this change was that its 'torsion-toggle' bogie had been thought adequate for the original design but that the Type 205's increasing weight demanded a change to ensure reliability of a bogie operating under strict space constraints. The new design used a volute spring, a circular tower comprised of a spiral of sheet steel, with rubber buffers at the limit of travel. This was not the work of a moment, Porsche designing and testing four design variants before settling on the final volute spring, which gave a gently rising rate with deflection. Krupp dealt with its exacting manufacture.

The spring of 1943 found Zuffenhausen's craftsmen building a full-size wooden mock-up of the Maus, both for demonstration purposes and to help co-ordinate the work of the project's suppliers. On 6 April Albert Speer inspected it during his snap visit to Zuffenhausen. After an on-off period in which the mock-up was first to be shown to Hitler at Berchtesgaden, packed up for that purpose and then

THE MOUSE THAT ROARED 221

Early May 1943 saw a wooden model of the Maus-to-be looming over Hitler and Porsche as they and colleagues reviewed it at the Wolf's Lair. Flamethrowers curved around the rear of its hull.

At the early May 1943 Maus presentation in Rastenburg Adolf Hitler and his retinue saw a wire-controlled scale model of the Maus demonstrate its agility. Even this failed to convince some critics of the merits of Porsche's concept.

unpacked again, it was shipped and assembled for a showing at the *Wolfsschanze* in East Prussia in early May. Hitler, who on his birthday the previous month had named Porsche a *Pionier der Arbeit* with a gold cluster, would not have the first running Maus that month, as he had requested, but could at least see what he had asked for.

'Nobody in the tank forces displayed any interest in the production of these monsters,' recalled Albert Speer, 'for each of them would have tied up the productive capacity needed to build six or seven Tiger tanks and in addition would present insoluble supply and spare-parts problems. The thing would be much too heavy, much too slow and moreover could only be built from the autumn of 1944 on. We – that is Professor Porsche, General Guderian, Chief of Staff Zeitzler and I – had agreed before the beginning of the inspection to express our scepticism at least by extreme reserve.

Special railcars were designed and built to transport the Maus. Even without the fuel and munitions that brought its battle weight to 207 tons, the huge Panzer's weight had to be distributed over fourteen pairs of wheels.

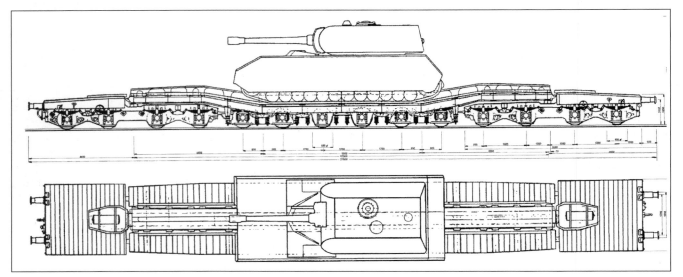

'In keeping with our arrangement,' said Speer of the presentation to Hitler, 'Porsche, when asked by Hitler what he thought of the vehicle, replied tersely in a noncommittal tone, 'Of course, mein Führer, we *can* build such tanks.' The rest of us stood silently in a circle. Otto Saur, observing Hitler's disappointment, began to rant enthusiastically about the chances for the monster and its importance in the development of military technology. Within a few minutes he and Hitler were launched on one of those euphoric raising-the-ante dialogues such as I had occasionally had with Hitler when we discussed future architectural projects.

'Unconfirmed reports on the building of super-heavy Russian tanks whipped them up further,' said Speer, 'until the two, throwing all technical inhibitions to the wind, arrived at the overpowering battle strength of a tank weighing 1,500 tons, which would be transported in sections on railroad cars and put together just before being committed to battle.'*

'Of course Hitler knew that the 1,500-ton tank was a monstrosity,' Speer added, 'but ultimately he was grateful to Saur for giving him such shots in the arm from time to time.' Saur's ideological commitment and 'booming voice' equipped him well for badinage with Hitler. 'Such viewings were always a big show,' said Walter Rohland. 'Adolf Hitler in the midst of his generals and his

---

* Incredibly, in 1942 Hitler ordered the building of such a tank, to be powered by four U-boat diesel engines developing a total of 10,000hp. With triple turrets carrying two 150mm cannon and one 800mm mortar, it would have had to scale 1,500 tons. An astonishing expanded version of the two vast tanks built at the end of the First World War, it did not progress.

closest collaborators. To the extent that Hitler was not naturally conceited, he inevitably was in the devoted demeanour of his coterie.'

Another feature of the display was a motorised $^1/_{15}$th-size model of the Maus. Controlled remotely by cable, it vividly illustrated the monster's potential agility. At the time, however, even this practical demonstration failed to silence the critics of Porsche's choice of track proportions.

During the review at Rastenburg, wrote John Milsom, 'complaints were made that the size of the tank made the 12.8 cm gun look like a child's toy and accordingly Krupp was ordered to prepare a new turret mounting a 15 cm gun, the coaxial 7.5 cm to be retained. It seems hardly probable that Hitler would have ordered a larger gun on purely aesthetic grounds but in point of fact it does appear that this is what actually occurred.'

Ferdinand Porsche could afford to be noncommittal when faced with the issue of the military viability of the Maus, said his son: 'My father, over whom the dabblers had long since gained the upper hand in strategic considerations, explained to those opposed to the project that he was not a soldier and was thus a spectator when it came to evaluating the military value of the super-heavy Panzer. He was a designer and as such had accepted the contract. He could give the assurance that he would resolve even this assignment.'

One factor that Ferry's father could not resolve was the Allies' intensifying bombing. The Essen works of Krupp, a prime target, was hit on 5 and 6 March 1943. 'All of our calculations and drawings for the Maus turret and armament were destroyed,' Krupp telexed to Porsche and the HWA. 'We have already started work on replacement drawings and are attempting in every way to make up for the loss. However, a significant delay in production start-up cannot be prevented. Hull delivery is on schedule so that there is no delay in completing the vehicles.'

In May the relevant office of the HWA confirmed production plans for the Maus. The number of units to be built was increased from 120 to 135. Their schedule called for the first two hulls and turrets for series vehicles to be delivered in November. Krupp was to supply five in December and eight in January. From the next month deliveries were to be at a rate of ten per month. An intervention by Guderian, who wanted to test the first units in battle before committing to more, led the Panzer Commission to maintain the launch schedule while curtailing the ultimate monthly rate to five.

For tinkerers with the specification of the Maus this was their last major opportunity. A welcome reduction in complexity came from abandonment of the flamethrower. Allied air activity led to a proposal to fit the turret with an anti-aircraft cannon. This too was given up on the grounds that deployment of the Maus would only take place with the big beast surrounded by a flotilla of smaller Panzers in the manner of a defensive fleet surrounding a convoy. They would shoot down attacking aircraft.

On 1 August, soon after the failure of the German offensive at Kursk, Alkett began assembly of the first Maus prototype in its Berlin-Spandau workshops. If this mega-machine lived up to expectations it looked like Hitler and his generals would have a big surprise, not only for the Russians in the east but also for the invaders expected in the west in the coming year.

No sooner did assembly begin than Krupp was hit by another heavy raid.

Karl Rabe amusingly but accurately drew himself at the centre of the complex web of alliances that built the Maus, showing Porsche, Alkett, Skoda, Siemens and Daimler-Benz as its creators. Zadnik was pictured down with the 'flu in Vienna.

Without compressed air and electricity, struggling in heaps of rubble and under shattered roofs, work on the first hulls and the turrets, running later in the schedule, had to be halted. 'On 18 August 1943,' wrote Panzer historian Thomas L. Jentz, 'Krupp concluded that due to the destruction from the bombing raid it will take seven months to restart hull production and eight months for turret production' to resume.

This was a heavy blow indeed. The 1944 combat season was all but certain to be denied the Type 205's mobile firepower. An interim schedule from the HWA saw production postponed to September 1944 at the earliest. The demands of production of other Panzers were such that Ferdinand Porsche and Karl Rabe were informed by Albert Speer, at a meeting in Berlin on 27 October, that the standing order to build 135 Maus tanks was being revoked.

Completion of the first prototype would continue, they were told. Later a second, with a diesel version of the Daimler-Benz aero engine, was approved. This V-12 was built from the outset to be upright rather than inverted. 'All of the manpower, machines and equipment for Maus production are to be immediately employed in increasing other armoured vehicle production,' said the cable sent confirming the meeting's decision. 'The armour is to be used for achieving the ordered increase in self-propelled attack-gun production.'

'It is a common belief that the Maus weighed too much and therefore the project was cancelled because of the lack of resources,' said Thomas L. Jentz after close study of the project's history. 'This is pure bunk. Production was well under way with armour plates already rolled for 30 Maus hulls and cut for nine hulls and turrets when disaster struck at Krupp. *This is the only incident in the war where a bombing raid succeeded in completely stopping mass production of a Panzer.*' (emphasis added)

At the end of 1943 Ferdinand Porsche's first Maus prototype was being completed at Alkett in Berlin. Although Albert Speer had left strict orders that the first Maus was not to be moved without his permission, the engineers were

Karl Gensberger was at the controls of the first Maus as it inched out of Alkett's Berlin works at Christmastime 1943, carrying its mass-substitute turret. It manoeuvred perfectly in response to its controls.

desperate to see how well it worked. At dusk on Christmas Eve Porsche engineer Walter Schmidt and Alkett works manager Seitz decided to risk an outing. Karl Gensberger drove the Maus out of its house through a gap that left only a few inches on each side. Its control precision was already evident. In view of the HWA's criticism of the Maus's track width and layout, nothing was more important to the engineers than a check to see how tightly the Maus could turn. It turned with astonishing sweetness, Gensberger found, on a diameter of only 26ft – as tightly as the best small cars. Maus trials continued on the adjoining loose-surfaced race track that Alkett used for its Panzer shakedowns.

Before joining his family for Christmas celebrations Schmidt informed Porsche, at Zell am See, of the good news by telex. But the Professor was not convinced. He ordered his flu-stricken electrical expert, Otto Zadnik, from his sick bed in Vienna to Spandau to take the helm. Zadnik found a fault in the assembly of the tracks which, when rectified, allowed the mammoth Maus to turn on its own axis – literally on a dime.* QED.

In January 1944 this prototype was ready to be shipped to the *Panzerkaserne* at Böblingen near Stuttgart for testing by the HWA.† It travelled by rail on a bespoke 14-wheel flat car. The trip took several days, from 11 to 14 January, not only to cope with bombing concerns but also because the Maus's width demanded caution through tunnels and sometimes required traffic on the oncoming rail to be sidelined. Porsche, who personally supervised its transport, decreed that its 3-plus miles from rail head to Böblingen be driven near midnight to escape the many air raids that Stuttgart was suffering.

At Böblingen the Panzer's test driver for the HWA was Karl Gensberger while Otto Zadnik continued to drive for Porsche. Its first movements after unloading were brief drives to the works halls, 'without incident'. Problems with shifting the two-speed gears in the motor drives meant that only the low ratio was usable

---

* During development of the Maus its general manoeuvrability would be further improved by redesigning the 'spuds', the bars across the tracks that grip the ground, so that they were on the outboard surfaces of the tracks only. British investigators after the war found this an innovation worth further evaluation.

† In 1958 the author's Signal Corps unit was based at the *Panzerkaserne*.

(From left to right) *Ferry, Rabe and Porsche stood in the slush at Böblingen early in 1944 in front of the first of their vast Type 205s. Tiny hammer and sickle motifs were intended to make spies think it was a captured Soviet machine.*

Arriving at Böblingen on 10 March 1944, the second Maus prototype was fitted with its correct turret during May. Both its curved turret face and frontal glacis were sloped to deflect incoming munitions.

at first. After further assembly some 8 miles of off-road driving were essayed on 31 January. This resulted in the failure of the rubber rings in the bogie wheels, a fault that had already been experienced in bench tests and for which new wheels were being made.

Porsche himself came aboard during 26 miles of off-road testing on 7 and 8 February. On the latter day he accompanied Fritz Nallinger and Otto Köhler of Daimler-Benz for a mile and a half of observation of their V-12 engine. Included was filming of the big Panzer's agility to add to cinema material already taken of its construction and concept. They showed it making figure-eight tracks in the snow. The Führer was well known to be susceptible to an impressive motion-picture presentation of one of his secret weapons; such a film had worked well for the builders of the V-2 ballistic missiles.

After these tests a new engine variant, the MB 509, was installed in the Maus. Requiring fuel of a minimum of 77 octane, the V-12 was fed with commercial petrol laced either with one-third aero fuel or with 0.09 per cent tetra-ethyl lead. Engine and drive-train functions were displayed on a special test instrument panel in the turret that was both filmed – with a frame every two seconds – and snapped by a fixed Leica when required. On 10 March the partly assembled second Maus, serial 205/02, arrived and was towed from its trailer by 205/01. The latter's towing power was so great that it moved the complete train, which had to be attached to a locomotive to hold it while Maus number two was pulled off.

By 1 April it was evident that the 205/01's 'steering ability on firm ground and slippery clay is good due to the separate drive for each track and the curved ends of the track links. Mechanical brakes are sufficient to halt the vehicle. It is necessary to make it easier to install the transmission, engine and generators.' Again the big tank's manoeuvrability was verified. The HWA remained unconvinced, however, even after Porsche sent them aerial photos of its convoluted trails.

When the proper turret for the Maus arrived on 3 May it was outfitted and installed on 205/02 by 8 June, ready for an inspection by Guderian who, however, was a no-show. Doubtless the Allied landings in France on 6 June required his attention elsewhere. In spite of the teething troubles inevitable in a new design both 'Mice' showed impressive reliability. A gearbox-bearing failure was seized upon by Porsche's critics as a sign of faulty design, but in fact it was a random fault that had not shown up on test rigs. Spring breakages were addressed with better volutes.

The Maus had a devastating effect on the stone-block paving of the Panzerkaserne *yards. This resulted from a combination of its sheer weight and the wiping effect during manoeuvres of its unusually wide tracks.*

In action the Maus was as fearsome as its specification suggested, said Richard von Frankenberg: 'When it drove down the street between houses the window panes burst and people thought the Last Judgement was at hand. The engine's bellowing and the clatter of the tracks on their 24 rollers was ear-shattering.' Böblingen's administration presented a bill for water pipes that were inadvertently broken. 'The high ground pressure', said a report, 'results in cobblestones being torn out in curves that are driven several times.' A more likely explanation was that the exceptional width of the tracks had a sweeping, twisting effect on the cobbles during turns.

All those who drove the Maus declared it 'child's play' to handle because everything 'went electrically'. Ferdinand Porsche asked test-driver Karl Gensberger about the Maus's maximum speed. It was 10km/h, he was told, about 6mph. Porsche ordered Gensberger out of the cockpit and took the controls himself. He made a series of top-speed runs that verified the prototype's achievement of its designed maximum speed of 20km/h or 12mph. 'My father drove the Maus himself a number of times,' said Ferry Porsche, 'and was absolutely satisfied with the design's performance.'

Encouraging as the Maus tests were, Ferdinand Porsche chanced his arm. During a meeting at the Berghof with Hitler on 4 March 1944 he updated the Führer on its progress and raised the idea of a resumption of production. This reached the ears of Krupp officials, who carried out a census of Maus components. Seven hulls had been built, they found, and enough material for eight more had already been worked. Material for several dozen more was on hand. In spite of the previous year's 27 October ukase, the Maus plating had not been recycled to make mobile attack guns after all.

'As related by Prof. Porsche,' read an internal HWA memo of 23 March, 'Hitler has ordered accelerated driving trials and to resume development of the Maus. In addition, Porsche has contacted Krupp for delivery of a second Maus I turret and the first Maus II turret. We request orders to clarify if the decision to complete only two chassis (one with a turret) has been rescinded.' A Mark II Maus had a revised turret with a more deflective frontal glacis and the 75mm gun mounted above its 128mm sister. It was part and parcel of Ferdinand Porsche's strategy to relaunch Maus manufacture.

Krupp made it clear that although in principle it could restart delivery of Maus hulls and turrets at mid-year of 1944, it could only do so with additional staff in the hundreds and displacement of some of its other current commitments. It also urged that preference in upgunning of the Mark II turret be given to the 150mm weapon agreed upon during the May 1943 Wolf's Lair review instead of a proposed 170mm cannon. All such speculation was ended by a 10 July letter from

Developed by Daimler-Benz from its aero V-12s, an MB 517 diesel engine was fitted to the second Maus, giving improved fuel economy. Both Type 205 prototypes were shipped to the Kummersdorf test grounds in October 1944.

Albert Speer 'with information that due to the current situation Hitler has ordered a halt to development of all armoured vehicles with heavy guns'. In his long and often frustrating battle with Porsche, Speer had finally prevailed.

Powered by the diesel version of the Daimler-Benz aero V-12, now called the MB 517, the second Maus showed its potential by consuming some one-third less fuel for a given mileage. At the end of October 1944, after their promising trials at Böblingen, both Maus prototypes were transported to the HWA's Kummersdorf proving grounds – graveyard of earlier Porsche tank designs. At Kummersdorf, however, 205/02 broke its crankshaft owing to misalignment with its generators. When a second diesel engine – the only available replacement – arrived in mid-March 1945 Porsche sent a team of mechanics to Kummersdorf to install it. Returning to Stuttgart on 3 April, they reported that they had successfully run the engine but the diesel Maus had not yet been driven.

First to Berlin, the Russians were also first to Kummersdorf. 'The defence was organised on the spot,' wrote Michael Thiriar, 'with the help of civilians, enlisted in the Volkssturm and armed with Panzerfausts. The Germans bravely took the offensive with a few tanks undergoing assessment including a Tiger, two American Shermans and an Italian P40.' The unarmed Maus 205/01, probably hampered by a breakdown, was immobilised by its operators near a Kummersdorf firing range. The 205/02 was sent to do battle in the Wünsdorf region where it too was disabled before it fell into Soviet hands. At Wünsdorf was Zossen, headquarters and communications centre of the Wehrmacht, only a few miles north of Kummersdorf towards Berlin.

The remains of the two Panzers were taken in hand by the Russians. Lacking heavy explosive material, German troops' efforts to destroy the massive machines were less than fully effective. Of the two Panzers the damage to 205/02 was the more severe. From the two prototypes the Soviets succeeded in making one Maus,

The crew of Maus 205/02 did a remarkably good job of disabling their charge near Wünsdorf at the war's end. Russian troops inspected its tail section (left) and turret. The latter topped a single Maus reconstructed by the Soviets.

carrying out the major tasks during March and April 1946 in situ at Kummersdorf, which was in their occupation zone. On one of the bespoke railcars the completed hybrid was transported to the armoured-vehicle proving grounds at Kubinka, some 35 miles from Moscow, where it arrived on 4 May 1946.

Uncertain is the extent to which the Russian technicians were able to bring their Maus to life. Although they analysed and reported on it in great detail in 1951–52, they did not vouchsafe performance findings. They were most critical of what they considered the excessive armouring of the Maus, especially the 100mm protection of its tracks, contributing significantly to its great weight. The final resting place of this sample Maus, marrying the 205/01 hull with the 205/02 turret, was the Museum of Armoured Forces at Kubinka, where it joined a surviving Ferdinand.

A typical *ex post facto* assessment of the Maus was that of 1969 by F.M. von Senger und Etterlin. 'It must be stressed that the Maus', he wrote in his review of German tanks, 'was a sorry example of misdirection of the German war effort at the highest level, which resulted in a waste of valuable personnel and material on projects whose tactical value was questionable. Despite this the Maus and E-100 prototypes represented remarkable achievements in design.'

Although judging it worthwhile for technology-transfer purposes to describe the Maus in immense detail, Britain's post-war analysts felt that 'much thought must have been given by the German Production Ministry as to whether it was worthwhile locking up tank production capacity to produce the vehicles, for every one of which it must certainly have been possible to build several Tigers or Panthers. It is thought that any commander in the field, even with his command in defence, would prefer the latter.' Here of course they echo the views expressed above by Guderian and others.

None of the critics of the Maus project begrudged the Porsche team the credit due for its mastery of a demanding brief. The Maus was able to demonstrate mobility and agility that confounded all its critics. Nevertheless opinions like those above have deeply coloured most subsequent appreciations of the Maus. For many observers it has gone down in history as a wasteful diversion of time, money, machinery and materials when the Reich was desperately short of all of these.

Largely overlooked has been the crucial role of the Maus as a *Geheimwaffe*, a secret weapon of the stripe that was close to the heart of Adolf Hitler. It ranked alongside the jet and rocket fighters, the electric submarines, the ground-to-air missiles and the V-1s and V-2s as a potentially devastating surprise for Germany's enemies. Moreover, much more than any of these it was a machine that Hitler felt that he understood, that he could tinker with and even help specify and design. And it was the work of his favourite engineer.

For better or worse – largely the latter for Europe – the Führer's knowledge of these weapons under development was his main sustaining force as the war progressed. When challenged by colleagues about the war's prospects he was quick to assure them that the *Geheimwaffen* being created by his brilliant scientists and engineers would soon turn the tide. This placated them while, tragically, underpinning Adolf Hitler's blind determination to continue the fighting to its bitter end.

CHAPTER 16

CHERRYSTONE IN PRODUCTION

'They were devilish weapons,' wrote Walter Cronkite about his days as a war correspondent in London. Starting in mid-June 1944, wrote the future icon of television news, 'hundreds of the so-called flying bombs were aimed at London from launching sites on the Channel coast. Air-raid sirens were screaming again, day and night. The people were told that they would probably have 15 seconds to seek shelter after the bombs/engines quit. The bombing was so random and frequent, however, that few people stayed for long in the shelters, despite the fact that the bombs exploded on impact and spread their death and destruction over wide areas.'

Just when Londoners were celebrating the end of the bombing Blitz and revelling in the news of an Allied landing in Normandy, this new menace darkened their days and ended many lives. Although the Allies had known since the end of 1943 that something of the sort was brewing, as a result attacking what seemed to be launching sites on the French coast, the actual arrival of the 'buzz bombs' or 'doodlebugs' was still a shock. This was an authentic secret weapon which the Germans expected to have a psychological as well as physical impact.

Officially the Fi 103, the new menace was code-named *Kirschkern* or 'Cherrystone'. Propaganda minister Josef Goebbels dubbed it the V-1, first of Germany's *Vergeltungswaffen* or Vengeance Weapons. Their launch was calculated to stiffen the backbone of Germany's military and population at a crucial stage of the war. The V-1 would soon be followed by the V-2, the ballistic missile developed by the Wehrmacht on the east coast of the Peenemünde peninsula on the Baltic.

On the west side of the same secret installation the Luftwaffe tested its V-1, which won the unofficial race to bomb Britain. Developed by the Georg Fieseler Werke under the direction of Robert Lusser, the V-1 was a pulse-jet-powered small mid-wing aeroplane that carried a 1-tonne explosive charge 150 to 200 miles, from its steam-catapult launch, to London from many sites on the European mainland and speedily enough, at some 430mph, to be hard to shoot down.

Ferry Porsche was derisive about the technology of the V-1, writing that it 'was original in concept and had some tactical value but in fact was made up of the most antiquated components – a pulse-jet engine operated by shutters that dated back to 1908 and an auto-pilot used during World War I. There was no other navigational device once the bomb was set on its course, which often proved very erratic. The trajectory of the V-1 was pre-set with a simple timing device which released the elevators and caused the bomb to dive steeply. Almost immediately gravitational force cut off the fuel supply, the pulse-jet stopped and that was the end of the run.'

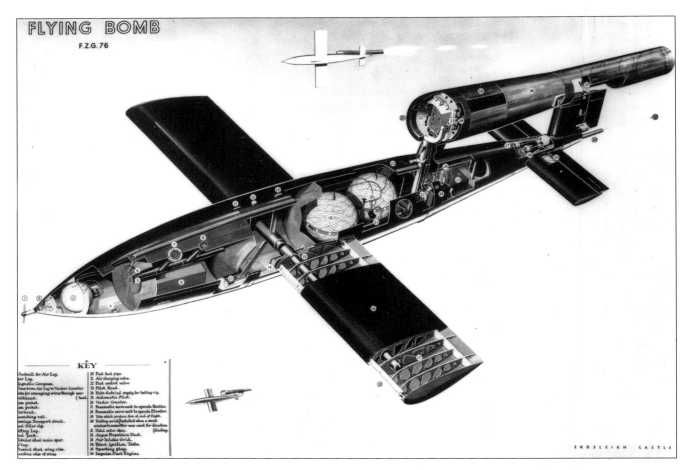

Working from the wreckage of surviving V-1s in Britain, aviation artist Peter Endsleigh Castle depicted the weapon's constituents for the War Office. To get to know its workings was to defeat it.

For Flight *Max Millar limned the Fi 103 or V-1. His view from the rear revealed the commendably simple mechanisms responsible for its direction-keeping, using gyroscopes, and altitude maintenance.*

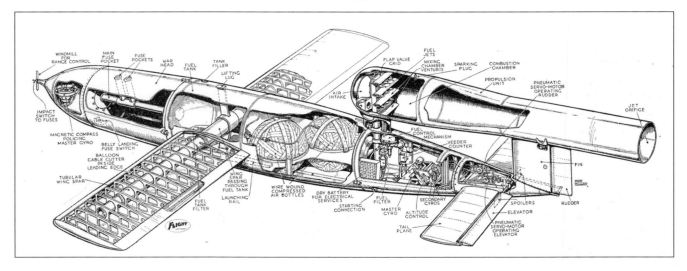

Giving 45 propulsive explosions every second, the practical pulse jet was the creation of Munich inventor Paul Schmidt, who had it running by the beginning of the 1930s. Its only drawback was thrust at rest that was too low to accelerate the V-1. This was solved by propelling it into flight from a ramp which boosted it to 200mph or air-dropping it from a bomber. Both methods forced more air through its entry shutters to bring it to maximum thrust, which built up to the V-1's maximum speed.

The V-1's fuel tank was at its centre, above the wings where consumption during the flight would have the least effect on its weight distribution. Between the tank and control mechanisms in the tail were two wire-wound spheres capable of holding air compressed at up to 225 atmospheres. Their air powered the control surfaces through servos and pressurised the fuel tank.

Its crudeness was part of the V-1's appeal to a resource-strapped Germany. Fabricated of mild steel, its wings, fuselage and warhead were easy to make. Its engine was happy to drink low-grade petrol, poorer than that suitable for cars. Cherrystone's developer Fieseler hoped to manufacture it, or at least much of it, but the sponsoring Luftwaffe had other ideas. It envisioned the mass production of these 'vengeance weapons' to stun Londoners with firing rates of 150 to 200 daily.

Although still behaving inconsistently, by mid-1943 the Fi 103 was promising enough for its production to be organised. This was a topic of a Berlin meeting on 17–18 June that included Hermann Goering and his deputy for production, Erhard Milch. Enthused by the weapon's potential, Goering asked for the 'ultimate goal of fifty thousand a month, if possible'. For the time being the target was an order of magnitude less, to reach 5,000 a month by May 1944.

Such ambitions demanded a big factory. One such, Germany's biggest in fact, was the KdF-Werke at Fallersleben. It was already tooled to manipulate mild sheet steel, the V-1's main structural material. Hoping to get valuable contracts for the production of aircraft fuselages, Ferdinand Porsche and his board colleagues had put their factory at the head of the queue for any work the Luftwaffe required. This had made them well known to Germany's airmen as the prime source of repair services for the Junkers Ju 88 and the making of Ju 88 parts, activity that would be the factory's dominant wartime task.

Goering and Milch easily identified the KdF-Werke as the ideal site to build flying bombs. This was welcomed by the plant's managers, whose ultimate goal was to bring the high-value production of complete aircraft to Fallersleben. This would be a small but promising start.

Getting the V-1 into production was another matter altogether. Porsche and Fieseler engineers went back and forth over problems and their solutions; spot-welding that worked for cars could not keep a V-1's fuselage intact under the stress of launch and flight. Yet the Nazis craved urgent mass production of this underdeveloped forerunner of the cruise missile. Its accuracy was so poor, thought one Luftwaffe officer, that 30,000 per month would be needed to cow the British into abandoning the war.

The KdF Works became the prime contractor for the V-1's production. It made the tail assemblies in its ancillary *Vorwerk* at Braunschweig, the hulls and engine nacelles on two parallel assembly lines in its main plant. It outsourced the wings, pulse-jet engines and guidance systems. Argus made the engine, its Type 109-014,

Production lines at Fallersleben both fabricated parts for the flying bombs and carried out their final assembly except for the wings, which were fitted near the point of launch for ease of transport.

producing 31,100 during the war. Much of the V-1's final assembly took place near each missile's firing point.

During 1943 the struggle continued to reach production-readiness for this radical new weapon, code-named '1114' by the KdF Works. On 19 August 1943 Ferdinand Porsche received the news that British bombers had attacked the development centre at Peenemünde. Damage to the Luftwaffe's west side of the peninsula was relatively light, however. By September a workforce of 1,453 was dedicated to V-1 production at Fallersleben, composed almost entirely of Germans for security reasons. In March 1944 series manufacture began. Just after the Allied landing at Normandy, V-1s began to be launched against England.*

A shift in oversight of the KdF-Werke took place in October 1943. Until then, wrote Simon Reich, 'the management board consisted of many who had played a major role in the company's pre-war development, including Lafferentz, Werlin, Schmidt, Porsche and Piëch. However, the authority of the board of directors expired on 13 October 1943, and any vestiges of corporate autonomy disappeared as the *Aufsichtsrat* (supervisory board) was succeeded by the head of the commercial division of the DAF,' Robert Ley's powerful national labour union. On-site general management remained the responsibility of Anton Piëch, Porsche's son-in-law.

The V-1 contract brought substantial business volume to Fallersleben. Of some 31,000 V-1s made during the war, Volkswagen was responsible for 19,500, the lion's share.† It also brought something much less desirable: the attention of the Allied forces' target selectors. A failed attempt in late 1943 to involve the Peugeot works in Montbéliard, France, in the production of components for the 1114 leaked some of its secrets to the Allied side, including the key role of Fallersleben in its production. This disclosure contributed to the previously obscure factory's designation as an important bombing target by the Allies. Earlier raids had been desultory. In mid-1940 one hall was hit by a few small bombs that caused little

---

* It is not our mission here to describe the sorry story of the lives lost to, and damage caused by, this deplorable yet at the same time undeniably innovative weapon in Britain and on the continent. For further reading on this subject numerous books are available. One was written by Bob Ogley, who said that 'The flying bomb was a unique, brilliantly conceived, indiscriminate and short-lived weapon that was launched by the Germans in a last-ditch orgy of terror, designed to turn the tide of war.'

† Another source refers to the production of 23,748 core fuselages under Fallersleben's direction in the year 1944.

A fragment of launching ramp showed the circular passage through which the piston travelled to launch the weapon. The V-1's steel hull was well suited to fabrication by its prime contractor, Porsche's Volkswagen Works.

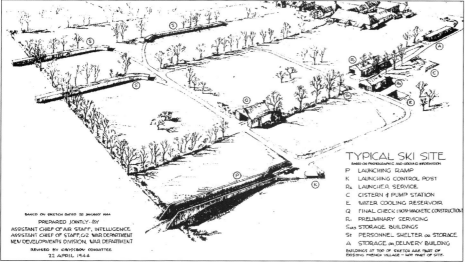

Called a 'ski site' for the curving ends of its storage buildings, so shaped to prevent shrapnel damage to the waiting V-1s, this feature fortunately made their launching sites more visible to photo reconnaissance and bombing.

damage and another raid that November did more damage to the nearby forest. Bombs that fell near the foundry in the summer of 1943 caused no damage.

In 1944 the scale and accuracy of attack escalated. Attended by the US Eighth Air Force on 8 April 1944, KdF-Stadt and its factory were visited by 56 aircraft which dropped 500 high-explosive and 450 incendiary bombs from 20,000ft. They damaged the roofs of three halls and hit the railway lines and a transformer. A supply of timber went up in flames. Casualties included 13 killed and 40 wounded.

Missed by this raid, the tool and die shop making Teller mines took a bizarre hit on 29 April when a disabled British Lancaster bomber crashed through its roof complete with its bomb load. The hall was comprehensively defenestrated by the resulting explosions and its floor penetrated in two places. The plane's RAF crew, who had baled out, were rounded up by members of the plant's anti-aircraft unit.

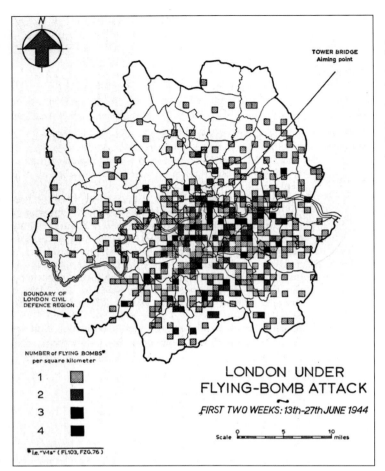

Although strikes on London in the V-1's first two weeks diverged widely from the aiming point at Tower Bridge, they nevertheless struck hard and randomly at Londoners who had thought air attacks a thing of the past.

More air attacks came on 20 and 29 June and 5 August, delivering in all 675 tons of explosives. The first of these was the largest, with 137 B-17s whose loading included 130 1-ton high-explosive bombs arriving over Fallersleben in four waves with their fighter escort. The second June raid, richly equipped with incendiaries, caused a temporary power cut by disabling one generator turbine. The June raids killed 35 and wounded 107. Nevertheless the V-1 attack on Britain was already well launched, the 2,000th firing taking place on the day of the second June raid.

In August a final attack was flown by 85 American B-24s. This was scheduled, wrote intelligence historian Harry Hinsley, after 'one plant, Fallersleben, had been firmly associated with the assembly of flying bomb fuselages. [Air Intelligence] recommended that this alone should be accorded the highest priority and that seventeen other plants believed to be associated with other stages of V-1 production should be treated as low priority targets.'

The August 1944 assault was precautionary because, as Hinsley wrote, after the June raids the Allies knew that 'the damage inflicted led the Germans to transfer [Fallersleben's] flying bomb assembly plant to safer quarters'. Production of powered fuselages for V-1s continued at the works in June and July with Rudolf Brörmann overseeing their quality, while the first steps were taken to move V-1 production to safety. This had the potential not only to protect the weapon's manufacture but also to reduce the KdF-Werke's attraction to Allied bombers. The vast plant was, after all, to be the post-war home of Germany's people's car, for which more than 300,000 buyers were already putting stamps in their savings books.

The air raids had already prompted Porsche, Anton Piëch and Bodo Lafferentz to launch a search for underground sites that could be not only safe havens for their production but also future space for expansion from Fallersleben. Looking best suited for this purpose early in 1944 was an iron-ore mine at Tiercelet across the Rhine to the west of Stuttgart and Strasbourg. Blandly code-named *Erz* or 'Ore', the tunnels there promised ample underground working space even though mining was continuing.

On the morning of 4 March 1944 Ferdinand Porsche and his son spoke with Heinrich Himmler of the SS about their need for a labour force to prepare the tunnels and later to help produce V-1s. Promising support, the bespectacled Himmler presented the Professor with the SS ring, bearing the death's-head emblem, as a talisman of their past and – Himmler hoped – future co-operation. Himmler's deputy for administration, Oswald Pohl, forwarded the request for labour.

At a meeting of the Luftwaffe Fighter Staff on 17 March Porsche and Lafferentz confirmed that they had the means to proceed at Tiercelet. Three days earlier, they said, the SS had confirmed the availability of 3,500 prisoners of war to assist the several thousand of the Todt Organisation already at work. Thus the workforce problem there was solved for the KdF-Werke, Porsche told the air officers, which would 'staff all the tunnels with concentration-camp prisoners'. Tiercelet, however, lost its lustre when it was seen to be too close to the invading Allies.

The administrator in charge of factory sequestering, Paul Figge, identified the *Mittelwerk* or 'Central Works' near Nordhausen, a mammoth underground factory bored deep beneath the rock of the Kohnstein mountain range, as suitable for Porsche's V-1 production. The Nordhausen site was under the complete control of Heinrich Himmler's SS, which dreamed of exploiting the war to build for itself a peacetime industrial empire. SS General Hans Kammler, a civil engineer who had no compunctions about using captive labour under the worst conditions imaginable, oversaw operations.

Figge's initiative was resisted by both Porsche and Anton Piëch, who hoped to keep control of V-1 production by shifting some of it to other underground sites directly affiliated with the KdF-Werke. In this aim they clashed with the highly placed Paul Figge. The latter solved this problem by lodging a complaint against Piëch that led to his dismissal from his post at Fallersleben. At news of this the DAF chief Robert Ley – who in effect had paid for the plant – was extremely angry. Ley was instrumental in Piëch's restoration to the Volkswagen factory's leadership.

Continuing their search for their own subterranean sites, Porsche and Piëch checked the alternatives. Porsche paid a visit on 27 June to abandoned railway

For Ferdinand Porsche an involvement with the V-1 proved to be a pact with the devil. In no other aspect of his wartime activities, both at Fallersleben and later at Nordhausen, did he come to rely so much on slave labour.

After the devastation wrought on the VW Works by bombing the damage was left unrepaired to imply that no more attacks were required. Production of the first vengeance weapon was moved south for safety to Nordhausen.

Their carving completed by slave labour, the Nordhausen tunnels were major production sites. Junkers occupied twenty lateral tunnels at the north and the V-2 rocket twenty more. Only three at the south were enough for the V-1.

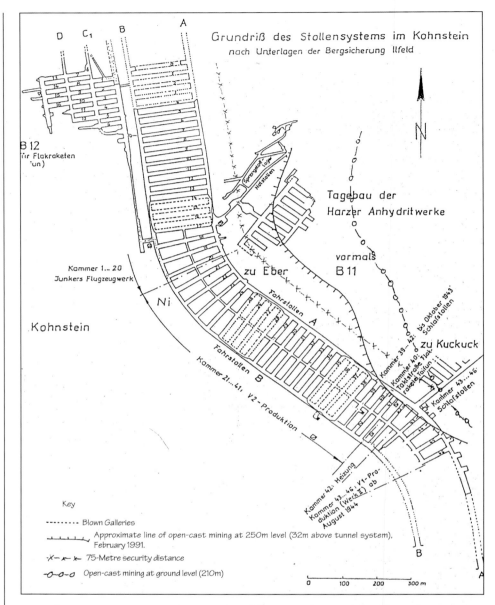

tunnels near Dernau, northwest of Koblenz. Code-named *Rebstock* or 'Vine', these looked ready to give up mushroom production in favour of Cherrystones. In fact Fallersleben sent 22 rail carriages of machinery there in mid-July.

With Dernau too looking threatened by a westerly invasion, attention turned to an asphalt mine at Eschershausen, south of Hanover and only 60 miles from Fallersleben. Assessing the site personally on 13 August, Ferdinand Porsche was pleased to find high underground spaces that could accommodate all but his tallest presses. Code-named *Hecht* or 'Pike', the Eschershausen tunnels were equipped with machine tools confiscated from French factories. Staffing was again by concentration-camp inmates, some transferred from Tiercelet, living under the most desperate conditions.

To bring VW's share of the Eschershausen mine up to a production standard, ten stamping presses from Fallersleben were arriving and delivery was accepted of

144 of the 329 machines en route from Peugeot, which was under VW management. Mommsen and Grieger wrote that somewhere to the south a 22-car train carrying 45 machine tools from Montbéliard was thought to be on its way but in Germany's internal chaos its location could not be verified. Stamping presses from Citroën's Paris factory, needed to make V-1 and V-2 parts, got no further than the border with Luxembourg by mid-1944 and were soon repatriated.

Useful though some of these initiatives were, Porsche and Piëch could not withstand indefinitely the pressure from on high to sequester their Cherrystone production at Nordhausen's Central Works. As they feared, the shift meant a reduction in the authority of the Porsche/Lafferentz/Piëch troika and their on-site engineer, Josef Aengeneydt, in favour of the SS and Albert Speer's designated czar for V-weapon production, porcine but demanding efficiency expert Gerhard Degenkolb. When one deals with the Devil, one must accept the consequences.

After a tussle with the SS, Fallersleben was obliged to relinquish its role as prime contractor for the V-1. Henceforth it acted as a subcontractor to the Central Works, now the nexus of the supply network of 16 widely dispersed subcontractors that the Porsche people had established. One was the Volkswagen plant, which in spite of its damaged condition could still account for 35.4 per cent of V-1 fuselage production in 1945.

Under SS supervision a workforce of captives needed fully six weeks to shift VW's key V-1 production operations to Nordhausen, 60 miles south of Fallersleben. In August 1944 the tooling for the main components was successfully transported. At the Central Works it was installed in a set of tunnels near the South Portal. The V-1 found itself being produced near its rival, the Army's rocket-

Swathed in camouflage netting, the southern entrance of the Nordhausen tunnel complex was near the V-1's assembly area. Stored outside were components including its wire-wrapped compressed-air tanks.

Much as it had at Fallersleben, assembly of V-1s continued in the tunnels of the benignly named Central Works. Although reluctantly, Fallersleben had to accept the role of a subcontractor.

powered V-2, Mercedes-Benz aero engines and other critical war materiel. By the beginning of November 1944 the Volkswagen team had relaunched V-1 production.

Compared to the elaborate V-2, the pulse-jet-powered V-1 was a weapon of moderate cost: its selling price to the Luftwaffe was RM5,000 the copy. Remarkably this was only 19 per cent more than the price of a Schwimmwagen. The V-1 was valuable trade for the Volkswagen Works. Its total business turnover in 1944 was RM290 million, of which the V-1 accounted for fully RM100 million. Conditions worsened in 1945, when subcontractor Fallersleben began complaining to the Central Works about its slow and incomplete rendering of payments due.

With all his products petroleum-powered, including the V-1, Ferdinand Porsche joined a group meeting with Adolf Hitler in the summer of 1944 to discuss the fuel shortage resulting from the Allied bombing of production centres. Others at the conference were Albert Speer, Field Marshal Keitel, Chief of Staff Zeitzler and arms maker Hermann Röchling. Hitler's only response to the discouraging fuel-supply figures he was given was to gloat over the coming terror attacks on London with the V-1 and V-2, saying that 'The British will suffer. They'll find out what retaliation is! Terror will be smashed by terror.' Those in attendance, including Porsche, could only take this irrelevant outburst as confirmation of the validity of the V-1's production.

Fieseler's Fe 103 became the basis of another secret weapon, a manned attack aeroplane with an integral warhead. In a project spearheaded by famed aviator Hanna Reitsch, this was none other than the V-1 with its air reservoirs resited to make room for a cockpit just below the pulse jet's inlet. Always considered a stable airframe, the Fe 103R or *Reichenberg*,* as the craft was known, proved its qualities when flown as a glider in pilot trials. Dual-control versions, both passive

---

* It could surely have been no more than coincidental that the Fi 103R was code-named after the town only 3 miles from Porsche's birthplace in which the young Ferdinand first studied engineering. Or was it?

CHERRYSTONE IN PRODUCTION 241

After the Allies rolled up the French sites that had first launched the V-1s, the weapons were air-dropped by Heinkel He 111 bombers. The last flying bombs fell on Britain during 29 March 1945.

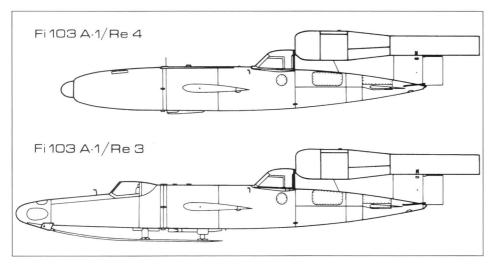

After the war the Allies discovered evidence of the Fi 103R or Reichenberg, a piloted version of the V-1. Although successfully test-flown, this suicide weapon was abandoned as unsuited to Germany's military traditions.

In addition to the standard Reichenberg, several were made for pilot tuition with a second instructor's cockpit replacing the payload in the nose. Its cushioned skid provided a reasonably smooth landing.

and powered, were built for instruction with another cockpit in the warhead's place.

Reitsch's efforts to develop such a semi-suicide craft languished until Adolf Hitler asked Otto Skorzeny to take an interest. Vienna-born Skorzeny, famous for his successful abduction of Benito Mussolini from captivity in Italy in September 1943,* had been investigating manned torpedoes for use against Allied shipping but changed course to take on the Reichenberg project. He brought fresh support to the initiative of Hanna Reitsch, who had convinced Hitler that such weapons were a good idea.

Ernst Heinkel offered Skorzeny a workshop to build trial Reichenbergs but Skorzeny found his project blocked by a lack of flying bombs to use as prototypes. None was available, an Admiral advised him. 'That's not what Professor Porsche, a friend of mine, told me,' Skorzeny related. 'There are several hundred V-1s in his Volkswagen factories waiting to be picked up. I can assure you that he'd gladly let me have a dozen.'

Ultimately Skorzeny received more than 14 dozen, for some 175 Reichenbergs were assembled at sites at Neu Tramm in Lower Saxony and Pulverhof near Hamburg. Recruited to pilot them were members of the specially formed Leonidas Squadron. Although its fliers had signed a declaration that death was a likely outcome of their mission, their aircraft's operating concept was that exit by parachute would be possible before impact. At least one test pilot succeeded in deplaning by parachute in spite of the cramped cockpit and nearby engine inlet.

Hanna Reitsch played a key role in flight tests that validated the Reichenberg's potential. Although the invading Allies in France presented attractive targets, Albert Speer recommended to Hitler in July 1944 that a better option would be Russia's power stations. The Leonidas flyers were just getting the hang of the Fi 103R when a new commandant in October 1944 had other ideas. When he and Speer met with Hitler on 15 March 1945 they won the dictator's assent to the proposition that suicide missions – although acceptable to Axis partner Japan – did not befit Germany's military traditions. Leonidas and its Reichenbergs were stood down.

When the Allies finally broke through from their beachheads in July 1944 they swept past the launching sites that suited the V-1. Although the brigades firing them moved to new locations while airborne launches began, the limitations of the weapon's range became problematic. While Argus worked on a more powerful version of its pulse jet, orders went out to jet-engine builder BMW and to a new player, Porsche, to create a suitable turbojet engine for the V-1.

For this prestigious assignment the Porsche men reached well ahead of their current project numbers to award the magic Type 300. Success in its achievement was important to them in order to keep their money-spinning V-1 viable against the growing threat to it from a steadily improving V-2 and its long-range derivatives being promoted by the Wernher von Braun design team.

Requirements for the little jet were to extend the V-1's range to 310 miles and to propel the flying bomb at a 500mph clip. Higher speed was needed because the Allied fighters, especially the Hawker Typhoon and soon the jet-engined Meteor, were getting fast enough to keep pace with it in level flight. Efforts were also to be made to improve the V-1's aim, because an extrapolation of the existing mean divergence from target would only mean even poorer results at longer range.

---

* Recent research has cast doubt on the importance of the role played by Skorzeny in Mussolini's release. Detained at a hotel in the Dolomites, the Fascist leader was not that well guarded. The tall and good-looking Skorzeny, it seems, was more a highly prominent onlooker at Il Duce's release than the prime mover of the rescue mission.

The jet-engine design itself 'presented very little difficulty,' Ferry Porsche recalled, 'but there was a condition attached which turned the whole project into a nightmare. Our design had to be of such low cost that the loss or destruction of the bomb after only one flight would not add much to the war budget.' In other words this was to be a disposable turbojet, the first of its kind, to be designed when the science of jet engines was still in its infancy. 'If it's come to this,' Ernst Piëch recalled his grandfather saying, 'things must really be getting a bit desperate.'

Fortuitously the existing German jet technology pointed the way for the Type 300, known to the Luftwaffe as the 109-005. Germany's shortage of heat-resisting alloys had already forced her engineers to find ways to use simpler steels for critical parts like nozzles and turbines. Hollow turbine blades were internally air-cooled to cope with temperatures up to 700 degrees C.

While Josef Mickl and Paul Netzker were Porsche's key men on the project, they had help from experienced gas-turbine engineer Max Adolf Müller. Positing a nine-stage axial compressor, their October 1944 studies forecast a thrust of 880lb at a speed of 12,500rpm. This was a one-third advance on the Argus pulse jet with fuel consumption forecast to be lower. Making studies of the trade-off between range and speed, a report at that time indicated that 'it seems not inconceivable that speeds approaching 500mph at a flight distance of just 375 miles will be reached. Definitive values can only be communicated after the presentation and evaluation of this powerplant project.'

Even with the support of greater resources, forecast the October 1944 report, realisation of such an engine would take at least eight months. The chaotic conditions of the weeks ahead, not to mention the other demands on Ferdinand Porsche's team, meant that no Type 300 jet engine was completed before VE Day. Some components were made and laboratory tests conducted before the invading Americans captured the drawings and pieces. BMW's efforts were even more in arrears.

Some 20 miles north of the Porsche demesne at Zell am See another engineer impatiently waited for the Type 300. This was Milan-born Mario Zippermayr, a remarkable example of the many entrepreneurs who succeeded in attracting the interest of the Wehrmacht, Luftwaffe and SS with their ideas for sensational new secret weapons. Of Austrian parentage, Zippermayr studied engineering in Germany before establishing development centres in Vienna and the Alpine village of Lofer, just over the border from Germany's Berchtesgaden.

In 1943 Mario Zippermayr seized the attention of Nazi authorities with his concept of an aerial torpedo whose fuel was a cargo of coal dust accompanied by liquid oxygen, the whole to be ignited during its dispersal. His plan envisioned aircraft firing these at ground targets at 1-mile distances and speeds of 400-plus mph. Tests of the bomb elements at the Starnberger See were promising enough that such weapons were beginning production at Nordhausen.

For the design of the torpedo itself Zippermayr conceived a sleek shape with wings that swept back along the sides of the hull, looking more like fins than conventional wings. In spite of the challenge of getting such a craft into the air, these wings promised efficiency at high speed that led to the idea of an interceptor of similar design.

The result was a new type of aircraft Zippermayr called the *Pfeil* or 'Arrow',

Verging on the fantastical was the Pfeil *interceptor concept of Mario Zippermayr, working in Austria near the German border. For its propulsion he was depending on Porsche's Type 300 turbojet engine.*

resembling a paper aeroplane with its short wingspan and nose-to-tail chord. With a cadre of evacuated Porsche engineers working near Lofer at Zell, it was natural enough that the Type 300 turbojet became the Pfeil's designated powerplant. But as with so many other undertakings, Zippermayr's concept ran into the sand, his chimerical project getting no further than partial construction of a full-scale trial glider.

Damage to the KdF-Werke contributed to the end of some of its car-production activities in August 1944. The 17th saw the last of 630 civilian KdF-Wagens produced. On the 26th the final Type 166 Schwimmwagen was built. The Kübelwagen was still being made; at the end of the month Ferdinand Porsche attended tests of a version of it equipped for artillery towing. Another variant, the Type 198, was equipped to drive starting shafts for Tiger and Panther tanks.

Production lost to Nordhausen was made up by Fallersleben in the last months of the war by adding the fabrication of subassemblies for Focke-Wulf's Ta 154 *Moskito* twin-engined night fighter. Plans to make parts for other aircraft, including the new Messerschmitt Me 262 twin-jet fighter-bomber, expired with the internal collapse of Germany in 1945.

Another opportunity for the KdF-Werke was a new device that came into its own at the mid-point of the war. Introduced in July 1943, the hand-held *Panzerfaust* or 'Panzer fist' anti-tank weapon proved sensationally effective. A creation of Leipzig's Hugo Schneider AG, it was recoilless by virtue of an open-ended tube with a black-powder charge that propelled a projectile at moderate speed over a 70-yard range. With a wooden tail from which fins sprouted when it left the tube, the projectile carried a shaped charge that was able to penetrate up to 200mm of armour.

Responding to the successes achieved by this one-man tank killer, on 18 August 1944 Adolf Hitler launched *Sonderaktion Panzerfaust* with a new top priority to have one million made in the next three months. Quick to learn of the initiative was Fallersleben, for which the Breslau branch of Hugo Schneider was the supplier of the KdF-Wagen's headlamp assembly. Soon the Panzerfaust was being made at Fallersleben, which gave it the '1138' designation.

Cheap to make and easy to distribute and carry, the 13lb Panzerfaust was wielded not only by dedicated teams of tank hunters but also by ordinary soldiers and members of the *Volkssturm* home guard, greatly bolstering both their menace and their confidence. Thrown onto the defensive by these weapons on both fronts, the Allies struggled to develop protection from the Panzerfaust for their armour.

Although not a 'secret' weapon, the Panzerfaust was an effective one, so much so that captured caches of the devices were exploited by the American 82nd Airborne Division in its campaigns in Sicily and France. In all Germany produced some eight million by the war's end. Of these fully three and a half million were made by Ferdinand Porsche's KdF-Werke in the 1944/45 winter. They hugely enhanced Hitler's ability to resist his attackers during the war's ultimate challenges.

CHAPTER 17

DISPERSAL UNDER PRESSURE

By early 1940 the pressing demands on Germany's infrastructure applied the brakes to completion of the Volkswagen plant and its adjoining town. Far from becoming a model city for the New Germany, the KdF-Stadt remained more a 'wild-west' village. In December 1941, when all construction there was halted, 2,358 dwellings had been erected, only 10 per cent of the intended number.

Matters were little better at the factory. Equipping of production sections in the main plant, such as the press shop, machining hall and toolroom, was approaching completion, although there were critical gaps: some of the special machine tools had been delayed in their delivery from America, their shipment blockaded by the state of war. These, including half the Gleason gear-cutting machines that VW had ordered, would never arrive. A still-neutral America could do no more to help the Allies.

Air assaults continued to make 1944 a turbulent year for the VW works and its machinery. The plant's sub-storey, designed to carry plant services, became a production site. Its 13ft height was adequate for machine tools and engine and vehicle assembly lines sheltering under the factory's heavy reinforced-concrete floor.* What could not be moved was protected. Heaps of sandbags surrounded the three vital turbines in the powerhouse. Heavy mortar-laid masonry walls rose from floor to ceiling as protective baffles around the big presses that had to remain upstairs. The plant's bombing affected Porsche's Panzer development, because his test rigs for suspensions and drive trains were sited at Fallersleben.

Under the bombsights of American Fortresses and British Lancasters, in 1944 trucks and trains struggled to keep rolling on the steadily shrinking territory of Greater Germany to help the VW Works meet the war-production commitments that kept it viable as a business. Some brought inbound machinery, for since mid-1943, as already mentioned, major factories of France's vehicle maker Peugeot were operating within the orbit of Ferdinand Porsche and Fallersleben.

After the German armies swept into France and occupied the northern part of that nation they commanded its major industrial areas. Industrialists in Germany soon weighed the ways they might exploit those assets. For Ferdinand Porsche a particular target was the Peugeot factory at Sochaux, a suburb of Montbéliard, in occupied France close to the Swiss border.

Although both Daimler-Benz and Auto Union viewed the plant and waved off, Porsche was attracted by Peugeot's 15,000-strong workforce and modern light-metal foundry, the latter filling a gap in Fallersleben's facilities. The planned non-ferrous foundry at Fallersleben had not been completed; a reliable source of aluminium castings was urgently needed. On 10 and 11 February 1943 Porsche

---

* Bizarrely this use of the sub-storey for production led some post-war Allied investigators to conclude that the factory's floor area was double what its designers intended.

A bereft Beetle lurked in the wreckage of the Volkswagen Works at Fallersleben after the Allied attacks. Its destruction led to the wide dispersal of production of many components, as much as possible to underground sites.

and members of his KdF-Werke team met in Paris with occupation officials to discuss Sochaux's casting capabilities in particular.

Having gathered the information he needed, 23 February found Porsche presenting a plan to Albert Speer that included the request to be granted the 'management of Peugeot'. Later this would be characterised as 'technical control' of the Sochaux plant's activities. A last-minute battle for control of the plant with Dessau's Junkers was resolved in favour of Fallersleben.

Getting the green light, the Professor and his colleagues met at Sochaux on 27 March with Peugeot's president since the end of 1941, Jean-Pierre Peugeot. Born in 1896, Jean-Pierre had joined the family firm in 1922. After preliminaries a one-hour main meeting took place at which the industrialists reached an understanding. Ferdinand Porsche invited Jean-Pierre Peugeot to visit Fallersleben at the end of April for further talks at what was now the mother factory.

In June 1943 the Peugeot factories officially came under the authority of the KdF-Werke. President Peugeot did visit Fallersleben,

Assigned oversight of the Peugeot works at Sochaux in occupied France, Porsche personnel dealt with Jean-Pierre Peugeot. He made clear his intention to have a good relationship with Porsche both during and after the war.

where he emphasised that he desired the best possible relations with the Porsches and the VW factory because he hoped to benefit from steady orders for parts from his plants from Volkswagen after the war. Further visits by Peugeot officials to the KdF-Werke took place in August and Porsche himself took part in a Paris meeting on 17 September. Wartime contacts were also made by Porsche and Anton Piëch with Renault in Paris, although their eventual involvement was less with that company than with Peugeot.*

While making the most earnest declarations of co-operation, the French deployed a wide range of delaying tactics to avoid producing the crankshafts, connecting rods, flywheels, transmission housings and sheet-metal parts that Fallersleben ordered. In fact they hoped in this way to be excused the attentions of Allied bombers so that their factories could remain intact. In this they were disappointed, for the Sochaux works were hit heavily on the night of 16 July. Further attacks on the plant early in 1944 meant that by April its work for the KdF-Werke virtually ended, while from June onwards the big factory was silent.

This brought a new phase in the Porsche/Peugeot relationship. If the machine tools could not operate there, with the Allies thrusting into France, they could be brought to Germany. While the Allied land attack hesitated in France, in a fateful period of ten weeks between early September and mid-November of 1944 all but one-tenth of the industrial potential of the Peugeot plant at Montbéliard was 'brought to safety'.

Shipped to Germany were 1,545 machine tools, 2,155 electric motors, 282 welders and 586 other items of factory equipment such as furnaces, hoists, compressors and measuring gauges. In addition the trains rolled north-east from Montbéliard with 5,500 tons of sheet steel, 3,600 tons of machining steel, 720 tons of cast iron, 117 tons of non-ferrous metals and 700 tons of tools and dies – materials Peugeot had been hoarding with the hope of soon resuming car production.

Ferdinand Porsche, whose 1943 itinerary found him in Prague with an experimental KdF-Wagen, also travelled to Paris to meet with Peugeot officials. They would be less than assiduous in delivering promised parts to Volkswagen.

---

* Later in 1943 the relationship that Porsche had forged with Peugeot was formalised by Albert Speer, who saw no sense in deporting French labourers and engineers to Germany if they could work even more productively in their home plants. In September 1943, wrote Joachim Fest, 'After several preliminary talks in Paris, Speer invited the young French production minister Jean Bichelonne to Berlin. They soon agreed on the mistakes and continuing resentments of the older generation, marked as it was by the First World War, and finally arranged to establish some so-called "protected enterprises" in France. These were to produce locally for the German economy and in return they were to be guaranteed exemption from all deportation measures. Within a few months more than 10,000 French factories were already delivering products for armaments and civilian needs to Germany.'

Some of Peugeot's machine tools went to Daimler-Benz. The bulk of them were delivered to Fallersleben's dispersal site in the asphalt mine at Eschershausen; none of significance was delivered to Fallersleben.

The United States Strategic Bombing Survey determined that the VW works itself had 2,476 machine tools at the time the US Air Force commenced its 1944 raids. This was a 38 per cent increase in the machine park from the 1,800 installed in 1939–40 – credible in light of the many machines added during the war for armament production. The Americans judged that their bombing had wrecked 10 per cent and damaged 20 per cent of these machines.

Hundreds of the VW factory's machine tools were dispersed to cottage-industry sites that were close enough to Fallersleben for their parts to be delivered to its production lines. Wrote Art Railton, '[Joe] Werner and 5,000 workers (many of them POWs) removed about 600 machines, including the precious Bullards from America. Taking over scores of barns and sheds, they used farm tractors to drag the machines on sheet-metal sleds to nearby villages where they were set up and kept in production.'

By 1945 components for the Kübelwagen were being made at Ahmstorff, Almke, Brackstedt, Gifhorn, Hehlingen, Lüneburg, Neindorf, Soltau and Tülau. Only Soltau and Lüneburg were further than 20km from the plant. Some 400 machines were evacuated to the latter and at the former 266 machine tools were making VW oil pans, cylinder heads, crankshafts and gearbox casings.

Through 1943 and the early months of 1944 Allied bombers roamed ever more freely above Germany's industrial centres. The spring of 1943 marked a sharp intensification of the bombing. The 11th attack on Stuttgart came on 6 September 1943, when the US Army Air Force made its first daytime bombing run over the industrial centre with its Daimler-Benz aircraft-engine factory. Nevertheless the home base of the Porsche team remained at Stuttgart-Zuffenhausen, even though many other firms with important wartime missions had already been dispersed, either underground or to strategically less important regions.

In April 1944, when the first bombs fell on the KdF plant at Fallersleben, Ferdinand Porsche became highly agitated. The company's archives, he complained to his son, were stored in the attics of the Stuttgart office buildings, where they were vulnerable to air attack. He insisted that they should be moved at once to the cellar where they would be better protected. The precaution had already been taken of triplicating the original drawings. One of the additional sets was stored in Porsche's Stuttgart villa and the other was at the residence of Ghislaine Kaes, Porsche's nephew and personal secretary, in the same city. All were packed in special locked containers of sheet steel. Dutifully on 13 March Ferry shifted the Zuffenhausen archives to the building's cellar.

As fate would have it, just eight days after the vital drawings had been moved to the cellar an Allied bomber released a single bomb that plummeted at such an angle that it bored straight into the cellar, missing everything else but incinerating the drawings. It was becoming evident that at least some of the Porsche team would have to move.

Richard von Frankenberg told the story of an industrialist who, on leaving Stuttgart after a heavy air raid in the autumn of 1944, passed on the road a Volkswagen parked high on the city's periphery where its houses and industries were spread out below as in an amphitheatre. Hundreds of fires were visible below pillars of smoke wavering in the wind. Near the Volkswagen the industrialist recognised Porsche, witnessing the devastation of his adopted city with tears in his eyes. 'Must it really be like this?' the engineer murmured, half to himself. 'Tell me, is this necessary? Is this war?'

Since 1943 Albert Speer had been urging the Porsche cadre to desert Stuttgart, a magnet for heavy bombers, for safer regions. 'Slippery as an eel,' said a colleague, Ferdinand Porsche resisted such requests, saying that he could hardly work remotely from his reliable suppliers. When the pressure increased, however, exploration of alternatives began.

Ferry Porsche took on the task of reconnoitring possible sites. Applying first to the authorities in Stuttgart, he was offered a property in Czechoslovakia. 'When I heard that,' said Ferry, 'I did my best to prevent it.' Though the senior Porsche was born in what was now Czech territory and both men had been Czech citizens after the First World War, the idea was deeply unattractive. They did not relish being abandoned there after the war among a populace who were less than thrilled about engineers who had helped the German war effort.

Ferry decided to bypass the Reich bureaucracy by inquiring in Salzburg about possible Austrian locations. He found one, a glider-equipped flying school at Zell am See. Superficially, at least, this was fortuitous. Ferdinand Porsche invested much of his substantial earnings in property. In 1939 he had bought a spacious property at Zell that had once been a copper-mining area.

'War is coming,' Porsche told his grandson Ernst Piëch. 'Money, forget it. We need property. We need a farm. This is a good investment, because land will always be there.' 'Father and mother knew from the last world war that the best thing is to have a farm,' Piëch added. The Zell property was a big advance on their acreage at Wiener Neustadt.

At Austria's idyllic Zell am See the Porsche/Piëch family property served for tractor testing, farming for sustenance and a bolt-hole at the end of the war. Storage of documents and equipment was an additional benefit.

Another justification supported the purchase. 'The idea was that we had an offer to construct a tractor,' Ernst Piëch recalled. 'We needed the area for testing.' He was referring to the pet project of Reich satrap Robert Ley, the *Volkspflug* or 'people's plough' mentioned earlier. The land at Zell would be ideal for trials of prototypes of the Type 110. From its residence the property extended 741 acres up the hillside and 124 acres down the valley.

On the Zell acres, known to the family as the *Schüttgut*, Porsche completely rebuilt the 150-year-old three-story chalet, installing seven bathrooms. It could house two dozen people in its 20 rooms. With conditions in Vienna worsening during the war, in March 1943 the Anton Piëch family decamped from Vienna to Zell am See to take up quarters in the chalet. At the end of 1943 Ferry's family also moved to Zell, which was equidistant between Vienna and Stuttgart and south of Berchtesgaden and Salzburg.

Another wartime bolt-hole for family members and close colleagues was a house at Unterdellach, on the south bank of the Wörthersee near the Carinthian capital of Klagenfurt. Located with the help of a cousin of Anton Piëch, the house was acquired by Ferdinand Porsche from Count Secy in the spring of 1939. Enjoying a 200-metre waterfront and a boathouse for four craft, the Unterdellach dwelling had 16 bedrooms and a three-car garage.

Showing the Austrian provinces of Salzburg and Carinthia, a map also picked out the Carinthian town of Gmünd, near which many of Porsche's engineers were relocated for safety in the autumn of 1944.

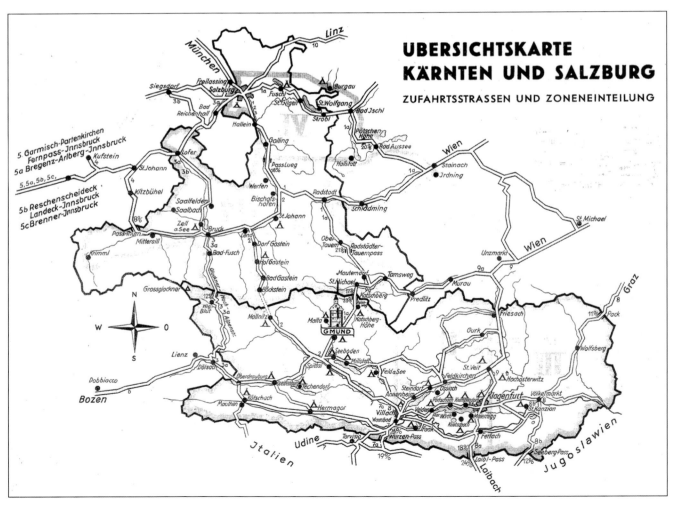

Although seemingly ideal because it adjoined the Porsche family property at Zell, the flying school's buildings were too small for the Porsche engineering team. 'One day,' Ernst Piëch recalled, 'an officer came to us with news. His name was Pischl; he was an American but in a German uniform.* "Could you organise a motorcycle for me?" he asked. Sure, why not? We were friendly with NSU, so we got him one of their 350cc.'

The mysterious but knowledgeable Pischl offered them a substantial sawmill at Karnerau 2 miles west of the town of Gmünd in Carinthia. The mill's remoteness was persuasive, in spite of the colossal effort needed to move men and machinery to its rural location – which lacked even a railhead – and prepare the primitive buildings to suit them.

On 7 May 1944 a family conclave on the move was convened at the Porsche villa among Ferry, his father, Karl Rabe and the company's business manager Hans Kern. A week later Ferdinand Porsche and his son were in Gmünd, after a stopover in Zell am See, to check out the facilities there of Werk Karnerau of the Berlin-based Holzgrossindustrie Willi Meineke, which was to be cleared for their use. They decided that the Gmünd buildings would suit as long as they could have the Zell flying school as well, chiefly for storage. Both were made available.

On 1 June came a thunderbolt: a telegram from armaments chief Albert Speer forbidding the dispersal! Speer, formerly a staunch advocate of moving, had changed his mind. On 13 June Ferdinand Porsche took up the matter with Speer, who reiterated his refusal. Two days later, however, after another meeting Speer gave his approval but with the proviso that no allowance would be granted for construction costs. The senior Porsche didn't give up easily. On the 25th he reported with some satisfaction that Speer had granted 80,000 marks towards the cost of the move.

Making the decision and acting on it were two different matters for an organisation under tremendous wartime pressure. Encouragement came on 25 July in the form of the heaviest air attack yet on Stuttgart. Ferry Porsche allocated the machine tools. 'Our operation was divided into three parts: the headquarters

Only after the war did Porsche erect a sign announcing its presence at the Karnerau sawmill to which its main staff were evacuated in 1944. Three offices in the building at the gate included one for Ferdinand Porsche.

---

* The name Östrom has also been mentioned in this connection.

remained in Stuttgart, production went into Gmünd, while the flying school at Zell am See became our storage depot.' Recalled Ferry, 'My idea was that we could always make a fresh start as long as we had one-third.' Vehicles also went to Zell.

On 8 August 1944 Karl Rabe and his wife drove to Gmünd via Munich, Zell, the Grossglockner Pass and Spittal, which had the nearest railhead 10 miles from Gmünd. Using a Type 82 Kübelwagen, they towed a trailer and took the Josef Mickls with them. On the following day Rabe reviewed the state of the facility:

The main hall has an area of 25 x 50 metres, behind which is a further uncompleted hall. There's still much construction under way; the main Meineke hall is not yet closed in. The carpentry work is still in full swing. Floor and roof are still lacking; the windows should soon be installed. We are informed that the hall can be occupied in about three weeks. In advance some 30 drafting boards can be installed. The question of quarters for a similar number of men can be solved.

The hall adjoining the main hall is open and would be used for storage of the machines already delivered and the drafting boards and the like. Shipment of the machine tools has caused no damage at all, save for a milling machine that toppled over but is practically undamaged.

A sufficiently large guest room with a spacious kitchen is available, needing only the slightest amelioration.

Nevertheless the present fire-safety arrangements leave much to be desired. Under the main hall is a cellar space with a wooden roof, filled with saw shavings, while barrels of tar are lying about outside. The fire-extinguishing equipment is largely inoperative and no adequate sprinklers are available.

The fire watch during the night is carried out by only a single man. Since as well the performance of the Russians, provided by Fallersleben, is very poor without constant supervision, I recommend the strengthening of the watch personnel to three men, who can be deployed as a reinforced guard during the night.

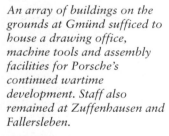

An array of buildings on the grounds at Gmünd sufficed to house a drawing office, machine tools and assembly facilities for Porsche's continued wartime development. Staff also remained at Zuffenhausen and Fallersleben.

The well-lit drawing office at Karnerau continued its support of Ferdinand Porsche's many commitments, including the Maus project, late-war Panzer proposals, the Type 300 turbojet and further development of VW variants.

The move to Gmünd involved many decisions that were human rather than political or technical. The Austrians who held leading positions on the Porsche staff were among those who least resisted the idea of returning to their homeland, especially – they admitted mainly to themselves – because conditions in Germany were worsening rapidly.

German-born employees tended to remain in Stuttgart. Many of those who did go to Gmünd returned home in the last months of the war to be nearer their relatives. Moreover, as Germany succumbed to the attacks from east and west the atmosphere in Austria grew increasingly inhospitable for German nationals.

Acting with his father's full authority, 35-year-old Ferry Porsche took command of operations at Gmünd. Karl Rabe reported to him as technical director and Hans Kern as financial director. Kern had joined Porsche in 1933 at the invitation of Hans von Veyder-Malberg, who was then sales manager and a shareholder in the firm. By 1942 Kern had become fully responsible for the financial affairs of the Porsche KG.

Holding the fort in Zuffenhausen was Karl Fröhlich, who maintained liaison with the team in Gmünd. New additions to the Austrian staff were engineers Wilhelm Hild and Gustav Vogelsang. Still on the move in the first months of 1945 was Ferdinand Porsche, indefatigable as ever at 69.

To provide homes for the emigrants ten two-family houses were put up on a gravelly area near the junction with the valleys of the Malta and Lieser Rivers. Of a standardised, modular design that the Germans were using widely at that time, their enclosed back gardens allowed vegetable growing and chicken keeping. Finished in December 1944, the houses were warmed by wood-burning stoves.

On concrete foundations poured in the sawmill buildings, machine tools for prototype manufacture were installed at Gmünd/Karnerau. They would come in handy for post-war production of a wide variety of retail products.

Overcoming the early reservations of the locals, the newcomers began to fit in at Gmünd. Porsche (left) could relax with both civilian and military colleagues at a dinner with musical accompaniment.

Three single-family houses near the Lieser River were gradually completed for the families of Hans Kern, Karl Rabe and Ferdinand Porsche.*

Tardy in taking up permanent residence at Gmünd, on 6 September Rabe was urged by the Porsches father and son to move there as soon as possible. He and his family were installed in Gmünd's Post Hotel by the 11th. On 20 September Rabe's office furniture arrived, brought like other shipments by the local Pfeifhofer company that had been contracted to help with the move.† On the 29th more staff arrived from Stuttgart, as did five vehicles carrying valuable machine tools from Italy. Porsche was co-operating on a transmission project with Turin truck maker OM, where engineer Rudolf Hruska was providing technical liaison.

A 30-wagon train brought supplies and equipment to Spittal to be transported to the works. Redundant wooden buildings at the KdF-Werke were dismantled and shipped to Gmünd, there to be erected as barracks or to be used as walls demarcating activities inside the sawmill's big open hall. Freshly poured concrete footings supported the incoming machine tools. Among the engineers arriving in July and August were Erwin Komenda, Leopold Schmid and Emil Rupilius. In September Ferry forwarded a request from his father and stepbrother that the entire Porsche staff be moved to Gmünd. This, said Karl Rabe with a sigh, would be 'a very difficult task'.

A link with the outside world was achieved on 10 October when the Karnerau telex machine clattered into life. It communicated instructions and findings on current projects, including the news from Ferdinand Porsche that the facilities of the VW Works at Fallersleben were substantially at the disposal of the Porsche designers. Work began at Gmünd with some 30 engineers, half of whom were assigned to the development of fuel injection for the Beetle engine, carried out with the help of a Munich engineer.

Other projects included the ongoing conversion of both the Volkswagen and its military sister, the Type 82, to operate on wood gas from on-board generators. Project 276 covered the creation of a version of the Kübelwagen with a towing hook for light artillery, tested by a climb to the courtyard of Gmünd's Burgholz, the ancient castle that loomed over the town. Most enigmatic was 'Project S' for the SS, covered by Type number 290. It required liaison with the Marbach facilities of electronics company Telefunken.

The refugees in Austria's Carinthia soon discovered that it was not the pastoral idyll they might have expected. Nearby Spittal had an important rail junction – just the kind of transport link that the Allies were attacking – so work was sometimes interrupted by air-raid alarms. Bombing runs on Graz and Linz overflew the area. No direct attacks on the Porsche compound were made, however. Its location was known only to the resident Austrians and Germans and ultimately to officers of America's Sixth Army who questioned Porsche's people at Zuffenhausen on 27 April as they made their way east.

Nor were the Porsche staff to be spared the indignity of formating as members of Hitler's *Volkssturm*, the home guard. Created by a Führer order of 25 September 1944, it obliged all able-bodied men between the ages of 16 and 60 on the home front to train as a local-defence militia. On 12 November Porsche executives were among those who marched in rows of three to a hall on Gmünd's main square to be sworn in by name. In his radio speech that evening SS chief

---

* 'While only the gatehouse exists from the former factory buildings,' wrote Karl Rabe's son Hans in 2002, 'some of the temporary homes are still in original condition, such as the houses of Joseph Goldinger (driver of Ferdinand Porsche), archivist Franz Sieberer and sales director Karl Volkert. Both former temporary homes of the Porsche family have been replaced by new construction. The Rabes' temporary home has been restored and is a holiday home owned by a German family.'

† Helmut, son of the senior Pfeifhofer, was a schoolmate of the children of the Porsche engineers. Capitalising on his collection of memorabilia from those days, in 1982 Helmut opened his Porsche Automuseum in Gmünd. Run by him and his son Christoph, the museum is a notable attraction in the area.

Included in Porsche's War Crimes dossier was a photo obtained by the American OSS from a special issue of Befreites Sudetenland *dated October 1938. Porsche was one of the Sudetenland's most notable sons.*

---

* Researchers from the American Sixth Army arrived at Zuffenhausen on 27 April to interrogate the Porsche engineers, only to be told that they had been 'evacuated' to Gmünd and Zell.

Heinrich Himmler read a proclamation by Hitler, accepting the loyalty oaths of the *Volkssturm*.

With the Third Reich reeling under the attacks of the Allies and doomed to imminent collapse, the Porsche clan gathered at the Zell estate in April 1945. The elder Porsche and his wife, based there since January, were soon joined by his son and daughter and their families. They had only to await the arrival of the Americans, the British or, at worst, the Russians.

With the Zell farm's produce at their disposal, Ferry related, 'our problem during that waiting period was not one of physical hardship but of psychological suspense, not knowing what was going to happen next. Germany was in a state of collapse, turmoil and utter confusion, and here we were awaiting the arrivals of our Allied conquerors. It was not a pleasant time, although personally it was something that I had foreseen for a long time as inevitable.'

At Zell they heard the news of the arrival of American forces from the west at Fallersleben on 10 April. By the following morning the Yanks had taken control of the KdF-Wagen works and town with 200 troops and their Sherman tanks. By the 15th they had formally occupied the KdF-Stadt and its controversial factory. On 13 April the Russians entered Vienna on their westerly drive. Russian and American troops met at Linz, fortuitously to the east of both Zell and Gmünd.

On 20 April, while muted celebrants wished Adolf Hitler a happy 56th birthday in his bunker deep below the bomb-battered Chancellery in Berlin, readers of *The New York Times* learned about the capture of the 'German "Willow Run"'. Accurately enough, the Associated Press reporter likened Fallersleben to the factory that Ford had built to produce the B-24 bombers that had done much of the damage to the VW Works.

Cowardly at the end, fully prepared to sacrifice his people as unworthy, Adolf Hitler committed suicide on 30 April. On 4 May an armistice was declared in north-west Germany and on the 7th Germany surrendered unconditionally in all theatres of war. The war against Japan was still on; America reverse-engineered the V-1 and planned to use it against the Japanese. But at last the guns fell silent in Europe.

One of the first to appear at Zell am See, to the families' astonishment, was the same officer Pischl who had helped them find their Gmünd quarters in 1943. Now, of course, he was in American rather than German uniform. 'How did this happen?' mused Ernst Piëch. 'We didn't know!' By the end of April American troops of James Van Fleet's III Corps, part of George Patton's Third Army, swept into Austria after overrunning the Bavarian Alps where Germany's propaganda had threatened a final defence in a 'national redoubt'.

Next to arrive at Zell am See, said Ernst Piëch, were 'one English and one American officer. They worked very hard to make copies of all our drawings.' 'The Englishman was an automotive specialist,' wrote Michael Thiriar, 'and the American was a Chrysler engineer who had met Ferdinand Porsche during his trip to the USA in 1936.' The investigating team of half a dozen men all told made Zell the centre of their blueprint research, moving drawings there from Gmünd.* A joint British–American military investigation team arrived at Zell on 13 May and worked there for two days before repairing to Gmünd for four days. They returned to Zell on the 20th for final debriefing of Ferdinand Porsche and his associates.

'Throughout the investigation every assistance was given to investigators by Professor Porsche himself and members of his organisation,' said the resulting 425-page intelligence report, bound as a hard-cover book, on many of his wartime projects.* 'It is perhaps significant,' its author added, 'that at no time did he refer adversely to other German tank design personalities. This is in contrast to the attitude towards him apparently adopted by such as Speer, Arnoldt, Heydekampf etc.'†

After another interrogation his examiners concluded that 'Dr. Porsche was either unaware of, or indifferent to, the personal and professional criticisms made concerning him by other German engineers. He certainly had no ill to say of others.' After all he had achieved in 45 years of advanced engineering in many spheres, the 69-year-old Porsche felt no need to rubbish mere tank designers.

'We were very strongly controlled by soldiers' at that time, Ernst Piëch recalled. 'Two or three days after the Americans the Russians came. Nobody expected them to come so far so quickly.' Although they presented themselves as members of a war graves commission, the Soviet officials were obviously on the hunt for military secrets. As far as Porsche was concerned, however, the Western Allies had beaten them to it.

Inevitably enough, Ferdinand Porsche was taken into captivity on 30 May. His profile had been among the highest of the technicians and industrialists who had contributed to the Third Reich's ability to wage war. After a long journey Porsche found himself lodged in the former servants' quarters of Kransberg Castle, near Bad Nauheim in the Taunus hills north of Frankfurt, which Albert Speer had rebuilt in 1939 to serve as Luftwaffe headquarters. Code-named 'Dustbin' by the British, it was the upmarket counterpart of the more Spartan 'Ashcan' in the Palace Hotel at Luxembourg's Mondorf-les-Bains.

'The prisoners were allowed to walk and talk freely,' wrote Richard Overy, 'to write occasional letters and to listen to the radio. The rooms were clean and the food, basic military rations, generous by the standards of defeated Germany.' 'We banished boredom by early-morning sports,' said Albert Speer, 'a series of scientific lectures, and once [Hjalmar] Schacht recited poetry, giving astonishingly emotional readings. A weekly cabaret was also conjured up.' 'In all we were about 40 people in Dustbin,' added Walter Rohland. 'Fluctuation among the camp's occupants was great, so we were kept very well informed about the situation in other camps.'

Ferdinand Porsche found himself associating again with such titans of the Third Reich as Speer, Wernher von Braun, Fritz Thyssen and Hjalmar Schacht. Nevertheless, said fellow detainee Gerd Stieler von Heydekampf, 'most distinguished among us fine fellows was Professor Porsche. He arrived in his own car with his chauffeur.' Von Heydekampf found himself interpreting for some of his colleagues, including Porsche who, 'taciturn as he was, kept them questioning quite nicely'.

Albert Speer saw one of his missions in Dustbin as identifying for the Allied inquisitors the men whose Third Reich involvement had been entirely technical and thus apolitical, contributing little to the regime's wartime decision-making. He put Porsche in this category, as did the engineer's interrogators.

Not for Porsche the haughty attitude of some of the German industrialists. On

The Völkischer Beobachter *of January 1939 was the source for the OSS photo of racing driver Hans Stuck. Although Stuck was no longer an Auto Union driver he had been supported by Himmler in the later pre-war years.*

* The author, who has seen many such Second World War reports, is unaware of any others that are of the magnitude of the book dedicated to Porsche's wartime activities.

† Small spelling corrections have been made by the author. We have already heard the criticisms of Speer and von Heydekampf; Kurt Arnoldt was Henschel's hard-working chief engineer. He disliked Porsche's torsion-bogey suspension but agreed with him that rear drive of the tracks was superior.

a 1945 trip to Germany ex-Bristol engineer Sir Roy Fedden met some of them, finding that 'for the most part these leaders of industry seemed quite unrepentant, and thought it was bad luck they had lost after so nearly winning'.

After his 70th birthday on 3 September, on the 11th Ferdinand Porsche was released into the care of Josef Goldinger for his return to Austria. He was given a *laissez-passer* signed on behalf of the US Director of Intelligence that allowed him to travel, stating as follows:

TO WHOM IT MAY CONCERN
1. Porsche, Prof. Ferdinand, has been detained at the Special Detention Camp DUSTBIN for exploitation by specialists.
2. This exploitation has been completed and subject is, therefore, being released in FRANKFURT.
3. He has been authorised to travel with his car and chauffeur from his place of interrogation to GMÜND and ZELL am SEE, where he is to continue to work for Allied Military Government.
4. Any assistance given to subject by Military Government for this trip is appreciated.

Allied interest in Porsche's wartime work did not end with the Kransberg interviews. Although his May interlocutors had given him a certificate stating that 'A complete investigation has been made of the activities, in connection with tank, automotive and aero engine design undertaken during the war by the development organisation of Professor Ferdinand Porsche,' the questioners kept coming.

After Porsche's pumping by the experts at Dustbin, small wonder that when another team of specialists turned up at Zell am See on 2 October they found 'a tired old man who would really rather not be interrogated unless there were some aspects of tank development remaining which had not already been fully covered. He warmed up somewhat during the interrogation but his answers often drifted from the questions.'*

At both Zell and Gmünd all design work by the Porsche engineers had been forbidden since 7 May. The Allies took no chances on their creation of new secret weapons. In the workshop at Zell am See Herbert Kaes, Porsche's nephew, once a liaison engineer between the Porsche KG and the German Army, now directed the repair of those vehicles that could make their way to his door.

Conditions were much the same at Gmünd, where the British occupying authorities put Karl Rabe in charge of the rump Porsche operation. For Porsche's men of extraordinary engineering skills and experience the outlook was bleak. 'If your father were a shoemaker,' one of the British authorities told Louise Piëch, 'he would surely make shoes, but he will never design automobiles again.'

---

* They were British experts interested in the way the German tank designers tackled the crossing of soft ground.

CHAPTER 18

OCCUPATION AND RESOLUTION

Ferdinand Porsche returned to a much-changed Austria. In their conference at Teheran, Iran, at the end of 1943 the Allies erected the structure of post-war Europe. At the initiative of Winston Churchill, Austria was singled out for special treatment. Like Germany, and unlike all other nations, it was to have zones of occupation by the Allies. Their borders were implemented in July 1945.*

Zell am See was in the American-ruled zone and thus cordoned off from Gmünd, which was in the southerly British zone. Innsbruck was in French-governed territory to the west and Vienna in the eastern Soviet sector. These were fine divisions for a country only fractionally larger than South Carolina, but the joint occupation succeeded in its goal of pre-empting the absorption of Austria into the Soviet Bloc.

The first breakthrough for the Porsche engineers at Gmünd came on 11 September 1945 when Major R. Andrews of the British Military Government signed a document giving the Porsche KG in Gmünd 'allowance to work on the development of an agricultural tractor. This is a very important task for Carinthia and the neighbouring countries. The design is almost finished. Experimental vehicles are to be constructed.' Their development was logged into the Porsche records on 7 December as Type 312 for a gasoline version and Type 313 for a diesel tractor.

In control of Carinthia, in which the main Porsche engineering force was employed at Gmünd, the British seemed well placed to capitalise on Ferdinand Porsche's legacy of technical concepts. By an accident of history the British were also the custodians of the Fallersleben factory. That they were not more opportunistic in the exploitation of either was owed to their view that they were trustees of these properties on behalf of the German people. While the Americans generally shared this perspective, the Russians emphatically did not.

Among those Britons who sought to raise awareness of the possible strategic value of Ferdinand Porsche and his ideas was Richard Parker of the Property Control Office in Berlin. From Parker the Control Office for Germany & Austria in London received a report dated 15 August 1946 compiled by the VW Works Board of Control. Over two single-spaced pages and 17 paragraphs it set out what was known of the relationships between the Porsche executives and businesses and the VW factory.

Pictured in the course of his post-war interrogations, Ferdinand Porsche adopted a stance that was readily recognisable from his earliest years at Austro Daimler. It said, 'This is who I am. Take me as you see me.'

* The zones remained in place until 1955, when a peace treaty declared Austria a free and sovereign nation.

Post-war prosperity meant that both Ferdinand Porsche (left) and Karl Rabe put on the pounds. In some cases, such as Anton Piëch's, this overly rapid process meant life-threatening health problems.

The thrust of the Parker document was that the Porsche people were busily endeavouring to exploit their patents and product ideas for the benefit of almost everybody but the British. It sought to make the case that Porsche's ventures were under British control with both Zell am See and Gmünd in the British Sector of Austria, even though the Stuttgart Porsche office was in the American Zone of Germany.

Setting out in an attachment the details of the funds received by Ferdinand Porsche and his family members before and during the war from various organs of the Third Reich, the Parker document added that 'Porsche himself was given the honorary rank of *Standartenführer* in the S.S., was deeply implicated in Nazi industry, and received large sums of money from it. He therefore qualifies for arrest.'* The latter point was reiterated in paragraph 17, which was action-orientated:

ACTION
17. Since various countries consider Porsche's inventions worth their active support, it is recommended that:
(1) It should be considered whether, in the light of the above, an immediate investigation into the possibilities of exploiting the Porsche inventions for the benefit of British industry should be undertaken.
(2) The personal position of Dr. Porsche should be considered so as
(a) to ensure that his knowledge and equipment be fully exploited in the interests of British industry,
(b) to secure his arrest if his past activities merit this.

Any need for a British arrest of Ferdinand Porsche was rendered redundant by actions already taken by another of the four occupying powers. At the beginning of November 1945 a French delegation arrived at Zell am See. Led by Lieutenant

---

* The document did make clear elsewhere that with regard to Porsche's honorary SS post 'the formalities of his appointment were not completed'. The payments and amounts described in the document included the following:
  • RM661,900 in the form of an interest-free loan granted in 1942–43 by the KdF-Werke account to Porsche KG, Stuttgart, for expansion of its buildings there.
  • RM18,144,450 paid over 1937 to 1945 to Porsche KG by the KdF-Werke for the development costs of the car.
  • RM6,533,651 paid over 1938 to 1945 by the KdF-Werke for the development costs of the tractor.
  • RM270,000 paid by the KdF-Werke to Anton Piëch on 18 September 1944 'as a fee for your services during the period from 1 June 1941 to 30 May 1944'.
  • RM3,565,000 in value of cheques alleged to have been taken from Berlin by Anton Piëch and to be held at Zell am See by his wife.
  • RM500,000 held by Anton Piëch, on which he claimed a lien because he was owed RM536,484 in expenses by the KdF-Werke.
  • RM4,200,000 thought to be held by Hans Riedel of the KdF-Werke in funds 'he got from Piëch when they left Berlin'.

Ernst Piëch recalled funds on the order of RM1.2–1.3 million that his father brought from the VW enterprise and deposited in Vienna's Spenglerbank until its fate was decided. 'But Austria froze all the German assets in Austria,' he said, 'and eventually seized the money.'

Lecompte, it was sent in the name of Marcel Paul, a leading member of the *Parti Communiste Français* (PCF). Paul was a newly minted minister in the French government with the portfolio for Industrial Production.* 'We soon discovered,' Ferry Porsche related, 'what Marcel Paul had in mind. It was nothing less than taking over the entire Volkswagen factory at Wolfsburg as a war reparations payment.'

Discussions continued after 16 November 1945 at the Hotel Mueller in the capital city of France's German zone, Baden-Baden, to which the Porsches, father and son, were invited. There they met Colonel Trevoux who, as Ferry Porsche recalled, 'wasted no time getting down to business. He at once told my father, as a *fait accompli*, that the French were to get one half of everything at Wolfsburg, including machinery and equipment. He also explained to our assembled group that the French wanted to start a nationalised automobile factory and that in due course proper use would be made of the equipment shipped from Germany.'

The French colonel added that he expected the Porsche organisation to be 'heavily involved in this affair'. The French wanted Porsche to oversee the transport of the factory to its new location – not yet decided – and to direct the launch of manufacturing there. They were also to restyle the car around the existing chassis, to reduce its resemblance to the dream car of the despised Hitler and give it 'a typical French look'.

For Ferdinand Porsche this was an exciting new challenge. 'I think that France will be the premier automotive nation in the Europe of tomorrow,' he told his hosts. 'I'll be happy to work for her.' The outlook was promising, as he said to his son, 'It seems to me that the French are handling this thing on a pretty big scale.'

'As soon as word got out about our visits to Baden-Baden,' Ferry Porsche related, 'the French auto industry began thinking up ways of blocking the whole project. The serious competition which a French version of the Volkswagen would offer them on the home market bothered them a great deal. In fact, it is fair to say that the entire French automotive industrial complex was dead set against the project.'† They had no need to look far to find a way to upset this applecart.

The wartime activities of Porsche and others from the KdF-Werke at the Peugeot plant in Montbéliard were soon brought into the frame. Those events seemed in the distant past in November 1945 when Porsche was being lionised by the French at the Hotel Mueller in Baden-Baden. On the 19th he was treated to a celebratory dinner and on the 21st he made a presentation on his KdF-Wagen project and its factory. Porsche returned to Zell, only to be invited back to Baden-Baden for another meeting on 13 December, when a contract was to be signed.

On Saturday, 15 December the French colonels and lieutenants fêted Ferdinand and Ferry Porsche and Anton Piëch at dinner, after which they were to return to Zell for the Christmas holidays. Always keen on racing, the Porsches were interested in hearing more about Marcel Paul's other pet project, the creation of a national racing car for France, the supercharged V-8 CTA-Arsenal. Lightning struck at a quarter past seven with the coffee. By the order of Justice Minister Pierre Henri Teitgen the Porsches and Piëch were arrested by two plainclothes officers of the Sûreté.

---

* Fernand Picard says that Paul became Minister for Industrial Production in March 1946.

† Some commentators, including Michael Thiriar, belittle this rationale for the French action, suggesting that it is insufficient motivation for their treatment of the Porsches and Piëch. They are not sufficiently aware of the deep fears that all West European auto makers had of the market-destroying potential of the KdF-Wagen project.

Point man for the French in their accusations against the detainees was none other than their erstwhile colleague Jean-Pierre Peugeot. Swiftly adapting himself to the new realities, Peugeot aimed a fusillade of accusations of the most heinous behaviour on the part of Ferdinand Porsche. 'No doubt to their intense satisfaction,' Ferry related, 'the French car manufacturers came up with "evidence" that my father was really a "war criminal".'

Ferdinand Porsche was held personally responsible by the French for the stripping out of equipment and materials from Montbéliard. Even French workers, Jean-Pierre Peugeot claimed, had been deported to Germany by Porsche. As well, said the Peugeot president, his company should have the 'right to help ourselves from the German manufacturers in order to replace the materials taken during the pillages suffered'. Porsche, his son and his son-in-law faced serious charges.

The accused were imprisoned to await trial in a jail that the Gestapo had originally established in Baden-Baden for high-ranking prisoners. Still at liberty was Herbert Kaes, who had decided to attend the cinema on the evening his colleagues were arrested. On 3 January he made representations on behalf of the detainees to General König, head of the French occupation authority, but they stayed put. His brother Ghislaine visited the prison on 30 January and made contact with his uncle. With a daily bribe of 20 marks to the German guards he was able to bring baskets to Ferdinand Porsche and take them away again.

On the way in the Kaes basket contained food and information, while on the way out it carried Porsche's instructions to his team. The instructions were reinforced by a document given to Ghislaine that restored his rights and duties as his uncle's official amanuensis. In this capacity Kaes contacted the French authorities and told them that the Professor was suffering from a progressive loss of sight. They transferred him to a hospital in Baden-Baden, a more comfortable situation that allowed Ghislaine Kaes to base himself in his uncle's sickroom.

On 27 February 1946 Ferry Porsche was released from Baden-Baden, only to be subjected to house arrest in the Hotel Sommerberg at Bad Rippoldsau. There he was supposed to be redesigning the KdF-Wagen to make it more suitable for the French market. Not until 24 July 1946 was he released to return to Austria, where he took charge of the activities at Gmünd.

At 6:00 a.m. on 3 May 1946 Ferdinand Porsche and Anton Piëch were transported by car to Paris and housed in the servants' quarters of a villa called Galice House in the Meudon Hills, not far from the Renault factories. The villa was one of the many properties expropriated from Louis Renault by the left-leaning French government. His mission, Porsche was told, was to advise on the design of the new 4CV model being designed by Renault.

Secretly during the war the Renault engineers, led by Charles Edmond Serre and his deputy Fernand Picard, had been developing a new rear-engined small car. Their effort, which Louis Renault personally encouraged, was not uninspired by the KdF-Wagen project. The 4CV's first prototype ran at the end of 1942 and in 1945 Renault had already decided to produce it in high volume.

The Renault engineers showed Porsche their prototypes and asked for his views about them. 'Clearly there was no spirit of co-operation on either side' of this forced marriage, Fernand Picard recalled. 'Often the conversation turned into a dialogue of the deaf. Nothing positive came out of these meetings, which were simply a long catalogue of complaints: requests by the Germans [sic] for the drawings and equipment to carry out research on a racing car and insistence on bringing over their families.'

Porsche and Piëch remained in Paris and at the beck and call of Renault and its engineers for more than nine months. On 27 January 1947 Renault's chief gained his regime's approval for the Austrians' departure from his company's orbit. Removed on 17 February to spend a night at Meudon Prison, the next day they were transferred, in manacles, to imprisonment in Dijon where 'the conditions of detention were atrocious', according to historian Michael Thiriar. 'The two men were separated and put in unheated cells.'

Not until 31 May 1947, 17 months after their arrest, was testimony taken against the two detainees in the presence of investigating judge Raymond. The initial views proffered by two Peugeot officials were impressively pro-Porsche. The engineer, they said, had successfully reversed deportations of Peugeot workers to Germany and fended off visits to the plant by snoopers from the Reich. At no time, said other witnesses, had Porsche allied with German security personnel against Peugeot or taken part personally in the expropriation of either men or machinery. In the wake of this testimony Jean-Pierre Peugeot decided not to make a personal appearance.

Especially helpful to Porsche were remarks about the role in the Peugeot factory of engineer Alfred Nauck. From his departmental job in Albert Speer's Ministry of Armaments, said Nauck, at Speer's personal request he was deputised to Fallersleben during the war expressly to maintain some control over the expensively hyperactive Porsche team. In the troubled year 1944 Alfred Nauck behaved so high-handedly at the Peugeot works that Bodo Lafferentz accused him of acting like a 'Russian commissar'.

Lafferentz and Porsche had been relieved when the pompous Nauck was reassigned to chase Arado bomber production in October 1944.* Although his stay there had been brief, Nauck's visibility at the Peugeot factory was so high that ample testimony was given that it was he who had taken the lead at Peugeot on behalf of Fallersleben in exploiting the French company and in uprooting its machine park for expatriation to Germany.†

So slight was the evidence against Porsche and Piëch in this preliminary hearing that one might have expected both to have been released forthwith. However, in a travesty which, wrote historian Hans Mommsen, was 'not a glorious page in the annals of French justice', they remained detained. Hints that the payment of a substantial caution might free the Dijon Two were taken up by the Professor's children, who in the meantime had gained a substantial contract from Italian industrialist Piero Dusio to design sports and racing cars for his Cisitalia company.

Freed from his French detention on 1 August 1947, as a 'displaced person' Porsche still had to reckon with controls on his movements within subdivided Austria. His birthplace was wrongly recorded.

---

* Well into 1945 Porsche and Lafferentz were importuning the engineer to return the Kübelwagen he had been assigned as a duty car by Fallersleben.

† Not until 2047 will we know exactly what happened at Porsche's hearing; the French have sealed the relevant documents for 100 years.

At Zell am See Porsche, with grandson Ernst Piëch, congratulated two woodsmen on handsome stags they brought in on a battered Schwimmwagen. At Zell and Gmünd the repair of such vehicles was a post-war moneymaker.

---

* Louise Piëch produced as evidence a box of golf balls – extreme rarities in wartime – given to Porsche by officials at Renault as a sign of gratitude. This was hardly a sign of malice towards the Professor.

† Charles Faroux later noted with some regret that as no case had ever been brought against Porsche or Piëch, the French by rights should repay the million francs. Ultimately they did, but by then, said Ernst Piëch, the currencies had depreciated so much that it was barely enough to buy a pair of shoes.

Dusio and his associates, with their relative freedom to travel and their numerous contacts in France, offered the best and indeed only hope for gaining the release of Ferdinand Porsche and Anton Piëch from prison. Through old friends in the French racing world, including drivers Louis Chiron and Raymond Sommer and journalist/editor Charles Faroux, Dusio and Louise Piëch arranged to have one million francs – part of the contract fee paid to Porsche and officially $8,380 at the time – turned over to French authorities. This was bail amounting to a ransom.*

At 4:30 p.m. on 1 August 1947, after imprisonment for 19 months and 14 days, Ferdinand Porsche and Anton Piëch were released on *Liberte provisoire*.† Ferry picked them up at Dijon and drove them back to the family holding at Zell am See. Much was new along the roads they traversed but one change was obvious above all: Volkswagens were visible, a new picture on Europe's roadways. Ferdinand Porsche was not only surprised but also moved to see so many Beetles scuttling about.

Ferdinand Porsche inevitably reflected on the random nature of the justice meted out to the losers by the Allies. His punishment – for so it must be viewed – was not unlike that suffered by Willy Messerschmitt, who was released in 1947 after two years' detention. Another recipient of the German National Prize, Ernst Heinkel, was tried for war crimes but released. By the end of 1949 he was denazified and able to return to work. Kurt Tank, the brains behind Focke-Wulf, re-entered civilian life with little delay.

With some of his colleagues, V-2 missile builder Wernher von Braun was whisked to America to resume his work on rocketry at White Sands, New Mexico. A designer of advanced submarines, Hellmuth Walter, soon found a home with the British, who were eager to tap his knowledge. Industrialists were treated more harshly, with Walter Rohland imprisoned for one year, Hermann Röchling sentenced to five years and Alfried Krupp von Bohlen handed a 12-year sentence, which, however, was commuted in 1951.

Another 'war criminal' released in 1951 was legendary Tatra designer Hans Ledwinka, who had been 67 at the war's end. As a supplier of military vehicles he was accused of collaboration with the Nazis by the Czech authorities, who tried and jailed this close contemporary of Porsche.* Falling foul of the French like Porsche, Hanns Trippel was not freed by them until 1949. Builder of the amphibian vehicles that the Wehrmacht had spurned in favour of the Schwimmwagen, Trippel had occupied the Bugatti factory at Molsheim during the war.

Less rigorous was the fate of Bremen vehicle maker Carl Borgward. During the war his plants made trucks, half-tracks, light Panzers, torpedoes and other apparatus for the Wehrmacht, not unlike the KdF-Werke which was also in the British zone of occupation. Borgward was detained in mid-September 1945 and released after only five and a half months, although his works used a higher proportion of forced labourers than any other German vehicle plant, a peak of 65 per cent. Denazification procedures kept him from leading his factories again until July 1948.

Porsche too had to wait before joining his engineering cadre at Gmünd. Only provisionally at liberty, after his release he was ordered to present himself at Kitzbühl, headquarters of the French zone of Austria. He arrived there on 4 August 1947 and encountered Ghislaine Kaes, whom he required to brief him fully on everything that had happened in the months he had spent near Paris and at Dijon. The Professor was initially confined to Kitzbühl, where he was booked into the Hotel Klausner by the French authorities.

Near their Gmünd residences Erwin Komenda (left) and the two Porsches posed with their first Porsche car, a mid-engined roadster whose design dated from mid-1947 but was only completed in April 1948.

---

* The Czechs had the cheek to suggest that Ledwinka return to Tatra after his release. He declined.

While Ferry Porsche oversaw development work on the Porsche sports cars, the first of which were produced in mid-1948, his father lent his expertise to their creation over the 1947/48 winter.

In front of the family villa in Stuttgart, in whose garage the first Volkswagens were built, Ferdinand Porsche bent for a word with his son. His prestige was the principal asset of the budding sports-car manufacturer.

For a former prisoner like Porsche, still burdened by the German citizenship awarded him by the Third Reich, journeys abroad were all but impossible. His only document was an Austrian travel pass that was 'only valid with a visa' and declared him to be a German national. The other Austrian zones were off limits to the Professor. Yet with the aid of his daughter and other members of the family, Porsche gradually was able to travel greater distances from Kitzbühl until, in the relatively relaxed occupation that was typical of Austria, he eventually regained a modicum of freedom.

Though most of the Porsche engineers would move back to Stuttgart and commence making cars there, the family's strong links with Austria were maintained. While son Ferry would become the guiding spirit of sports-car making, daughter Louise set up the Salzburg company that became the Austrian importer of Volkswagens and, later, Porsches. Thanks to high demand for their products and services, both arms of the Porsche family would enjoy prosperity.

The man whose unstinting dedication to his profession created the foundation of this prosperity would not live to see its realisation. On 3 September 1950 Ferdinand Porsche celebrated his 75th birthday in the midst of many friends and admirers at Castle Solitude near Stuttgart. One of Germany's leading auto clubs, the ADAC, awarded him, 'on the final day of his 75th year and after 50 years of great success in the field of motoring, an honorary membership'. It was the last of the many encomiums he received.

In November 1950 Ferdinand Porsche visited the VW factory in Wolfsburg and met with its chief, Heinz Nordhoff. Soon after his return to Zell am See the engineer was felled by a stroke that left him partially paralysed. Those close to him were certain that the harsh treatment given Porsche by the French authorities contributed to his declining health. Hospitalised from the beginning of 1951, too weak to shake off pneumonia, the senior Porsche died in hospital in Stuttgart on the morning of 30 January.

In a world that yearned for peace, preferring to turn its back on the war's horrors, Porsche had quickly regained his standing as an engineer of immense talent and achievement. He was fêted as the creator of the Volkswagen, the success machine of the post-war world. Early biographers skated lightly over the Second World War years, majoring on the benign Kübelwagen and seldom if ever mentioning the V-1.

If Ferdinand Porsche reached conclusions about his role in the Third Reich, he said little about them. Hearing afterwards about the horrors of the concentration camps, he told his nephew that 'the Führer knew nothing about all that'. For Porsche, Ghislaine Kaes recalled, Hitler's arrival on the scene had been 'fateful' and 'God-given'. He had done his duty, as he saw it, as much to the men whom he had asked to join his new engineering company as to Greater Germany. He and they were designers, and if there were designs to be made, they would make them.

In August 1950 the two Porsches conversed in the course of that year's German Grand Prix as photographed by Rodolfo Mailander. Notably thinner, Porsche would succumb to pneumonia the following January.

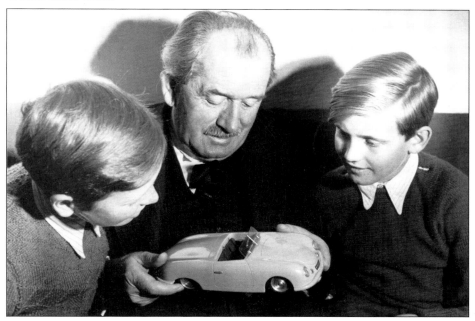

On his 70th birthday Porsche posed with grandsons Ferdinand F.A. Porsche (left), future stylist of the 911, and Ferdinand K. Piëch, future chief of Porsche engineering and chairman of Volkswagen.

It is fitting that our final image of Ferdinand Porsche is at the wheel of a VW, easily his most famous creation. But those eyes witnessed far more over the four decades during which he made historic military contributions.

Although Porsche had been a member of the Nazi Party since 1937, he shunned its pomp and furbelows. Said Wolfgang Porsche, 'My grandfather was made a member of the NSDAP because he had a great reputation as an engineering developer.' Wolfgang's father Ferry had successfully fended off Party membership. Nor had members of the Porsche and Piëch families manifested anti-Semitism. The ranks and honours distributed like confetti by the SS could not reasonably be avoided.

As arguably one of the world's most conspicuously renowned automobile engineers, Porsche's prestige rubbed off on that of the Nazi regime, said Richard von Frankenberg:

> For the Third Reich Ferdinand Porsche was not only a man who had helped to fashion the technical and economic foundation of this era, but also he resembled an advertising placard for this era and its political leadership. For solely on the basis that Ferdinand Porsche actually worked in the Third Reich, he the widely known and oft-honoured Volkswagen designer, the Third Reich had the ability to say: Look here, this man is working wholeheartedly with maximum effort for us; this man commits himself to us; how then can there be any doubt about our regime if such outstanding men as Ferdinand Porsche openly support it?

Porsche had contributed to the war effort of the Third Reich and had been rewarded, financially and in honours, for that contribution. He had not only designed the Kübelwagen and Schwimmwagen but also established and run a huge factory both to produce them and to manufacture offensive weapons that included the Ferdinand and the V-1, which although technically a Luftwaffe device was one of the least discriminating yet damaging artillery pieces in recorded history.

The Porsche role in the Third Reich reflected the most astonishing serendipity. When the engineer established his independent company in 1930–31, in the depths of the Depression, its outlook could not have been worse. Auto makers were collapsing, not ordering up new models. Paying his small cadre of designers was a hand-to-mouth affair. It was not Porsche's fault that the dictator who took charge of Germany in 1933 had long admired his work and was a fellow Austrian who had every confidence in his ability. Manifestly Hitler never lost that confidence.

As for Porsche's relationship with Hitler, said his nephew and secretary, 'The professor was a completely unpolitical person. It was later said of him that he snubbed Hitler, in that he abjured the German greeting "Heil mein Führer" and said "Guten Tag" instead. That is not so. He would even have greeted Hitler by rubbing noses if it were useful for his work. And just a year after the so-called seizure of power Hitler was extraordinarily useful to him.'

The feeling was mutual, for Porsche was a favourite of Adolf Hitler. Although best known for his monologues, Hitler enjoyed dialogues 'under four eyes', as the Germans say, with Porsche about Panzer matters. In February 1942 the Führer mused to friends that 'Dr. Porsche is the greatest engineering genius in Germany today – although one wouldn't think it, seeing him so modest and self-effacing. He has the courage to give his ideas time to ripen, although the capitalists are always urging him on to produce for quick profit.' In Hitler-speak 'capitalists' meant 'Jews', for the supreme leader had formed the view that Porsche was a man who had struggled under Jewish financial demands and constraints, only to prevail. In fact there was little evidence that this had been the case.

'One could say,' wrote Richard von Frankenberg, that 'this Ferdinand Porsche has been implicated in everything of which the National Socialists have been accused, implicated in that particular degree in which leaders must always be held to a higher responsibility than followers.' Certainly some of the British occupiers took this line. But hidden away in Austria as Porsche was, out of the line of fire, he was able to wait out the surprisingly short period during which Germans and Austrians mutated from despised enemies to valuable allies in the commencing Cold War.

A team of academics sought to sum up the role of Ferdinand Porsche in the Hitler years.* Their conclusions included the following:

> Properly to evaluate his role in the Third Reich, one must consider Porsche's absolute, almost naïve belief in technology. In his view technology was an absolute, free of good and evil, remote from any emotion. For him there were only tasks to be achieved which were to be solved by all available means.†
>
> He simply did not see the moral reprehensibility that was possibly associated with his products. Nor did he question the means for realising his ideas, as naturally he took advantage of his good contacts with Hitler to silence those opponents who could endanger him; as naturally he demanded from Himmler 10,000 prisoners to maintain his factory's

* Andreas Falkenhagen, Christoph W.O. Matthies and Maik Ziemann in a January 2000 paper at the Braunschweig Technical University.

† This part of the academic assessment paints Porsche as a dry technocrat devoid of emotion, which as we have seen was far from the case. His fondness for motor racing was only one of the many ways in which Porsche the man asserted his great enthusiasm for life.

production. And he did not shrink from subjugating the French Peugeot factory.

To be sure his son described him as non-nationalistic and his wife saw him as unpolitical. Also with his conspicuous wearing of civilian clothes he eschewed the staged trappings of the Nazis and ignored the National Socialist awards given him. But nevertheless he found himself in an unholy alliance with a system that gave him the scope of action that he needed.

On the one hand he used the National Socialist leadership in his own interest like no other; they did not give Porsche orders, but he them. On the other hand in Porsche the Third Reich had one of its most loyal and able servants.

It could be argued that while Porsche's Panzer inventions and designs had been among Germany's most spectacular, few of them had actually waged war. His Mixte-drive Leopard and Tiger prototypes and his Maus expended time, money and materials that the Third Reich could ill afford. Only his Ferdinand anti-tank Panzer was fielded in force, and that controversially with its struggles in the cruel environment of Kursk. Contradicting his reputation for profligacy, Porsche often argued for simplicity and practicality in the design of Germany's weaponry – usually in vain.

Although not Porsche's direct responsibility, the Nibelung Works, which had been at his disposal for new Panzer developments, proved anything but a paragon of productivity. His Fallersleben factory had built the V-1, to which he only belatedly began to make a design contribution, as well as many warplane components and his VW-based vehicles for the Army and the SS. As a catalogue of malefaction it was underwhelming; his net contribution to the Reich's armaments could well be assessed as neutral if not negative.

As mentioned earlier, in their secret 'wonder weapon' role both the Ferdinand and the Maus underpinned Adolf Hitler's determination to continue waging war. The Führer cited both in his pep talks to his generals, persuading them that these were battle-winning machines that would turn the war's tide in Germany's favour. That he was able to make reference to such new developments bolstered Hitler's own confidence in the future. Accordingly they deserve some credit – or blame – along with the other *Wunderwaffen* for Germany's endurance and persistence in battling on to the very end of an unwinnable war.

The years between the wars, first with Daimler-Benz and then with his own engineering company, found Ferdinand Porsche much more involved with military matters than is generally perceived. Although shot full of fascinating detail solutions, his GT I prototype tanks foundered on outdated engines and unrealistic demands, such as seaworthiness, imposed by the regime. Porsche's attractive eight-wheeled command and reconnaissance vehicle was a solution ahead of its time, as the generals freely admitted.

Most conflicted during those inter-war years were Porsche's aero-engine initiatives. A leader in the design and construction of aviation engines during the First World War, Porsche had ended that activity with a remarkable W-configuration prototype that could have pointed the way to a revolution in efficient design at lower cost. Leaving this in his turbulent wake, Porsche again addressed aero engines at Mercedes-Benz. Although remaining largely prototypes, his V-12s provided a bridge for the company to the MB 600-series twelves that would serve valiantly through the coming war.

Ferdinand Porsche was frustrated in his fitful efforts to contribute to aviation-engine design in the Second World War. His 1930s proposals for aero engines were unconvincing to the Luftwaffe, for they resembled hasty adaptations of auto-engine ideas. In any event by then the Junkers-Daimler-BMW troika had a firm grip on aeronautical engines with minor interventions from the likes of Bramo and Argus. Porsche and his colleagues could only hope that their Fallersleben works would gain aero-engine contracts, but this expectation was in vain.

Ferdinand Porsche's support of the Central Powers in the First World War was substantial and consequential. In harness with Skoda his Austro-Daimler was relied upon by the dual monarchy's panjandrums for motive power of all categories including his astounding C-Trains, which soldiered on in the service of various powers straight through the Second World War. No application, save perhaps the Maus, made better use of his Mixte drive system.

An active pioneer in aviation and a close colleague of Austria's innovators in this new field, Porsche and his team quickly grasped the essential imperatives of the aero engine. This was at a time when the aeroplane was desperately in need of more power from less weight with reliability. Rejecting the auto-derived designs of most, they brought new production techniques and design features to these engines. Both rival designers and engine experts acknowledge Porsche's key role in advancing the state of the art.

Deeply embedded as it was after four decades of service, the military gene easily survived the death of Ferdinand Porsche. For the last two of those decades

The placard 'And he swims!' on the prototype Type 597 Jagdwagen or Hunter at the 1955 Geneva Salon could better have read 'And he floats!' Porsche aimed for both governmental and commercial success for its post-war off-roader.

As completed with its ribbed bodywork the T597 was both more 'military-looking' and agile with its five-speed gearbox. Porsche spent some DM1.8 million on its development and body shells sourced from Karmann.

Of Porsche's output of seventy-one Type 597s, forty-nine were for the civilian market. Needing more such vehicles than Porsche could supply at short notice, the Bundeswehr picked a cheaper alternative from volume producer DKW-Auto Union.

both Ferry Porsche and Karl Rabe participated intimately in the company's work for the armed forces, as did many of their senior engineers. Building up its strength after the war, becoming a partner in NATO, Germany's Bundeswehr naturally turned to Porsche to meet some of its vehicular needs.

In fact Porsche was an outsider when the rebuilding German Army sought a post-war counterpart to the legendary Kübelwagen. When it officially launched work on its Type 597 on 19 December 1953, Porsche did so without the government backing that was sustaining its rivals: Goliath, part of the Borgward group, and DKW, famed for its two-stroke engines. Ferry Porsche hoped that adoption of its 597 would lead to in-house production that would supplement the still-uncertain demand for sports cars.

In charge of work on the Type 597 was none other than Franz Xaver Reimspiess. A veteran of negotiations between Porsche and German military authorities, he was chief designer at the tank-building Niebelung Works during the war. Known as the *Jagdwagen* or Hunter, the four-wheel-drive Porsche was first exposed to the public at the Geneva Salon in March 1955.

Using all the familiar Porsche design elements plus the four-wheel drive the Kübelwagen had lacked, the purposeful-looking Hunter had a 1.6-litre Porsche flat four with gearing that allowed it to mount grades as steep as 60 per cent. In a 1955 brochure the company also offered its Hunter to the public as 'a light cross-country car', pointing out that 'it will carry four passengers over hill and dale to hunting lodges or fishing spots'.

This proved premature. Germany's ordnance experts were less resistant than expected to the smoky two-stroke engine of the DKW Munga, the simpler and less expensive vehicle that was finally adopted by the Bundeswehr. No military contract meant that there was no civilian production either for the go-anywhere Type 597 Hunter.* Porsche's hopes of having a second string to its bow at Zuffenhausen were dashed.

Overlapping the end of the Hunter project was another competition, this time for the design of a new home-built Panzer to replace the Bundeswehr's American

In the summer of 1963 Germany's Defence Ministry chose Porsche's Type 714 as its new battle tank. Production preparations included twenty-eight prototypes and fifty of a pilot series – time for perfection that Porsche was denied during the Second World War.

---

* Of the first version and a post-1957 revised model, Porsche made 71 Hunters in all through 1958 for service evaluations. Although some were sold, this project was recorded on its books as a costly waste of money for the still-small company.

The Porsche-designed Type 714 Leopard was acquired and fielded by numerous nations. Although able to reach 40mph, in spite of its torsion-bar suspension 30mph was the maximum its crew could tolerate on uneven terrain.

tanks. After the initial requirements were released in November 1956 the Porsche men began their preliminary designs on 15 January 1957 as their Type 714. They were one of three rival teams. Porsche was part of Team A, Rheinmetall leading Team B and Borgward constituting Team C. Except for the last, which quickly fell out, each of the teams was to build two prototypes of its design.

After several phases of prototype build, during which the French withdrew as possible customers, Porsche's design was chosen for manufacture by Munich's Krauss-Maffei. Dubbed the 'Leopard', it had in common with Porsche's 1939–40 design of the same name a V-10 powerplant, this one made by MTU. A multi-fuel engine, it produced 830bhp and drove through a conventional transmission. Lightly armoured for agility and capable of 40mph, the Leopard was respected for its deadly British-built 105mm cannon.

Deliveries of Leopards began in late 1965. Other NATO members and allies soon joined the queue, operators eventually including Belgium, Brazil, the Netherlands, Norway, Denmark, Australia, Canada, Turkey and Greece. Italy bought some and negotiated home production of more. Through 1979 6,485 in all were produced, 4,744 being battle tanks and the rest of lesser configurations.

Porsche designed and built its Leopard prototypes at the remote southern end of its engineering centre and proving ground at Weissach, south-west of Stuttgart. When in 1981 it stood down its military activity the vast sheds and offices were taken over by its motor-sports department to build racing cars. As an example of swords into ploughshares it could hardly be bettered.

ACKNOWLEDGEMENTS

I have been tilling the fields of Porsche since I first tackled the history of the company of that name in the 1970s. This came about when editor and friend Dean Batchelor asked me write such a history, a project which was finally realised as *Porsche – Excellence Was Expected* by L. Scott Bailey under the auspices of his creation, *Automobile Quarterly*. In recognition of the bold initiative and trust in this author manifested by both men this book is dedicated to their memories.

At that time I had only a generalised appreciation of the pre-Porsche-car activity of Ferdinand Porsche. My first step in remedying this was researching and writing a book about the creation of the Volkswagen factory and its car and the subsequent post-war fate of both. The result, *Battle for the Beetle*, was published in 2000 by Bentley Publishers of Cambridge, Mass.

I confess to a partiality for Bentley because I was a frequent customer of Robert Bentley's Boston book shop when he was an early importer and publisher of automotive books and I was a student in and near Boston. Robert's initiatives have been accepted and amplified by his son Michael, who now heads Bentley Publishers.

I have drawn upon chapter 1 of *Battle for the Beetle* for some paragraphs in this book's chapters 8 and 9 about the role of Hitler and his early assignments to Porsche.

Chapter 2 of *Battle* contains text that appears in chapters 10 and 11 of this work, dealing with the wartime versions of the VW Beetle, Types 82, 87 and 166, and also in chapter 17, concerning the bombing and dispersion of the VW factory at Fallersleben. Some text from the same chapter, about wartime engineering work on the Beetle and related projects, appears here in chapter 14.

My next deep dive into the Porsche story was the researching and writing of a book about the career of Ferdinand Porsche up to and including the establishment of his own design office in 1930-31. An initiative of Ernst Piëch, this resulted in *Porsche – Genesis of Genius*, published by Bentley in 2008. The first full accounting of Porsche's early creativity, it is inevitably drawn upon for the present work.

This book's chapter 1 benefits from my findings as published in the first four chapters of *Genesis of Genius*. Chapters 2 and 4 here about military vehicles are greatly expanded from that book's chapter 7, while its chapter 8 provided a basis for the much more detailed chapters 3 and 5 about aviation in this volume. Part of chapter 12 of *Genesis* is a kernel of this book's vastly expanded chapters 6 and 7 concerning Porsche's work at Daimler-Benz. Elements of its chapters 13 and 14 contribute to this volume's chapter 8 about Porsche's post-Daimler activity.

Most recently for Bentley I have written *Porsche – Origin of the Species*. Published in 2012, it explores in depth the origin and creation of the first

Porsche car, the Type 356. For that book's chapters 3 and 4 I researched the circumstances of the dispersal to Austria of the Porsche engineering team and the immediate post-war fates of Ferdinand and Ferry Porsche. Some of their text is mirrored in this book's last two chapters.

For this, my first venture into another genre of authorship, I want to express my deepest appreciation to Bentley Publishers and to Michael Bentley personally for permission to make use of the above-mentioned material in this entirely new context. Its use here in no way vitiates the rights held by Bentley for the use of my copyright text in the volumes referenced above.

My idea of concentrating on Ferdinand Porsche's military activities serves as a way to integrate the story of this great engineer from his birth to his death in a manner I have not previously achieved. The mind boggles at the size of the book that would be needed if this were also to include all of his work on passenger cars and commercial vehicles. While I greatly respect the latter, Porsche's military work commands immense respect for its breadth, depth and variety.

From my first thoughts of a work of this kind the publisher I had in mind was Pen & Sword Books with its long history of volumes in the military field. By the happiest of chances an editor with whom I had previously worked, Rupert Harding, had joined Pen & Sword. Many thanks are due to Rupert for his close attention to this project, to Sarah Cook for her thoughtful edit, to Jon Wilkinson for his striking jacket design, to Sylvia Menzies-Earl for her diligent typesetting and attention to detail, and to the rest of the team at Barnsley for their efforts on behalf of an unusual title.

Incredibly tolerant of my researches into Porsche history have been the custodians of the Porsche Archive in Stuttgart-Zuffenhausen. I am indebted to Klaus Parr for invaluable material on Ferdinand Porsche's early years. His successor Dieter Landenberger has been no less supportive. In both regimes Jens Torner has provided invaluable continuity and assistance.

The years from 1923 to 1928 when Porsche was Daimler and then Daimler-Benz board member for product development required special attention for this study. In this I was ably assisted by Gerhard Heidbrink and Dennis Heck of the Mercedes-Benz Classic Archive. Both Maria Feifel and Michael J. Jung assisted with many hitherto-unpublished images of the little-known but remarkable contributions made by Porsche in these years.

Speaking of contributions, I could never do justice to this or any other work without the support and encouragement that I receive from my wife Annette. It goes far beyond mere care and feeding of this author. Our 30 years together have been profoundly fulfilling in every respect. Annette assists me with administration, as does Mike Holland in the running of my own archive and the conduct of special research. My debts to both are not repayable.

Another source of first-hand information has been the Duxford archive of the Imperial War Museum. There Stephen Walton has been of great assistance. His findings and my others are integrated into the text where sources are cited in connection with quotations. All such sources are detailed in the bibliography. I hope to be excused for the absence of academic notes in a book that is intended to appeal to the general reader. Needless to say, any errors of fact or interpretation are to the account of the author alone.

BIBLIOGRAPHY

Anon, *The Times History of the War*, Woodward & Van Slyke, New York, ND.
Allen, Michael Thad, *Hitler's Slave Lords – The Business of Forced Labour in Occupied Europe*, Tempus, Stroud, 2004.
Beck, Alfred M., *Hitler's Ambivalent Attaché – Lt. Gen. Friedrich von Boetticher in America 1933–1941*, Potomac Books, Washington, 2005.
Bellon, Bernard P., *Mercedes in Peace and War*, Columbia University Press, New York, 1990.
Bingham, Victor F., *Major Piston Aero-Engines of World War II*, Airlife, Shrewsbury, 1998.
Bittorf, Wilhelm, 'Die Geschichte eines Autos', special edition, *Der Spiegel*, Limbach, 1954.
Bornemann, Manfred, *Geheimproject Mittelbau – Vom zentralen Öllager des Deutschen Reichs zur grössten Raketenfabrik im Zweiten Weltkrieg*, Dörfler, Eggolsheim, 1994.
Boschen, Lothar and Barth, Jürgen, *Porsche Specials*, Patrick Stephens, Wellingborough, 1986.
Bruce, J.M., *British Aeroplanes 1914–18*, Putnam, London, 1957.
Bullock, Alan, *Hitler – A Study in Tyranny*, Penguin Books, London, 1962.
Cronkite, Walter, *A Reporter's Life*, Ballantine Books, New York, 1996.
Dabrowski, Hans-Peter, *Flying Wings of the Horten Brothers*, Schiffer, Atglen, 1995.
Dallas, Gregor, *Poisoned Peace – 1945, The War that Never Ended*, John Murray, London, 2005.
Dooly, William G., Jr., *Great Weapons of World War I*, Army Times Publishing, New York, 1969.
Engelmann, Joachim, *V1 – The Flying Bomb*, Schiffer, Atglen, 1992.
Etzold, Hans-Rüdiger, Rother, Ewald and Erdmann, Thomas, *Im Zeichen der vier Ringe 1873–1945. Band 1*, Edition quattro GmbH, Ingolstadt, 1992.
Etzold, Hans-Rüdiger *et al*, *Der Käfer – eine Dokumentation II*, Motorbuch Verlag, Stuttgart, 1992.
Fedden, Sir Roy, 'Inquest on Chaos – What Shall be Done with Germany's Research Equipment?', *Flight*, London, 29 November 1945.
Fest, Joachim, *Speer – The Final Verdict*, Weidenfeld & Nicolson, London, 2001.
Forty, George, *World War Two Tanks*, Osprey, London, 1995.
Fowler, Will, *Kursk – The Vital 24 Hours*, Spellmount, Staplehurst, 2005.
von Frankenberg, Richard, *Porsche – the Man and His Cars*, G.T. Foulis & Co., London, 1961.
von Frankenberg, Richard, writing as Quint, Herbert A., *Porsche – Der Weg eines Zeitalters*, Steingrüben Verlag, Stuttgart, 1951.
Fröhlich, Michael, *Kampfpanzer Maus – Der überschwere Panzer Porsche Type 205*, Motorbuch Verlag, Stuttgart, 2013.
Fürweger, Wolfgang, *Die PS-Dynastie – Ferdinand Porsche und seine Nachkommen*, Überreuter, Vienna, 2007.
Gander, Terry, *The German 88 – The Most Famous Gun of the Second World War*, Pen & Sword Military, Barnsley, 2009.
Georg, Friedrich, *'Unternehmen Patentraub' 1945 – Die Geheimgeschichte des grössten Technologieraubs aller Zeiten*, Grabert Verlag, Tübingen, 2008.
Glanfield, John, *The Devil's Chariots – The Birth and Secret Battles of the First Tanks*, Sutton Publishing, Stroud, 2001/2006.
Glantz, David M. and House, Jonathan M., *The Battle of Kursk*, University Press of Kansas, Lawrence, 1999.
Guderian, General Heinz, *Panzer Leader*, E.P. Dutton, New York, 1952.
Gudgin, Peter, *The Tiger Tanks*, Arms and Armour Press, London, 1991.
Hanfstaengl, Ernst 'Putzi', *Hitler: the Missing Years*, Arcade, New York, 1994.
Hauke, Erwin, Schroeder, Walter and Tötschinger, Bernhard, *Die Flugzeuge der k.u.k. Luftfahrtruppe und Seeflieger 1914–1918*, H. Weishapt Verlag, Graz, 1997.
Hitler, Adolf, *Mein Kampf*, Pimlico, London, 1992.
Holmes, Richard, *The World at War – The Landmark Oral History from the Previously Unpublished Archives*, Ebury Press, London, 2007.
Hopfinger, K.B., *The Volkswagen Story*, G.T. Foulis, Henley-on-Thames, 1971.
Horch, August, *Ich baute Autos – Vom Schmiedelehrling zum Autoindustriellen*, Schützen Verlag, Berlin, 1937.
Hurst, Kenneth A. MIMechE (SA), *William Beardmore – Transport is the Thing*, National Museums of Scotland, Edinburgh, 2004.

James, Harold, *The German Slump – Politics and Economics 1924–1936*, Clarendon Press, Oxford, 1986.
Jenks, William A., *Vienna and the Young Hitler*, Columbia University, New York, 1960.
Jentz, Thomas L. and Doyle, Hilary L., *Panzerkampfwagen VI P (Sd.Kfz.181) – The History of the Porsche Typ 100 and 101 also known as the LEOPARD and TIGER (P)*, Darlington Productions, Darlington, 1999.
Jentz, Thomas L. and Doyle, Hilary L., *Schwere Panzerkampfwagen Maus and E 100 – development and production from 1942 to 1945*, Panzer Tracts, Boyds, 2008.
Kaes, Ghislaine, *Ferdinand Porsche – Einhundert Jahre nach seiner Geburt*, unpublished manuscript, Stuttgart, 1975.
Keimel, Reinhard, *Luftfahrzeugbau in Österreich – von den Anfängen bis zur Gegenwart – Enzyklopädie*, Aviatik Verlag, Oberhaching, 2003.
Kershaw, Ian, *Hitler – 1936–1945: Nemesis*, Allen Lane, London, 2000.
Kershaw, Ian, *Hitler – 1889–1936: Hubris*, Penguin Books, London, 2001.
Kilduff, Peter, *The Red Baron – Beyond the Legend*, Ted Smart, London, 1994.
Kirchberg, Peter, *Grand-Prix-Report – Auto Union 1934 bis 1939*, Motorbuch Verlag, Stuttgart, 1982.
Knopp, Guido et al, *Hitlers Manager*, C. Bertelsmann Verlag, Munich, 2004.
Liddell Hart, B.H., *The German Generals Talk*, William Morrow, New York, 1948.
Loubet, Jean-Louis, *Citroën, Peugeot, Renault et les autres*, Le Monde-Èditions, Paris, 1995.
Ludvigsen, Karl, *Mercedes-Benz – Quicksilver Century*, Transport Bookman, Isleworth, 1995.
Ludvigsen, Karl, *Battle for the Beetle – The untold story of the post-war battle for Adolf Hitler's giant Volkswagen factory and the Porsche-designed car that became an icon for generations around the globe*, Bentley Publishers, Cambridge, 2000/2013.
Ludvigsen, Karl, *Porsche – Excellence Was Expected*, Bentley Publishers, Cambridge, 2003/2008.
Macksey, Kenneth, *Why the Germans Lose at War – The Myth of German Military Superiority*, Greenhill Books, London, 1996/2006.
Milsom, John F., *German Super-Heavy Tanks* in *Armoured Fighting Vehicles of Germany – World War II*, Barrie & Jenkins, London, 1978.
Mitcham, Samuel W., *Why Hitler? – The Genesis of the Nazi Reich*, Praeger, Westport, 1996.
Mommsen, Hans and Grieger, Manfred, *Das Volkswagenwerk und seine Arbeiter im Dritten Reich*, ECON Verlag, Düsseldorf, 1996.
Mönnich, Horst, *The BMW Story – A Company In Its Time*, Sidgwick & Jackson, London, 1991.
Morrow, John Howard, Jr., *Building German Airpower, 1909–1914*, University of Tennessee Press, Knoxville, 1976.
Nahum, Andrew, *The Rotary Aero Engine*, Her Majesty's Stationery Office, London, 1987.
Nelson, Walter Henry, *Small Wonder*, Little Brown, Boston, 1970.
Njipe, George M., *Blood, Steel and Myth – The II. SS-Panzer-Korps and the road to Prochorowka. July 1943*, RZM Publishing, Stamford, 2011.
O'Connor, Dr Martin, *Air Aces of the Austro-Hungarian Empire 1914–1918*, Flying Machines Press, Mountain View, 1994.
Olley, Maurice, *The Motor Car Industry in Germany During the Period 1939–1945*, His Majesty's Stationery Office, London, 1949.
Osteroth, Reinhard, *Ferdinand Porsche – Der Pionier und seine Welt*, Rowolt Verlag, Hamburg, 2004.
Overy, Richard, *Goering*, Phoenix Press, London, 2000.
Overy, Richard, *Interrogations – Inside the Minds of the Nazi Elite*, Penguin, London, 2002.
Pfundner, Martin, *Austro Daimler und Steyr – Rivalen bis zur Fusion*, Böhlau Verlag, Vienna, 2007.
Picard, Fernand, *L'Épopée de Renault*, Albin Michel, Paris, 1976.
Pomeroy, Laurence, *The Grand Prix Car*, Motor Racing Publications, Abingdon-on-Thames, 1949.
Porsche, Dr. Ing. h.c. Ferry with Bentley, John, *We at Porsche*, Doubleday, Garden City, 1976.
Porsche, Professor Dr. Ing. h.c. Ferry with Günther Molter, *Ferry Porsche: Cars Are My Life*, Patrick Stephens, Wellingborough, 1989.
Post, Dan R., *Volkswagen – Nine Lives Later*, Horizon House, Arcadia, 1966.
Prášil, Michal, *Skoda Heavy Guns*, Schiffer Military/Aviation History, Atglen, 1997.
Railton, Arthur, *'The Beetle'*, Eurotax, Pfäffikon, 1985.
Rauck, Max and von Lengerke, Friedrich B., *Die Renngeschichte der Daimler-Benz Aktiengesellschaft und Ihrer Ursprungsfirmen*, Daimler-Benz, Stuttgart, 1939.
Rauscher, Karl-Heinz and Knogler, Franz, *Das Steyr-Baby und seine Verwandten*, Weishaupt Verlag, Gnas, 2002.
Reeves, Lt. Col. G.C., *The War-Time Activities of Dr. Ing. h.c. F. Porsche, K.G.*, Combined Intelligence Objectives Sub-Committee, London, 1948.
Reich, Simon, *The Fruits of Fascism – Postwar Prosperity in Historical Perspective*, Cornell University Press, Ithaca, 1990.
Roberts, Stephen H., *The House That Hitler Built*, Harper & Brothers, New York, 1938.

Rohland, Walter, *Bewegte Zeiten – Erinnerungen eines Eisenhüttenmannes*, Seewald Verlag, Stuttgart, 1978.
Schreier, Konrad F., Jr., *VW Kubelwagen Military Portfolio*, Brooklands Books, Cobham, c. 1991.
von Senger und Etterlin, Dr F.M., *German Tanks of World War II*, Galahad Books, New York, 1969–1973.
Smelser, Ronald, *Robert Ley – Hitler's Labour Front Leader*, Berg, Oxford, 1988.
Smith, Michael, *The Secrets of Station X – How the Bletchley Park Codebreakers Helped Win the War*, Biteback Publishing, London, 2011.
Snyder, Dr Louis L., *Encyclopedia of the Third Reich*, Paragon House, New York, 1989.
Speer, Albert, *Inside the Third Reich*, Book Club Associates, London, 1971.
Speer, Albert, *Infiltration – How Heinrich Himmler Schemed to Build an SS Industrial Empire*, Macmillan, New York, 1981.
Speer, Albert, *Spandau – The Secret Diaries*, Phoenix Press, London, 2000.
Spielberger, Walter J., *Motorisierung der Deutschen Reichswehr 1920–1935*, Motorbuch Verlag, Stuttgart, 1979.
Spielberger, Walter J. and Feist, Uwe, *Panzerkampfwagen VI – Tiger I and 'Königstiger'*, Feist Publications, Berkeley, 1968.
Spielberger, Walter J. and Milsom, John, *Elefant and Maus (+E-100)*, Profile Publications, Windsor, 1973.
Starr, Fred, 'Development of the Poppet Type Exhaust Valve in the Internal Combustion Engine,' *The History of Engineering and Technology*, The Newcomen Society, London, Volume 82, Number 2, July 2012.
Strache, Dr Wolf and Schrader, Halwart, *100 Years of Porsche Mirrored in Contemporary History*, Porsche, Stuttgart, 1975.
Stuck, Hans, *Tagebuch eines Rennfahrers*, Moderne Verlag, Munich, 1967.
Svirin, Mikhail, *Das Schwere Sturmgeschütz Ferdinand*, Armada, Moscow, 2002.
Thiriar, Dr Michael, *Porsche Epic – Volume 1*, Editions Eder, Drogenbos, 1999.
Tooze, Adam, *The Wages of Destruction – The Making and Breaking of the Nazi Economy*, Allen Lane, London, 2006.
Trevor-Roper, Hugh R., preface and introduction, *Hitler's Table Talk 1941–1944 – His Private Conversations*, Phoenix Press, London, 2000.
Tubbs, D.B., 'Ferdinand Porsche', in *Automobile Design: Great Designers and Their Work*, edited by Ronald Barker and Anthony Harding, Robert Bentley, Cambridge, 1970.
van der Vat, Dan, *The Good Nazi – The Life and Lies of Albert Speer*, Weidenfeld & Nicholson, London, 1997.
Wik, Reynold M., *Benjamin Holt & Caterpillar Tracks & Combines*, American Society of Agricultural Engineers, St Joseph, 1984.
Zetterling, Niklas and Frankson, Anders, *Kursk 1943 – A Statistical Analysis*, Frank Cass, Abingdon, 2000,
Zumpf, Peter, Geissl, Gerhard and Weinzettl, Gerhard, *Austro Daimler Wiener Neustadt*, merbod-verlag, Wiener Neustadt, 2003.

INDEX

Page numbers in *italics* refer to illustrations

A-Train 46, 47
A7V tank 88–9
AAZ, Vienna 35, 36, 47, 48, 104
Adler 78, 136, 214
Aëro-Daimler 35, 36, 37, 38, *42*
airships 28–30, 102
Alkett 184, 219–20, 223, 224–5
Ambi-Budd 136, 146, *152*, *153*, *155*
Argus 98, 200, 233–4, 242, *243*, 271
Austria
    Allied occupation 259
    German annexation 16, 132
    independent state 70
    slave labour 205
    World War II 255, 256
Austria-America Rubber Factory 28
Austria-Hungary
    army, motorisation of x, 1, 2–3, 11–12, 17
    aviation, early 28–44
    aviation, World War I 60–9, 96, 270
    Bosnia-Herzegovina 19, 24
    disintegration of 70
    food shortages 69–70
    Habsburg monarchy 1, 2, 24, *24*, 70
    World War I 40, 45–9, 58–69
Austrian Aircraft Factory AG (ÖFFAG) 41, *63*, 66, 69, 72
Austrian Motor Aircraft Company 29, 37, 41
Austro Daimler 3, 9, *13*, 14–17
    aviation, early *28*, 29, 32–7, 38, 39–44
    aviation, World War I 60–9, 271
    Castiglioni, Camillo and 71, 73
    demise of 55
    land trains *45*, 52, 53–5, 271
    M 06 artillery tug *18*
    M 08 'Robbe' artillery tug *18*, 19, 21
    M 09 'Titan' artillery tug 21
    M 12 'Hundred' artillery tug *23*, 25, 27
    M 17 'Goliath' artillery tug 25–6, *27*, 72, 77, 211
    Maja cars 15, 90, 109
    Pferd tractor 56–8, 72
    Prince Heinrich Trials viii, 20–1, *31*, 32
    railway locomotives 58
    Skoda control of 22–3, 41, 73
    see also Aëro-Daimler
Austro Daimler-Puch 104
Auto Union AG
    formed 107
    P-Wagen 111
    racing programme 112, 113, 124, 131, 137, 182
Aviatik aircraft *40*, 41, 64

B-Train *46*, 47, 50, 58, 70
Barbarossa, Operation 168
Beardmore Aero-Engines Ltd *43*, 44

Becht, Walter 145, 205
Becker, General Emil 131, 134, 160–1
Béla Egger & Company *see* United Electrical Corporation (VEAG)
Belgium
    fortifications 22, 24–5
    World War I 24–5, 27, 47, 48
Benz 21, 40, 77, 96, 101
Benz, Karl 2
Berge, Ernst 74–5
Berger, Alfred 96, 99
Bierenz, Josef Eduard 2, 3
Blaicher, Ernst 187, 196, 199
Blomberg, General Werner von 92–3, 117
BMW
    inter-war period 95, 124, 131
    motorcycles 149
    V-1 turbojet 242, *243*
    World War II dominance 271
Borgward 265, 273, 274
Bormann, Martin *151*, 197
Bosnia-Herzegovina 19, 24, 45, 47
Boxan, Walter 104, 106
Brandenburg aircraft *41*, *42*, 63
Brauchitsch, Manfred von *110*, 111
Braun, Wernher von 242, 257, 265
Britain
    Porsche, post-war 259–60
    V weapons 231, 234, *236*
    World War I 44, 60, 88–9, 164–5
    World War II *142*, 143, 155, 177, 190, 207, 235, 245
Bulge, Battle of the (1944-45) 188
Büssing-NAG 81, 85, 87

C-Train 50–5, 218, 271
Caracciola, Rudolf 109
Castiglioni, Camillo 28–9, 71–3, 95, 104
Churchill, Winston 43, 138, 259
Clerget engines 32
Cody, Samuel Franklin 43
Czechoslovakia 53–5, 70, 132, 146, 249, 265

Daily Mail Circuit of Britain Air Race (1911) 42–3
Daimler, Gottlieb 2, 3
Daimler, Paul 3, 7, 14, 16–17, 23, 74, 75, 77, 79, 93
Daimler-Benz AG
    Allied bombing 248
    aviation, inter-war period 93–102, 124
    BF2 marine engine 97, 99, *100*, 101
    D IIa aero engine 95
    DB 602 aero engine 102
    DB 603 aero engine 124, 217
    DB 605 aero engine 204
    F1 aero engine 94

F2 aero engine 96, 97, 98
F3 aero engine 98–9
F7502 aero engine *93*, 94, 98
formation of 77
GT I tank 89–93, 98, 162, 170, 270
MB 502 marine engine 102
MB 503 tank engine 217, 219, 226
MTW 1 reconnaissance vehicle 81–3, *84*, 85, *86*, 87
OF2 aero engine *100*, 101–2
tank proposals 175, 176
war production 131
World War II dominance 271
see also Mercedes-Benz
Daimler-Motoren-Gesellschaft
  A7V tank 88–9
  Benz, merger with 77
  DZ I tug 77
  DZ III tug 77
  DZVR armoured car *76*, 77–8
  G 1 78–9
  post-World War I 74–87
  purchases Mercedes 14
  separates from Austro Daimler 22
de Dion 7, 9
Derbuel, Victor 208, 209
désaxé cylinder arrangement 32, 33, 61
Diesel, Rudolf 3
Dietrich, Lieutenant General Josef 'Sepp' 118, 132
Dirmoser, Oswald 23, 48, 50
Dirmoser, Richard 23, 48, 50, 52
DKW *89*, 90, 98, 107, 136, 273
Dr. Ing. h.c. F. Porsche GmbH 106
Drauz coachbuilder *147*, 148
Dusio, Piero 263–4
Dustbin, Camp, Frankfurt 257, 258

Egger, Béla 3–4, 6
Egger-Lohner electric car 4
Eschershausen underground factory 238, 248
Etrich, Igo 30–1, 32, 35, 37–8, 41, 42
Exelberg hillclimb *1*, 11

Fallersleben factory *123*, *129*
  Allied bombing of 234–6, *237*, 245, *245*, 248
  bombing of 153
  construction *119*, 120, *122*, 130, 161, 245
  DAF control of 234
  dispersal of 248
  Kübelwagen production *136*, 138, *139*, 146, 244, 248
  Panzerfaust production 244
  prisoner labour 151–2, 237, 239, 248, 269–70
  Schwimmwagen production 152, 244
  US occupation 256
  V-1 production 233–4, 236, 239, 270
  Volkswagen production 132, 270
  war production 123, 125, 127–8, 132, 207, 208, 233, 244, 245, 263, 269
Fiat 34, 73, 96, 98

Fichtner, Lieutenant Colonel Sebastian 134, 169
Fischamend, Austria 29, 41, 60, 68–9, 72, *106*
Fischer, Eduard 2, 14, 16, 17, *18*, 21, 23, 28, 74
Fischer Brothers machine shop 2, 13
Focke-Wulf 169, 244, 264
Ford 108, 138, 146, 256
France
  Allied bombing 247
  fortifications 22
  Porsche, detention of 261–2, 263–4, 265–6
  Volkswagen, interest in 260–1, 262
  World War I 47, 48, 88–9, 164
  World War II 146, 207, 227, 234, 244, 245, 246–8
Frankenberg, Richard von *5*, 108, 131, 132, 207, 227, 248–9, 268, 269
Franz Ferdinand, Archduke
  army, motorisation of x, 1, 11–12
  assassination of 24
Franz Josef I, Emperor 2, 11, 36, 59
Friedmann lubricator 34, *39*, 40, 61, 67
Fröhlich, Karl 104, *106*, 116, 253

Gensberger, Karl 225, 227
Georg Fieseler Werke 231, 233
German Workers Front (DAF) 118, 120, 125, 134, 234, 237
Germany
  Army, motorisation of 1–2, 17, 116
  aviation, early 37
  aviation, inter-war period 93–102, 270–1
  Bundeswehr 273
  German Grand Prix 109
  inter-war period 55, 106–7
  Kriegsmarine 99, 101, 102, 169, 175, 184
  post-World War I 74, *75*
  Reichswehr 77–8, 81, 116
  Soviet invasion of 146, 195, 208, 229–30
  Soviet Union, invasion of 55, 140, 143, 168, 169, 183, 186, 211
  war production 160–1, 168, 174, 207
  Wehrmacht 55, 116, 160
  World War I 24–5, 27, 40, 48, 64, 69, 70, 88–9
  World War II 120, 140–3, 168–9, 177, 183, 184, 186, 188–93, 194–5, 207–8, 245
  see also Luftwaffe
GEZUVOR 115, 117, 132
Ginzkey brothers mill *5*, 6
Glanfield, John 27, 164–5
Gmünd, Austria 250, *251*, 252–3, *254*, 255, 256, 258, 259–60, 262
Goebbels, Josef *128*, 129, 231
Goering, Hermann
  V-1 flying bomb 233
  Volkswagen *121*
  war production 125, 160, 161, 168, 178, 195, 219
Goldinger, Josef 207, 255, 258
Greece 140, *143*, 167, 274
Guderian, General Heinz

Inspector General of the Armed Forces 188
Panther tank 191
Ram-Tiger tanks 183
T-34 tank 169
tank development 197, 199
Type 130 *Ferdinand* tank destroyer 184, 189, 191–2
Type 205 *Maus* super tank 216, 219, 222, 223

Hacker, Oskar 125, 168
Hansa-Brandenburg aircraft *41*, 63
Heereswaffenamt (HWA)
   artillery tugs 211–13
   Kübelwagen 138, 140
   passenger cars 78–81, 134
   Porsche, relations with 131–2, 162, 168, 179
   reconnaissance vehicles 81, 87
   Schwimmwagen 147, 148, 152, 153
   super heavy tanks 215, 219, 223, 224, 227, 228
   tank destroyers 167, 184
   tank engines 201–5
   tanks 132, 161, 163, 167–8, 169, 199–200
   Volkswagen 116, 117, 118
Heinkel, Ernst 41, 127, *128*, 242, 264
Henschel
   Jagdtiger tank destroyer 187
   VK3001(H) tank 163
   VK4501(H)/Type 101 Tiger tank 173, 176, 177, 178, 179, 183, 189, 197, 215
   VK4503(H) Tiger II tank 199
Hess, Rudolf 111
Heydekampf, Gerd Stieler von 179, 187–8, 197, 207, 219–20, 257
Hiero engines 40, 42, 60
Himmler, Heinrich 149, 151, *152*, 206, 236, 237, 255–6, 269–70
Hindenburg, General Paul von 58, 69
Hitler, Adolf
   Albert Speer, relationship with 175, 214
   armaments, interest in 130, 163, 166, 167, 175, 215
   artillery tug 213
   assassination attempt 208
   death of 256
   early life 109
   Germany, defence of 146
   Mercedes-Benz AG G 4 81
   motoring enthusiast 109, *110*, 111
   Panzerfaust 244
   Porsche, first meetings 108–9, 112
   Porsche, relationship with 123, 163, 175, 178, 197, 207, 269
   rise to power 109
   Schwimmwagen 149, 151, *152*
   super heavy tanks 214, 219, 220, *221*, 222–3, 228–9, 230, 270
   tank destroyers 186, 187, 189, 190
   tank development 167, 173, 175, 177–9, 180–1, 183–4, 195
   V weapons 240, 242
   Volkswagen programme 113–14, 116, 117, 118, *119*, 120, 140, 159
   war economy 160–1
Hobart, Major General Percy 155
Holt caterpillar tracks 27, 88, 164
Horch 74, 78, 107, *112*, 142
Hötzendorf, General Franz Conrad von 23, 30, 31
Hruska, Rudolf 182, 255
Hugo Schneider AG 244
Hühnlein, Adolf *110*, 117, 137
Hungary
   independence 70
   see also Austria-Hungary

Illner, Karl 32, *35*, 36–7, 41, *42*, 43
Italy
   Austria-Hungary, alliance with 2
   fortifications 22, 23, *55*
   World War I 48, 52, 53, *54*, *55*, 69
   World War II 186, 194, 195, 244

Jagdtiger tank destroyer 187–8, *204*, 206–7
Jakob Lohner & Co. 3, 8, 37–8
   Pfeilflieger 42, 43
Japan 130, 177, 242, 256
Jeep 146, 155, 157
Jellinek, Emil *10*, 13–15, 90
Junkers
   Ju 88 125, 127, *128*, 233
   Ju 188 127–8
   Ju 388 128
   Nordhausen factory *238*
   Peugeot 246
   World War II dominance 124, 129, 271

Kaes, Ernst 20
Kaes, Ghislaine *20*, 68, 105, 106, 174–5, 216, 262, 265, 267
Kaes, Herbert 258, 262
Kales, Josef 104, 106
Kama proving grounds 85, 92
Kamm, Wunibald 205, 206
Karl I, Emperor 70
Kaulla, Alfred von 75
KdF-Stadt 123, 130, 235, 245, 256
KdF-Wagen *see* Volkswagen
KdF-Werke *see* Fallersleben factory
Keitel, Field Marshal Wilhelm 149, 151, 240
Kern, Hans 251, 253, 255
Kissel, Wilhelm 95, 131, 162
Kitzbühl, Austria 265, 266
Klemm, Hanns 94, *95*, 98
Kniepkamp, Ernst 132, 169, 180, 199, 218–19
Knoller, Richard 68–9
Köhler, Otto 15, *37*, 58, 72, 75, 81, 89, 226
Komenda, Erwin 106, 116, 255, *265*
Kraft durch Freude (KdF) (Strength through Joy) movement 118, 138, 147
Krobatin, General Alexander von 49, 53
Krupp
   88mm cannon 167
   420mm howitzer 25, 48

Allied bombing of 207, 223–4
E-100 super heavy tank 214, 230
passenger vehicles 79
Porsche tank components 168, 170, 171, 173, 178, 181, 182, 218, 219, 223, 228
tank prototypes 92, 93
war production 161
Kübelwagen 78, 116, 267
development 132–8, 255
production *136*, 138, *139*, 140, 146, 244, 248
Kursk, Battle of (1943) 188–93, 223, 270

Lafferentz, Bodo *117*, *119*, 120, 121, 125, 213, 234, 236–7, 263
land trains 45, *46*, 47–8, 50–5
Landwehr von Pragenau, General Ottokar 45, 69, 70
Lang, Hermann *110*
Lang, Richard 74, 75, 95
Lebaudy brothers 29
Ley, Robert *117*, *118*, *119*, 120, *121*, 129, 132, 134, 237, 250
Libya 38, 143
Liddell Hart, Basil H. 175, 176
Liège, Belgium *24*, 25
Lohner, Ludwig *1*, 3–4, 7–9, 11, 13, 14, 37
Lohner-Porsche Mixte x, 1, 3, 7, 8, 9–12, 13
Luftwaffe 123, 124, 125, 175, 257, 271
V-1 flying bomb 231, 233–4, 240, 243, 269
LZ 129 *Hindenburg* 102
LZ 130 *Graf Zeppelin* 102

Magirus 81, 82, 85
MAN 101, 169, 175, 176
Mannsbarth, First Lieutenant Franz 29–30
Manstein, General Field Marshal Erich von 183, 189, 190, 191, 193
Manteuffel, General Hasso von 175–6
Maria Theresa, Empress 16
Matilda tank 167
Maybach engines
airships 102
Henschel Tiger tank 176, 183
Magirus reconnaissance vehicle 85
Panzerkampfwagen IV 163
Porsche Tiger tank 182
Type 130 Ferdinand tank destroyer 184
Type 205 Maus super heavy tank 217
Type 245 tank 200
Mercedes
Daimler purchases 14
Lohner-Porsche engine 10
World War I 40, 65
Mercedes-Benz AG
170V 146
G 1 78–9, 87
G 3 79, *80*, 81
G 4 *80*, 81, 92
racing programme 112
Stuttgart 200 78
Type 150 *114*
see also Daimler-Benz AG

Messerschmitt, Willy 127, *128*, 132, 264
Mickl, Josef 41, 61, 72, 106, 116, 131, 205, 213, 252
Milch, Erhard 123, 233
Mixte powertrain
aero engines 67–8
land trains 45
Lohner-Porsche x, 1, 3, 7, 8, 9–12, 13
Type 100 'Leopard' tank 164, *165*
Type 130 Ferdinand tank destroyer 184
VK4502 (P)/Type 180/181 Tiger II 199
Model, General Walter 189, 191–2
Möwe (Gull) aircraft 32, 35, 61
Müller, Alfred 206–7

Nallinger, Fritz 100, 101, 217, 226
National Socialist Motoring Corps (NSKK) 117, 133, 134
NATO 273, 274
Netzker, Paul 164, 202, 210, 243
Neubauer, Alfred 52, 72, 74, 75
Nibel, Hans 96, 99, 101
Nibelung Works, St Valentin 161–2, *166*, *174*, 177, 181, 184, *187*, *189*, 195–6, 270, 273
Nice Week motoring competitions 10–11
Nordhausen underground factory 237, *238*, 239, 240
North Africa 140, 141–2, 143, 167, 177, 181, 186
NSU 107, 111, 113, 131, 149

Oertzen, Baron Detlof von 111–12, 113
Opel 21, 128, 136, 149, 197
Österreichische Daimler-Motoren-Gesellschaft *see* Austro Daimler

Panzerfaust 229, 244
Panzerjäger 183, 184, 191
Panzerkampfwagen IV 163, 166, 181, 183
Panzerkampfwagen V Panther 175–6, 181, 189, 191, 193, 206, 207
Parker, Richard 259–60
Parseval, Major August von 28, 29
Paulal, Karl 37
Pechan, Professor Joseph 6
Pemberton-Billing PB 29E 43–4
Petróczy von Petrócz, Major Stephan 68
Peugeot 146, 207, 234, 239, 245–8, 261, 262, 263
Peugeot, Jean-Pierre 246–7, 262, 263
Pfeil (Arrow) aerial torpedo 243–4
Picard, Fernand 261, 262
Piëch, Anton (son-in-law of FP) *105*
French detention of 261–2, 263–4
health 260
legal adviser 105
Renault 247, 262–3
war production 125, 128, 207, 234, 236, 237
Zell am See 250
Piëch, Ernst (grandson of FP) *viii*, 264
on Ferdinand Porsche vii–ix, 11
gas turbines 207

property purchases 249–50, 251
V-1 flying bomb 243
wind turbines 213
Piëch, Ferdinand K. (grandson of FP) 267
Piëch, Louise (née Porsche) 258, 264, 266
Pischof, Alfred Ritter von 32, 39
Porsche, Aloisia 'Louise' Johanna (née Kaes, wife of FP) 11, 20
Porsche, Anna (née Ehrlich, mother of FP) 4, 5–6
Porsche, Anna (sister of FP) 4
Porsche, Anton (brother of FP) 4
Porsche, Anton (father of FP) 4–6, 10, 20
Porsche, Ferdinand 161, 237, 256, 260, 267
    Albert Speer and 123, 173–4, 177, 195, 197, 214, 257
    Allied bombing 248–9
    Allied debriefing of 256–7, 258, 259
    Archduke Franz Ferdinand and x, 1, 11–12
    Austro Daimler 14, 15–16, 17, 20–1, 31, 71–3, 271
    Auto Union racing cars 112, 131, 137, 182
    aviation, early 28–30, 31, 32–7, 38, 39–44
    aviation, inter-war period 93–102, 124, 270–1
    birth and early life 4–6
    Daimler-Benz 77–87, 89, 92, 93–102, 103, 131, 270
    Daimler-Motoren-Gesellschaft 74–6
    death of 267, 271
    detention of 260, 261–2, 263–4, 265–6
    early career 4, 6–7, 12
    education 4, 5, 6
    factory dispersal 249, 251, 253, 254, 255
    France, Volkswagen interest in 260–1
    Fritz Todt and 161, 168, 174
    gas turbine engines 205, 206
    Heereswaffenamt and 131–2, 162, 168, 179
    Hitler, first meetings 108–9, 112
    Hitler, relationship with 123, 163, 175, 178, 197, 207, 269
    honours 59, 70–1, 76, 123, 127, 128, 178, 207, 222, 267
    independent engineer 105–8
    internment of 257–8
    Jagdpanzer E25 (Arta) 200–1
    Jagdtiger tank destroyer 187
    Kübelwagen 133, 134–5, 137, 141–2, 244, 267
    land trains 45, 46, 47–8, 50–3, 271
    Ludwig Lohner and 1, 4, 7–9, 11
    M 06 artillery tug 18
    M 08 'Robbe' artillery tug 18, 19, 21
    M 09 'Titan' artillery tug 21
    M 12 'Hundred' artillery tug 23
    M 17 'Goliath' artillery tug 25–6, 27, 72
    marriage 20
    Mixte powertrain x, 1, 3, 7, 8, 9–12, 45, 67–8, 164, 165, 184
    Nibelung Works, St Valentin 161–2, 270
    Panzer Commission of the Armaments Ministry chairman 168, 175, 181, 195
    Peugeot 245–6, 261, 262, 263
    Pferd tractor 56–8, 72
    Porsche 356 265–6
    prisoner labour 151–2, 237, 269–70
    Renault 262–3
    Schwimmwagen 147, 149, 264
    self-confidence ix, 12
    Soviet offer 107–8
    Steyr 103–5, 125–6
    Type 100 'Leopard' tank 162, 163–4, 165–6, 170, 270
    Type 101 Tiger tank 173, 174, 176, 177–8, 179, 182, 183, 270
    Type 130 Ferdinand tank destroyer 186, 189, 270
    Type 175 *Radschlepper Ost* artillery tug 212–13
    Type 205 *Maus* super heavy tank 215, 216, 219, 222, 225, 226, 227–8, 270
    Type 212/220 Sla 16 tank engine 204, 217
    Type 245 tank 200
    V-1 flying bomb 237, 238, 240, 242, 243, 267
    Volkspflug 129–30
    Volkswagen 113–17, 119, 120, 121, 197, 264, 267, 268
    war production 123–4, 125, 128–9
    World War I 27, 45–59, 60–9, 70–1
Porsche, Ferdinand Alexander (grandson of FP) 267
Porsche, Ferdinand Anton Ernst 'Ferry' (son of FP) 20, 103, 107, 267
    factory dispersal 249, 251, 253, 255
    France, Volkswagen interest in 261
    French arrest of 261–2, 264
    Guderian, Heinz 197, 199
    Hitler and 197
    Kübelwagen 135, 138
    Porsche archives 248
    Porsche KG 106, 207
    Porsche Type 356 car 265–6
    post-war military work 273
    Schwimmwagen 147, 148, 149, 151, 152
    Steyr-Werke 104
    Tiger tanks 178, 183
    Type 205 *Maus* super heavy tank 226
    V-1 flying bomb 231, 243
    Volkswagen 117, 118, 121, 209
    World War I 70
    Zell am See 250
Porsche, Hedwig (sister of FP) 4
Porsche, Louise Hedwig Anna Wilhelmine Marie (daughter of FP) 20, 70
    see also Piëch, Louise (née Porsche)
Porsche, Oscar (brother of FP) 4, 6, 10
Porsche, Wolfgang (grandson of FP) 268
Porsche KG
    Allied bombing 251–2
    income and expenditure 146
    Jagdpanzer E25 (Arta) 200–1
    Jagdtiger tank destroyer 187–8, 204, 206–7
    Kübelwagen 132–3, 136, 244, 255, 267
    Porsche 356 265–6

Schwimmwagen 147–8, 150, 151, 152–3, 244
Type 101 tank engine 163–4, *201*
Type 130 Ferdinand tank destroyer 167–8, 182–7, 189–90, 191–3, *194*, 195, *196*, 217, 270
Type 136 wind turbine 213
Type 175 *Radschlepper Ost* artillery tug 211–13, 217
Type 205 *Maus* super tank 214–30, 270
Type 212/220 Sla 16 tank engine 201–5, 217
Type 245 tank *199*, 200
Type 293 personnel carrier 208
Type 300 turbojet engine 242–3, 244
Type 305 gas-turbine engine *205*, 206
Type 312/313 tractor 259
Type 597 *Jagdwagen* (Hunter) 271–2, *273*
Type 714 Leopard tank *273*, 274
VK3001 (P)/Type 100 Leopard tank 163–4, 165–6, *172*, 270
VK4501 (P)/Type 101 Tiger tank 170–1, *172*, 173, *174*, 176, 177–81, 183, 215, 217, 270
VK4502 (P)/Type 180/181 Tiger II tank *198*, 199
Volkswagen *122*, 145–6, 208–10
powered observation platforms 68, 69
Prince Heinrich Trial *viii*, 20–1, *31*, 32
prisoner and slave labour 151–2, 205, 237, 239, 248, 269–70
Puch 73, 104, 113

Rabe, Hans 147, 178, 181, 255
Rabe, Karl *224*, 226
    Allied occupation 258
    Austro Daimler 72, 73
    factory dispersal 251, 252, 253, 255
    joins Porsche KG 105, *106*
    Kübelwagen 133
    Pferd tractor 57–8
    post-war *260*
    post-war military work 273
    Schwimmwagen 147
    Type 100 Leopard tank 162
    Type 130 Ferdinand tank destroyer 167
    Type 175 *Radschlepper Ost* artillery tug 211
    Volkswagen 116
Rapp, Karl 63, 64
Reichenberg manned V-1 240, *241*, 242
Reimspiess, Franz Xaver 106, 116, 162, 165, 208, 273
Reitsch, Hanna 240, 242
Renault 95, 247, 262–3, 264
Rheinmetall 91, 92, 93, 200, 274
Ribbentrop, Joachim von 55, 132
Röchling, Hermann 240, 265
Rohland, Walter
    Allied bombing 207
    on Ferdinand Porsche 163
    imprisonment 265
    internment of 257
    tank production 166, 169, 178, 183, 186
    Tiger tank 180

Rolls-Royce 65, 132, 202
Rommel, Field Marshal Erwin 16, 141–2, 177, 181
Rosenberger, Alfred 105, 106, *110*, 111
Royal Aircraft Factory 43, 44
Rumpler, Edmund 37, 72, 105
Rupilius, Emil 182, 255
Russia
    World War I 53, 58, 70
    see also Soviet Union

Saint-Chamond tank 164
Saint-Germain, Treaty of (1919) 70, 71
Salvator, Archduke Leopold 17, 18–19, 20–1
Saur, Karl-Otto 174, 176, 187, 222
Schacht, Hjalmar 74, 257
Schlieffen, Field Marshal Alfred von 24
Schmid, Leopold 215–16, 255
Schnellboot 101, 102
Schönfeld, Count Heinrich *14*, 17, 21, 35
Schwimmwagen 150, *154*, 155, 264
    development 147–9, 150–1, 152–3
    production 149, 151, 152, 153, 155, 159, 244
    US evaluation of 156–8, 159
Seeckt, General Hans von 78, 81, 91, 92
Seelöwe, Operation 167, 168
Selve 78, 79
Semmering hill climb 7, 37
Semper Vivus ('Always Alive') hybrid vehicle 9
Serbia 19, 24, 45
Sherman tank 229, 256
Siemens-Schuckert 164, 171, 184, 186, 217, 219, *224*
Simmering-Graz Pauker AG 164, 182, 201–5, 217
Skoda
    artillery tugs *17*, 18, 211–13
    M 11 305mm mortar 23, 25, *26*, 27, 47–8
    M 14 420mm howitzer 49–50, *51*
    M 16 380mm howitzer 49–50, *52*
    M 98 240mm howitzer 21, 53
    Controls Austro Daimler 22–3, 41, 73
    Type 205 *Maus* super heavy tank 219
Skoda, Emil, Baron von 22
Skoda, Karel, Baron von 22–3, 41, 47, 48–9
Skorzeny, Otto 242
Sopwith Aircraft 43, 44
Soviet Union
    Austria, invasion of 256, 257
    German experimental stations 85, 91–2
    German invasion 55, 140, 143, 168, 169, 183, 186, 211
    Germany, invasion of 146, 195, 208, 229–30
    Kursk, Battle of (1943) 188–93
    Porsche, offer to 107–8
    tank development 214
    see also Russia
Speer, Albert
    artillery tug 213
    factory dispersal 251
    and Ferdinand Porsche 123, 173–4, *177*, 195, 197, *214*, 257

fuel shortages 240
Hitler, relationship with 175, 214
Internment of 257
Peugeot 247
super heavy tanks 215, 220, 222, 224–5
tank development 175, 177, *178*, 180, 181, 182, 183
V-1 flying bomb 242
Volkswagen 132
war production 125, 174, 181
Spielberger, Walter J. 87, 92, 132, 165
Stagl, Hans Otto 29–30
Stahl, Otto 16, 17–18, 44, 74
Stalin, Josef 107
Stalingrad 183, 186
Steinfeld aerodrome 31, 32, *38*
Stern, Colonel Albert 'Bertie' 164–5
Steyr-Daimler-Puch AG 104, 125–6, *128*, 161, 168, 196, 204–5
Steyr-Werke AG 103–5
Stoewer 78, 140
Stuck, Hans 111, 112, 113, *257*
Svirin, Mikhail 168, 189, 193, 195

T-34 tank 167, 169–70
Targa Florio, Italy 76
Taube (Pigeon) aircraft 32, *35*, 37–8, 41, 42
Teller mines 128, 235
Thomale, Colonel 181, 182
Thomas, Major General Georg 123, 125, 160
Todt, Fritz
  armaments production 160–1, 166, 168, 174
  death of 173
  and Ferdinand Porsche 161, 168, 174
  honours 127, *128*
Todt Organisation 205, 237
Trippel, Hanns 148–9, 157, 265

Udet, Ernst 124, 125
Uniform Personnel Car 137–8, 139, 140, 148
United Electrical Corporation (VEAG) 3–4, 6, 7, 8, 12, 20
United States
  rocketry programme 265
  Schwimmwagen evaluation 156–8, 159
  World War II 130, 146, 177, 184, 195, 207, 235, 236, 244, 245, 248, 256

V-1 flying bomb 230
  description 231, *232*, 233
  turbojet-engined version 242–3
  launching of 233, *235*, *241*, 242
  production 233–4, 236, 237–40, 267, 270
  Reichenberg manned V-1 240, *241*, 242
V-2 ballistic missile 208, 226, 230, 231, *238*, 239–40, 242, 265
Versailles, Treaty of (1919) 77, 81, 92, 93
Veyder-Malberg, Hans von 111, 131, 253
Voith hydraulic transmission 170–1, *172*, 199, 200, 201
Volkspflug (people's plough) 129–30, 250
Volkssturm 229, 244, 255–6
Volkswagen
  development of 115–17, *122*, 145–6, 197, 208–10
  French interest in 260–1, 262
  Hitler proposes 113–14
  military variants 144–5
  post-war 264, 267, *268*
  production 118, *119*, 120, 132, 136, 138, 159, 270
  see also Kübelwagen; Schwimmwagen
Vollmer, Joseph 88–9, 105

Walb, Willy *111*, 112
Wanderer 78, 107, *133*
Warchalowski, Adolf 32, 35, 36, *38*, 42
Wels, Franz 30–1, 32
Werlin, Jakob *109*, 111, 114, 115, 132, 163
Werner, Joseph 138, 248
Wiersch, Bernd 137, 148
Wolf, Captain Robert 2, 3, 17, 19
wood-gas generators 130, 140, *145*, 255
World War I
  end of 69, 70
  outbreak 24–5, 27, 44
World War II
  end of 256
  outbreak 55, 120

Zadnik, Otto 65, 106, 172, 178, *184*, 206, 225
Zeitzler, General Kurt 222, 240
Zell am See, Austria 249–50, 251, 252, 256, 258, 259–60, 264
Zeppelins 102
ZF transmissions 81, 83, *84*, 137
Zimmerit anti-magnetic coating *194*, 195
Zippermayr, Mario 243–4
Zitadelle, (Citadel) Operation 188–93, 223, 270
Zündapp 107, 113, 149, 150